Green Nanotechnology

Green Nanotechnology

Oleg Figovsky
Dmitry Beilin

Published by

Pan Stanford Publishing Pte. Ltd.
Penthouse Level, Suntec Tower 3
8 Temasek Boulevard
Singapore 038988

Email: editorial@panstanford.com
Web: www.panstanford.com

British Library Cataloguing-in-Publication Data
A catalogue record for this book is available from the British Library.

Green Nanotechnology

ISBN 978-981-4774-10-9 (Hardcover)
ISBN 978-1-315-22928-7 (eBook)

Printed in the USA

Contents

Preface

Green nanotechnology has two goals: producing nanomaterials and products without harming the environment or human health and producing nanoproducts that provide solutions to environmental problems. It uses existing principles of green chemistry and green engineering to make nanomaterials and nanoproducts without toxic ingredients, at low temperatures using less energy and renewable inputs wherever possible, and using life cycle thinking in all design and engineering stages.[a]

Green nanotechnology aims to develop clean technologies to minimize potential environmental and human health risks associated with the manufacture and use of nanotechnology products and to encourage the replacement of existing products with new nanomaterials that are more environmentally friendly.

There are two key aspects to green nanotechnology. The first involves nanoproducts that provide solutions to environmental challenges. These green nanoproducts are used to prevent harm from known pollutants and are incorporated into environmental technologies to remediate hazardous waste sites, clean up polluted streams, and desalinate water, among other applications. The second aspect of green nanotechnology involves producing nanomaterials and products containing nanomaterials with a view toward minimizing harm to human health or the environment.[b]

Green nanotechnology involves the following:[c]

- Use of less energy during manufacture
- Ability to recycle after use
- Use of ecofriendly materials

---

[a] https://en.wikipedia.org/wiki/Green_nanotechnology
[b] B. Karn & L. Bergeson, *Natural Resources & Environment*, **24**(2), 2009
[c] http://www.azocleantech.com/article.aspx?ArticleID=330

The most important component of nanotechnology is nanomaterials, that is, materials with the ordered structure of their nanofragments having sizes from 1 to 100 nm. The production and process aspects of green nanotechnology involve both making nanomaterials in a more environmentally benign fashion and using nanomaterials to make current chemical processes more environmentally acceptable. A 2003 estimate by the NanoBusiness Alliance identified nanomaterials as the largest single category of nanotech start-ups.

According to the recommendation of the 7th International Conference on Nanostructured Materials, Wiesbaden, 2004, nanomaterials are classified as:

- Nanoporous structures
- Nanoparticles
- Nanotubes and nanofibers
- Nanodispersions (colloids)
- Nanostructured surfaces and films
- Nanocrystals and nanoclusters
- Nanocomposites

There are two basic ways to create nanoobjects:

- Reduce the size of macroscopic objects (dispersing, disintegrating, or grinding to the cluster level using a ball mill or using the mechanochemical synthesis)
- Create nanostructures from atom and molecule (crystallization) clustering, nanostructuring, nucleation, condensation, coagulation, polymerization, etc.

The prospect of a new materials technology that can function as a low-cost alternative to high-performance materials has, thus, become irresistible around the world. By this means nanotechnology presents a new approach to material science and engineering as well as for design of new devices and processes.

According to the Congressional Research Service USA the world industry uses nanotechnology in the production of 80 groups of consumer goods and more than 600 kinds of raw materials, component items, and industrial equipment. The figure below can give some suggestion of the global business segments of nanotechnology.

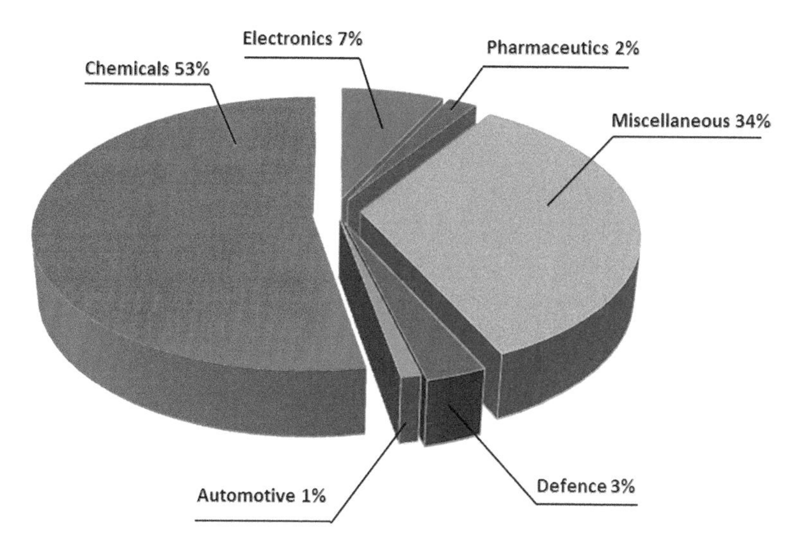

Figure 1 Global nanotechnology.

BCC Research provides an updated analysis of the nanotechnology products market in its report "Nanotechnology: A Realistic Market Assessment" (NAN031F). The global market for nanotechnology products was valued at about $26 billion in 2014. This market is expected to reach about $64.2 billion in 2019, with a compound annual growth rate of 19.8% from 2014 to 2019.

The American Association of National Science Foundation predicts, in the next 10–15 years, the market growth of nanogoods and services up to $1 trillion:

This book contains information about how advanced nanomaterials can be produced without harming the environment or human health. This encompasses the production of nanomaterials without environmental toxicity, at room temperature, and with the use of renewable energy sources. The book contains the descriptions and results of theoretical and experimental research in the field of environment-friendly nanotechnology carried out over the past decade by the scientific team of the company

Polymate Ltd. International Nanotechnology Research Center (www.polymateltd.com, Israel) under the leadership of Professor O. Figovsky. Developments of the company have been used in industry and agriculture and protected by more than 25 patents of the US, Germany, and Russia.

Let's summarize the contents of the monograph.

Chapter 1 is concerned with the interpenetrating polymer networks (IPN) principle in the production of composite materials and provides a unique possibility to regulate their both micro- and nanostructures and properties. The chapter discusses principal features and characteristics of IPN composites. Formation of rubberizing ebonite coatings on samples of oligobutadienes is examined. Recent advances in chemistry and technology of nonisocyanate polyurethane (NIPU) materials based on cyclic carbonate oligomers are reviewed in this chapter. The use of NIPU materials as coatings, adhesives, and foams is described.

Chapter 2 presents a few methods of sol-gel synthesis: alkoxide, nonhydrolytic, and colloidal. The sol-gel technology of nanocomposites based on the use of soluble silicates as precursors is discussed. Different types of nanophases used for producing the nanocomposites are examined. The various models of packaging of nanoparticles (spherical, fibrous, and layered) introduced into the nanocomposite structure during its preparation are studied. Polymeric materials for structures and coatings are increasingly dominating the corrosion-protection technology.

Chapter 3 describes the most effective method of improving protective properties by the use of additional components reducing the rate of diffusion of electrolytes in polymers and anticorrosive silicate compounds. Proposed is the set of inorganic substances of composite polymeric materials that selectively interact with the water or water solutions of acids, salts, and alkalis in order to decrease their penetrability and increase their chemical resistance simultaneously.

Chapter 4 contains a description of the new "green" manufacturing process of the nanostructured composite materials based on using the physical phenomenon superdeep penetration (SDP). Synthesis of a skeleton and formation of nanostructure is realized in metals, polymers, and ceramics. Physical anomalies at the impact, which appear in conditions when the relative depth of a crater exceeds 10, determining the size of the striker, are considered. SDP is used for manufacturing of special composite metal materials with an unusual complex of properties. The SDP method of polymer tracking membranes production was developed.

Chapter 5 is devoted to the creation of a new bioactive composite on the basis of silver nanoparticles. The biocidal effect of nanoparticle-modified paints and coatings is investigated. The structure and technology of biologically active nanocomposite preparation is offered.

Chapter 6 presents a brief overview of the work in producing and studying of environment-friendly nanostructured polymeric composites. Preparation technology and main applications of the nanocellulose is described. Novel environment-friendly hydrophobic polymer composites were developed. Various types of layer composites and their applications in the production of packaging materials are described. The proposed biodegradable nanocomposite coating increases the strength of the natural packaging materials and serves as an effective barrier against water and grease. Waste of the novel polymer materials can be utilized in two ways: by repulping and by biodegradation.

Chapter 7 is concerned with the problem of improving of seed germination conditions and development of plants and protecting plants from anticipated and averaged adverse conditions with the help of biologically active nanochips.

The major results of the works presented in this monograph were published mainly in the journal *Scientific Israel-Advanced Technology* (www.sita-journal.com) during the period 2005–2016.

The book will be useful to specialists in the field of chemical technology and materials engineering.

Acknowledgments

The authors are happy to express profound gratitude to the research teams of Polymate Ltd.-INRC (http://polymateltd.com), which include N. Blank, O. Birukova, N. Kudryavtsev, A. Leykin, R. Potashnikova, L. Shapovalov, and S. Sivocon and their colleagues E. Gotlib, E. Egorova, M. Ioelovich, B. Kudryavtsev, P. Kudryavtsev, L. Moshinsky, J. Owsik, Yu. Pushkarev, I. Ruban, S. Usherenko, and N. Voropaeva, for their invaluable contribution to the development of an environment-friendly technology and creation of advanced nanostructured composite materials. These research works formed the basis of the presented monograph.

A special thanks to the CEO of Polymate Ltd.-INRC A. Trossman (Israel) and the CEO of Nanotech Industries J. Kristul (http://www.nanotechindustriesinc.com, USA) for their assistance in the industrial application of the scientific research results.

The authors would like to acknowledge Mr. Sarabjeet Garcha for his help and highly professional editing.

Chapter 1

Nanostructured Composites Based on Interpenetrated Polymer Networks

1.1 Kinds, Classification, Properties, Synthesis, Application

1.1.1 Introduction

Interpenetrating polymer networks (IPNs) are unique "alloys" of crosslinked polymers in which at least one network is synthesized and/or crosslinked in the presence of the other. Unlike chemical blends, there are ideally no covalent bonds between components in IPNs [1–5]. Thus, there is some type of "interpenetration." However, the term *interpenetrating polymer network* was coined before current aspects of phase separation and morphology were understood. Now we know that most IPNs do not interpenetrate on a molecular scale; they may, however, form finely divided phases of only tens of nanometers in size. Many IPNs exhibit dual phase continuity, which means that two or more polymers in the system form phases that are continuous on a macroscopic scale [3, 5].

Green Nanotechnology
Oleg Figovsky and Dmitry Beilin

ISBN 978-981-4774-10-9 (Hardcover), 978-1-315-22928-7 (eBook)
www.panstanford.com

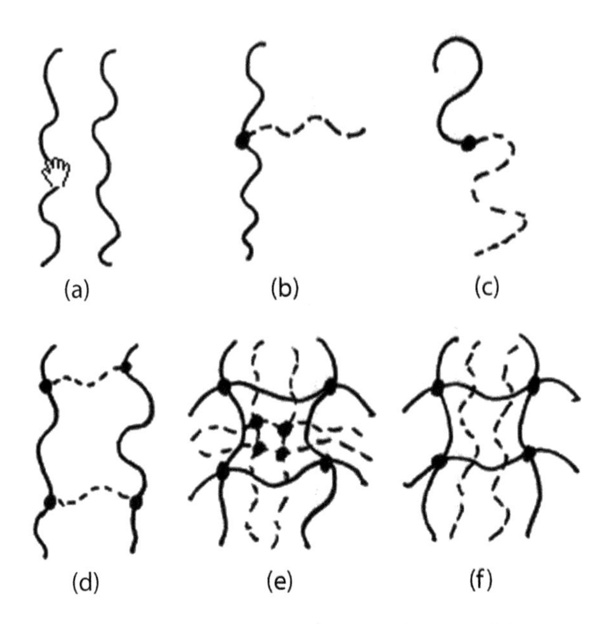

Figure 1.1 Six basic combinations of two polymers: (a) polymer blend, no bonding between chains; (b) graft copolymer; (c) block copolymer; (d) AB-graft copolymer; (e) IPN; and (f) SIPN. Structures (a–c) are thermoplastic; structures (d–f) are thermoset [1, 3]. Reprinted from Ref. [1] with permission from *Engineering Journal of Don*.

IPNs are often created for the purpose of conferring key attributes of one of the components while maintaining the critical attributes of another. In some cases, entirely new, and sometimes surprising, properties are exhibited by the IPN that are not observed in either of the two single networks alone [5].

When two or more polymers are mixed, the resulting composition can be called a multicomponent polymer material. There are several ways to mix two kinds of polymer molecules (Fig. 1.1). Simple mixing, as in an extruder, results in a polymer blend. If the chains are bonded together, graft or block copolymers result. When there is bonding between some portion of the backbone of polymer I and the end of polymer II, the result is called a graft copolymer; chains bonded end to end result in block copolymers. Other types of copolymers include AB-crosslinked copolymers, where two polymers make up one network, and the IPNs, and semi-IPNs (SIPNs) [3].

1.1.2 Kinds of IPNs

From a synthetic standpoint, IPNs come in two varieties:

(a) sequential IPN, in which one network is swollen and polymerized in the presence of the other
(b) simultaneous IPN, in which both of the network precursors are synthesized at the same time by independent, non-interfering routes

When one of the two polymers is linear (uncrosslinked), a semi-IPN results and when both of the polymers are identical, a homo-IPN results. An important parameter that must be considered in the fabrication of IPNs is the mutual miscibility of the interpenetrating polymers; in general, polymers do not mix well with each other, resulting in the phase separation of the resultant blend. However, because crosslinking provides a way to "enforce" mixing between two otherwise immiscible materials, there are virtually endless combinations of polymers worthy of exploration as IPNs for a variety of applications [5].

IPNs can be made in many different ways. Brief definitions of some of the more important IPN materials are as follows [3]:

- *Sequential IPN.* Polymer network I is made. Monomer II plus crosslinker and activator are swollen into network I and polymerized in situ (Fig. 1.2A). The sequential IPNs include many possible materials where the synthesis of one network follows the other.
- *Simultaneous interpenetrating network (SIN).* The monomers or prepolymers plus crosslinkers and activators of both networks are mixed. The reactions are carried out simultaneously, but by noninterfering reactions. An example involves chain and step polymerization kinetics (Fig. 1.2B).
- *Latex IPN.* The IPNs are made in the form of latexes, frequently with a core and shell structure. A variation is to mix two different latexes and then form a film, which crosslink both polymers. This variation is sometimes called an interpenetrating elastomer network (IEN).

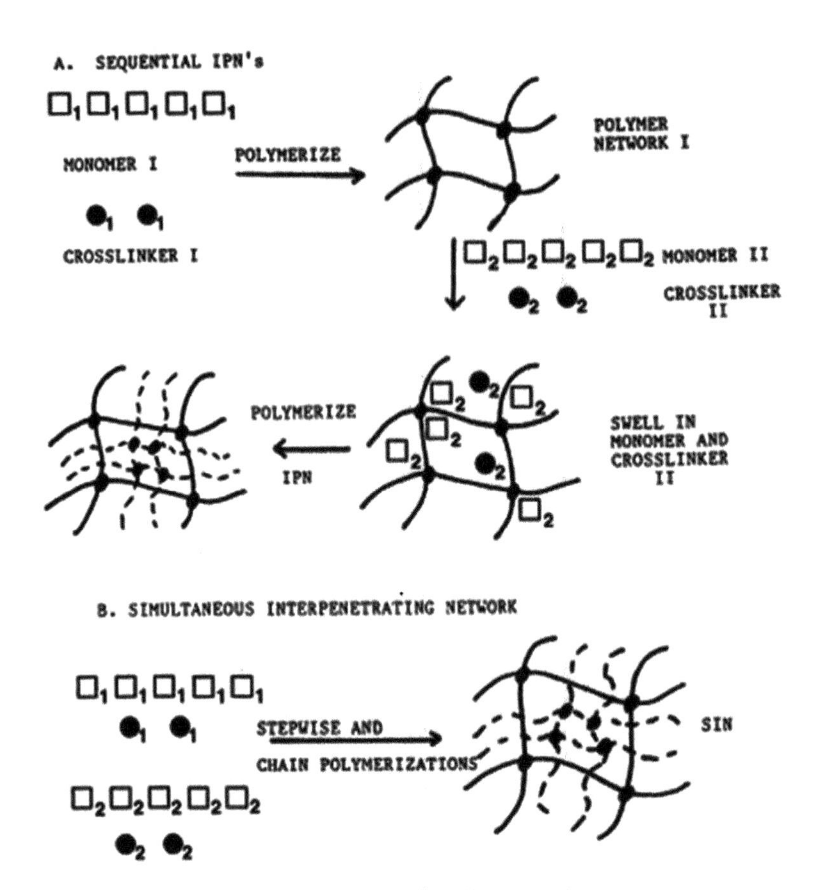

Figure 1.2 Basic synthesis methods for IPNs: (A) Sequential IPNs and (B) simultaneous interpenetrating polymer networks (SINs) [1, 2]. Reprinted from Ref. [1] with permission from *Engineering Journal of Don.*

- *Gradient IPN.* Gradient IPNs are materials in which the overall composition or crosslink density of the material varies from location to location on the macroscopic level. For example, a film can be made with network I predominantly on one surface, network II on the other surface and a gradient in composition throughout the interior.
- *Thermoplastic IPN.* Thermoplastic IPN materials are hybrids between polymer blends and IPNs that involve physical crosslinks rather than chemical crosslinks. Thus, these materials flow at elevated temperatures, similar to the thermoplastic elastomers,

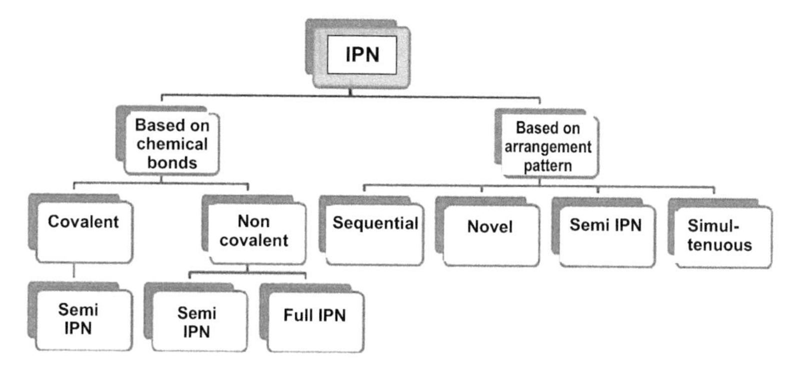

Figure 1.3 Different forms of IPN [1, 4]. Reprinted from Ref. [1] with permission from *Engineering Journal of Don.*

and at use temperature, they are crosslinked and behave like IPNs. Types of crosslinks include block copolymer morphologies, ionic groups, and semicrystallinity.

- *Semi-IPN.* Compositions in which one or more polymers are crosslinked and one or more polymers are linear or branched are semi-IPN (SIPN).

1.1.3 Classification of IPN [1, 4]

Different forms of IPN are illustrated in Fig. 1.3.

1.1.3.1 IPN based on chemical bonding

Covalent crosslinking leads to formation of hydrogels with a permanent network structure, since irreversible chemical links are formed. This type of linkage allows absorption of water and/or bioactive compounds without dissolution and permits drug release by diffusion.

- *Covalent semi-IPN.* A covalent semi-IPN contains two separate polymer systems that are crosslinked to form a single polymer network.
- *Noncovalent semi-IPN.* A noncovalent semi IPN is one in which only one of the polymer system is crosslinked.
- *Noncovalent full IPN.* A noncovalent full IPN is a one in which the two separate polymers are independently crosslinked.

1.1.3.2 IPN based on arrangement pattern

- *Sequential IPN.* In sequential IPN, the second polymeric component network is polymerized following the completion of polymerization of the first component network.
- *Novel IPN.* Polymer comprising two or more polymer networks which are at least partially interlocked on a molecular scale but not covalently bonded to each other and cannot be separated unless chemical bonds are broken.
- *Semi-IPN.* If only one component of the assembly is crosslinked leaving the other in a linear form, the system is transferred as semi IPN.
- *Simultaneously IPN.* Simultaneously IPN is prepared by a process in which both component networks are polymerized concurrently, the IPN may be referred to as a simultaneously IPN.

1.1.4 Properties of IPN [1, 4]

- A gel composed of two interpenetrating networks by crosslinking a polymer (or polyelectrolyte) into a pre-existing highly crosslinked network of a polymer (or polyelectrolyte) of a different kind increases elastic and mechanical properties, which were measured by the stress–strain behavior and comparing their elastic moduli and breaking points.
- According to US Patent data, the calculated true stress per unit solid and strain shows that PGA/PAA IPNs are much stronger than either the individual polymer networks or copolymers. The effect of IPN formation on tensile strength is nonlinear, as the maximum strength is many times higher than that of PEG-PAA copolymer. The elastic moduli and tensile strength can be modified by changing the molecular weight.
- *Oxygen permeability.* IPN hydrogels composed of PEG as the first network and a second network of poly acrylic acid had oxygen permeability of 95.9 ± 28.5 Barrer.
- *Shape memory.* Materials are said to show shape memory effect if they can be deformed and fixed into a temporary shape and recover their original permanent shape only on the exposure of external stimuli, like heat, light, etc.

- *Equilibrium water content.* IPN can swell in solvent without dissolving. The water content of hydrogels was evaluated in terms of the swollen weight to dry weight ratio. The dry hydrogel was weighed and then immersed in water as well as phosphate buffered saline. At regular intervals, the swollen gel was lifted, patted dried, and weighed until the equilibrium was attained. The percentage of equilibrium water content (WC) was calculated from the swollen and dry weight of hydrogel:

$$\mathrm{WC} = \frac{W_s - W_d}{W_s} \times 100,$$

  where, W_s and W_d are weight of swollen and dry hydrogels, respectively.

- IPN systems are known to increase the phase stability of the final product.
- Thermodynamic incompatibility can be overcome due to the permanent interlocking of the network segments.
- High thermostability.
- Good dielectric properties.
- Radar transparency.
- Nutrient permeability.
- Optical clarity—the percentage of light transmittance was found to be 90% and the refractive index was found to be 1.35.

1.1.5 Synthesis of Some IPN

The scheme of mechanisms for the formation of hydrogels is shown in Fig. 1.4.

1.1.5.1 Synthesis of IPN [1, 4, 6]

Figure 1.5 shows steps in IPN preparation.

The starting material to make the hydrogel was a solution of telechelic macro monomers (**1**) with functional end groups (**a**). They were polymerized to form a polymer network followed by the addition second polymer (**4**) of hydrophilic monomers (**3**) which were polymerized and crosslinked in the presence of the first polymer (**2**). This resulted in the formation of IPN hydrogel (**5**).

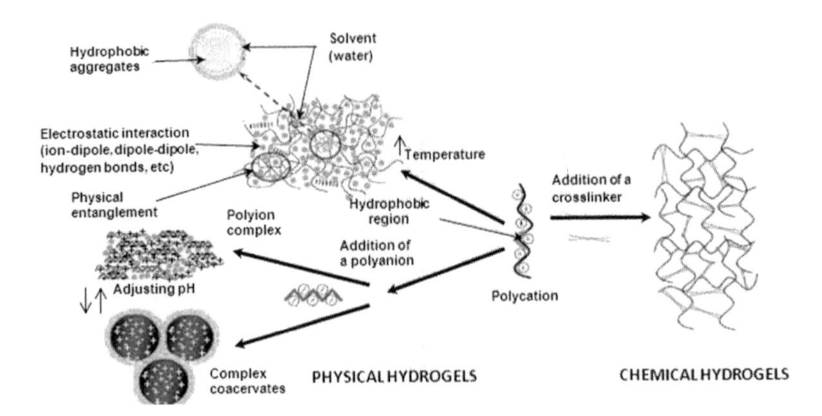

Figure 1.4 Overview of mechanisms for formation of physical and chemical hydrogels [1, 4]. Reprinted from Ref. [1] with permission from *Engineering Journal of Don.*

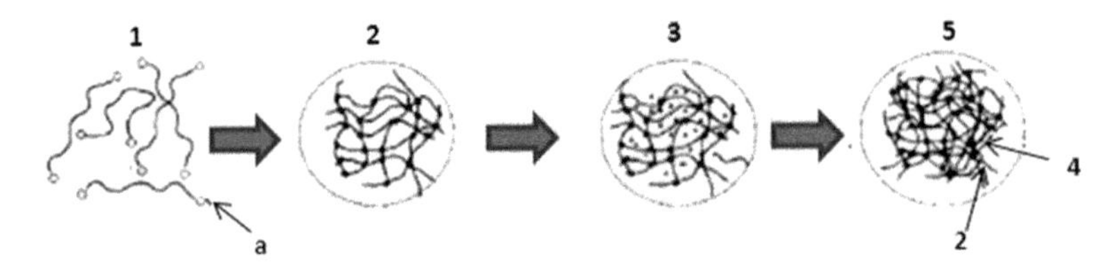

Figure 1.5 IPN hydrogels composed of poly(ethyleneglycol) macromere (PEGM) and chitosan [1, 4]. Reprinted from Ref. [1] with permission from *Engineering Journal of Don.*

Polyethylene glycol (PEG) can also be used a first polymer as it is biocompatible, soluble in aqueous solution and gives wide range of molecular weights and chemical structures. IPN hydrogel was also prepared by UV initiated free radical polymerization. It used the first network as PEG diacrylate (PEG-DA) or PEG dimethacrylate (PEG-DMA) dissolved in phosphate buffered saline (PBS).

We shall consider in detail the problem of polyurethane and epoxy based IPN.

The importance of polyurethane-based IPNs needs special emphasis. Such materials are relatively easy to synthesize and have outstanding properties. The polyurethane elastomers may serve as network in a sequential IPN synthesis or, in prepolymer form, may serve as one component in simultaneous interpenetrating networks.

Several methods of synthesis between the classical sequential and simultaneous methods have been worked out [3].

Meyer and co-workers [7–11] investigated the composition *cross*-polyurethane-*inter-cross*-poly(methyl methacrylate). In general, the synthesis involved an aromatic triisocyanate and a polyether glycol catalyzed by stannous octanoate. The poly(methyl methacrylate) (PMMA) network resulted from an AIBN-initiated (azobisisobutyronitrile-initiated) free radical polymerization with a trimethacrylate crosslinker. All of the components were mixed together. The polyurethane (PU) network was allowed to form first at room temperature, followed by heating to initiate the polymerization of the methyl methacrylate (MMA) monomer. The resulting IPNs exhibited two loss peaks in tan δ-temperature studies, but the glass-transition temperatures were shifted inward significantly and broadened. Thus, Meyer et al. concluded that these materials exhibited incomplete phase separation [3].

Jin and Meyer [12] studied the kinetics of reaction of these IPNs via Fourier transform infrared spectroscopy (FTIR). The PU-network was formed at room temperature again; then the PMMA was free radical polymerized at 60°C. The authors adopted the term "in situ" sequential IPNs for such materials to emphasize that all the reagents are introduced simultaneously, but that the networks are formed sequentially. This formulation is an example of kinetics partway between sequential and simultaneous syntheses [3].

Jin et al. [3, 13] showed that the polyurethane has two effects on the formation of the acrylic phase:

(1) By conferring a very high viscosity on the reaction medium from the very beginning of the polymerization process, an initial high reaction rate and early gelation effect is induced.
(2) The polyurethane clearly acts as a diluent and keeps the T_g of the reaction medium below the T_g of the PMMA; hence, complete monomer to polymer conversion at 50–70°C is allowed. Polymerization usually stops at or just beyond the point of glassification, due to slowed diffusion processes [2, 13].

Allen et al. [3, 15–20] carried out an interstitial polymerization of vinyl monomers within PU gels. The reaction scheme was as follows:

$$\text{Elastomer precursors} + \text{MMA} \xrightarrow{\text{PU catalyst}} \text{PU gel swollen with MMA} \xrightarrow{\text{MMA Initiator}} \text{SIPN}$$

Allen et al. pointed out that the maximum swelling capacity of the gel should not be exceeded (taking into account χ_1) or macrosyneresis will occur. On the other hand, the crosslink density must be high enough to prevent macroscopic phase separation on polymerization of the MMA. For all of the systems considered by Allen et al., the PU was gelled at room temperature and the MMA was initiated at elevated temperatures, in a manner similar to the work of Meyer et al. [3, 12, 13].

Although the focus of these papers has been on the polyurethane (PU), note that each of these papers also has poly(methyl-methacrylate) as the mate polymer. This pair develops excellent mechanical behavior over a wide range of compositions.

PUs are among the most used polymers in many modern technologies [21]. However, the use of toxic components, such as isocyanates, in the manufacturing process can render PU production extremely toxic and dangerous [22]. Nonisocyanate sources for PU production have been sought for a long time.

Isocyanates are critical components used in conventional polyurethane products such as coatings and foam. However, exposure to isocyanates is known to cause skin and respiratory problems and prolonged exposure has been known to cause severe asthma and even death. Isocyanates are also toxic to wildlife. When burnt, isocyanates form toxic and corrosive fumes including nitrogen oxides and hydrogen cyanide. Due to these hazards, isocyanates are regulated as by the EPA and other government agencies.

Non-isocyanate polyurethane (NIPU) based on the reaction of polycyclic carbonates and poly-amines are known for more than 50 years. Fundamentals for the practical application of NIPU on the basis of five-membered cyclic carbonates (1,3-dioxolan-2-ones) in coatings, sealants, adhesives, etc., were largely developed by O. Figovsky in the 1970–1980s [23]. Recently, some reviews dedicated to the synthesis of cyclic carbonates and NIPU have been presented [24–26]. In these works the advantages of NIPU have been described in detail.

We had been created materials on the basis of hydroxyurethane and epoxy NPI, which are described in a number of patents of the authors and their collaborators [27–34]. It should be emphasized the entirely new developments contained in patents [27] and [29]. The first of these provides information about nonisocyanate polyurethanes (NIPU) on the oligomers' basis with terminal cyclocarbonate and epoxy groups; nanostructured NIPU are described in the second part of this chapter.

1.1.6 Dendripolymers Based on the Epoxy-Amine Reactions

Of special interest are the highly branched dendritic structure polyaminopolyamides (PAMAM) based on the epoxy-amine reactions. The dendripolymers constitute a new class of substances, which combine the properties of polymers and small discrete molecules.

L. Moshinsky [35] has offered a new method for dendripolymer preparation. Using this method, rows of new dendripolyamide PAMAM kind were prepared and studied. The method was based on the epoxy-amine reactions, which were organized as dendrisynthetical succession of the exhaustive reactions.

Characteristics of the new highly branched polyamino-polyamides and results of the dendrimers study as epoxy hardeners are discussed below.

Two different methods in principle for the preparation of highly branched polymers are known. One method is based on the principles of dendrisynthesis that ensure the formation of polymers having well-ordered star-shaped or tree-like structure. The other method uses the principle of "stopped polycondensation," which allows preparing conventional hyperbranched polymers.

A ramification of the hyperbranched polymers depends on topological characteristics of the monomers used and upon the polycondensation extent. For example, permissible extent of the component conversion constitutes approximately 50% in a case of polycondensation of bi- and trifunctional monomers. So these polymers represent polydisperse products, having highly branched but chaotic undigested structure. Nevertheless, a comparatively simple technology together with quite interesting properties of hyperbranched polymers have defined some possibilities for their technical application, for example, as components of coatings, adhesives, putties, and modifiers for thermoplastic polymers.

The dendrisynthesis may be realized in two variants: a divergent scheme of dendripolymer preparation and a convergent outline for tree-like polymer formation. Divergent synthesis begins with the selection of a multi-functional compound (core maker), which serves as a center of a dendripolymer-to-be.

The following scheme can serve as an example of such type of reactions [35]:

$$\text{-CH}_2\text{CH}_2\text{NH}_2 + 2\text{CH}_2\text{=CH-}\overset{\overset{\text{O}}{\|}}{\text{C}}\text{-OCH}_3 \xrightarrow{t^\circ\text{C}} \text{-CH}_2\text{CH}_2\text{N}\left\langle \begin{array}{l} \text{CH}_2\text{CH}_2\text{-}\overset{\overset{\text{O}}{\|}}{\text{C}}\text{-OCH}_3 \\ \text{CH}_2\text{CH}_2\text{-}\overset{\overset{\text{O}}{\|}}{\text{C}}\text{-OCH}_3 \end{array} \right. \xrightarrow{120\text{-}200^\circ\text{C}}$$

$$\xrightarrow[\text{-2CH}_3\text{OH}]{2\text{H}_2\text{NCH}_2\text{CH}_2\text{NH}_2} \text{-CH}_2\text{CH}_2\text{N}\left\langle \begin{array}{l} \text{CH}_2\text{CH}_2\text{-}\overset{\overset{\text{O}}{\|}}{\text{C}}\text{-HNCH}_2\text{CH}_2\text{NH}_2 \\ \text{CH}_2\text{CH}_2\text{-}\overset{\overset{\text{O}}{\|}}{\text{C}}\text{-HNCH}_2\text{CH}_2\text{NH}_2 \end{array} \right.$$

The growth of dendripolymer branches is realized owing to exhaustive reactions of amine group acrylation (first step) and ester-group amidation (second step). To build a dendritic molecule, the reactions of the scheme must be repeated several times. This process

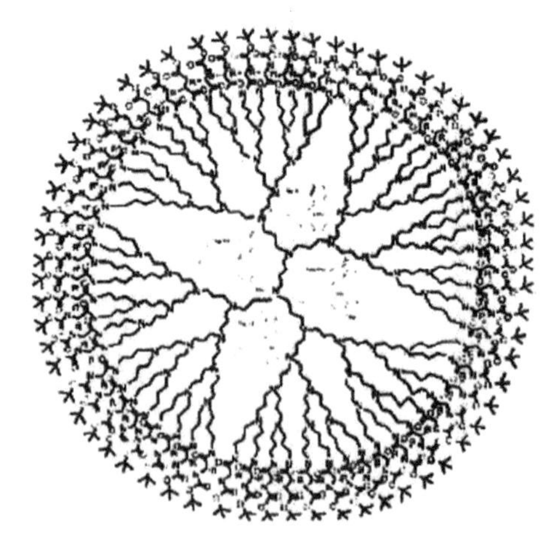

Figure 1.6 Schematic representation of a dendripolymer structure, prepared by divergent scheme [35].

gives at the end a divaricate product with expected molecular weight and branching extent. This reaction sequence results in the formation of the star-shaped molecule possessing spheroidal structure, wherein its functional groups are placed on periphery of the spheroid (Fig. 1.6).

Convergent method of dendrisynthesis consists in joining the beforehand-prepared oligomer segments of the dendripolymer-to-be. Further, analogously to divergent scheme, a core maker, an intermediate segment, and a multi-functional oligomer segment are exhaustively joined into the dendritic structure designed.

As it was mentioned above, convergent dendrisynthesis using amine-epoxy reactions has been realized for the first time. This method was used for the preparation of amine-functional highly branched polyamides. The chemical scheme of such kind of dendripolymer formation (right-hand half of core maker and the reactants) can represent as follows [35]:

a) $-R-CH_2-NH_2 + 2CH_2CH + CH_2O-C_6H_4-C(CH_3)_2-C_6H_4-OCH_2CH-CH_2$ (epoxy, O) →

→ $-R-CH2-N$ ⟨ $CH_2CH(OH)CH_2O-C_6H_4-C(CH_3)_2-C_6H_4-OCH_2CH-CH_2$ (epoxy, O) ⟩$_2$

b) $-R-CH2-N$ ⟨ $CH_2CH(OH)CH_2O-C_6H_4-C(CH_3)_2-C_6H_4-OCH_2CH-CH_2$ (epoxy, O) ⟩$_2$ $+ 2R'[C(=O)(NHCH_2CH_2CH_2)_3-NH_2]_2$ →

→ $-R-CH2-N$ ⟨ $CH_2CH(OH)CH_2O-C_6H_4-C(CH_3)_2-C_6H_4-OCH_2CH(OH)CH_2(NHCH_2CH_2)_3-NHC(=O)-R'-C(=O)(NHCH_2CH_2)_3-NH_2$ ⟩$_2$

Three essential moments distinguish the process according the scheme from other conventional methods for dendripolyamide preparation:

(1) In contrast to other methods, intermediate product (**a**), the multi-epoxy dendripolymer is an unstable substance, which can be spontaneously polymerized at room temperature. So, this dendripolymer should use immediately after its preparation because it goes to gel for some hours whereupon it does not fit for any polymer-analogous transformations.
(2) In contrast to other methods, both steps of the scheme may be realized at moderate temperature without by-product separation. So, the formation process of epoxy-amine dendripolymers is technologically very simple and ecologically clean.
(3) In contrast to many other methods, the regulation of molecular weight and molar functionality of the epoxy-amine dendripolymers does not demand several generation steps. A multifunctional core maker, as well as epoxy resins and diepoxy compounds with molecular weight in the range of 200–5,000 g/mole, and polyamine-polyamides with molecular weight of 500–2,000 g/mole may be used in this process. Selection of components from this variety makes it possible to get dendripolymer with given structure and necessary properties.

These epoxy-amine dendrimers in the absence of a solvent represent rubber-like substances with a softening point in the range of 170–210°C. The same dendripolymers, comprising comparatively small amount of suitable solvent, represent resin like products suitable for preparation of thermosetting epoxy compositions, including adhesives, coatings, sealants, etc. Some epoxy-amine dendrimers, containing a small amount of benzyl alcohol, have been produced at present as trademarks Dendrepox™ Polyamide. Physicochemical properties of the four typical dendripolyamides are shown in Table 1.1 [35].

The preliminary testing showed Dendrepox™ Polyamides cure the epoxy resin quicker than standard polyamide resins do it. For comparison, the curing rate constants of the systems comprising

Table 1.1 Properties of dendripofyamdes type Dendrepox™

Characteristics	Trademark of dendripolyamide			
	IB-100	**HB-101**	**AD-102**	**AH-103**
Appearance	Light-brown resin-like substance			
Molecular weight (g/mole)	6,500	8,300	12,000	13,300
Branching extent ($mole^{-1}$)	4	4	6	5
H-functionality ($mole^{-1}$)	30	30	45	37
Kind of functionality	As Versamide 125			
Viscosity (Pa·s)	130	120	240	250
Density (g/cm^3)	1.02	1.00	1.01	1.00
Solid residual (%)	75.0	77.0	73.5	72.0
Gel time at 22°C (minutes)	60	60	65	60
H-equivalent weight (g) (without solvent)	215	280	270	300
Mix ratio with epoxy resin (g/100 g)	150	200	200	220

Source: [35]
Note:
- Core maker of the IB-100—isophoronediamine;
- core maker of the HB-101—Hycar' ATBN 1300X42—the low-molecular amine-containing co-polymer of butadiene with acrylonitrile, BF Goodrich, USA (MW = 900 g/mole);
- core maker of the AD-102—adduct of the isophoronediamine with aliphatic epoxy resin;
- core maker of the AH-103—adduct of the Hycar* ATBN 1300X42 with aliphatic epoxy resin;
- the mix ratio was calculated on the epoxy resin with EEW = 190 g.

DGEBA and dendripolyamide (polyamide resin) are shown in Table 1.2 [35].

Dendrepox™ Polyamide IB-100 was tested as a component of simple epoxy adhesives. The adhesive compositions comprised the bisphenol A-based epoxy resin Epon® 828 (MW = 380 g/mole, density = 1.16 g/cm^3, viscosity = 15.6 Pa·s) and dendripolyamide, or its mix with other amine hardener. Two other polyamines—triethylenetetramine (TETA) and polyamide resin (Versamide® 140)—were used to make more integrate appreciation of the Dendrepox™ effectiveness. Strength characteristics of the adhesives depending on the hardener ratio are shown in Fig. 1.7.

According to these data, some blends of the Dendrepox™ with the conventional polyamide resin make a synergistic effect that is displayed in a maximum of the adhesive properties. Interestingly, the combined effect of the dendripolymer with other amine hardeners was absent.

Table 1.2 Curing rate constants of the Polyamide resins

Polyamide resins	Curing rate constants (min^{-1})
Dendrepox™ Polyamide IB-100 (114.0 g/100 g Epoxy)	2.00–10^{2}
Dendrepox™ Polyamide HB-101 (145.6 g/100 g Epoxy)	2.08–10$^{\sim 2}$
Dendrepox™ Polyamide AD-102 (140.4 g/100 g Epoxy)	1.23-T0^{2}
Dendrepox™ Polyamide AH-103 (189.2 g/100 g Epoxy)	1.64–10$^{\sim 2}$
Polyamide resin Versamide® 140 (86.7 g/100 g Epoxy)	7.78–10$^{\prime\prime 3}$

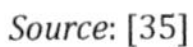
Source: [35]

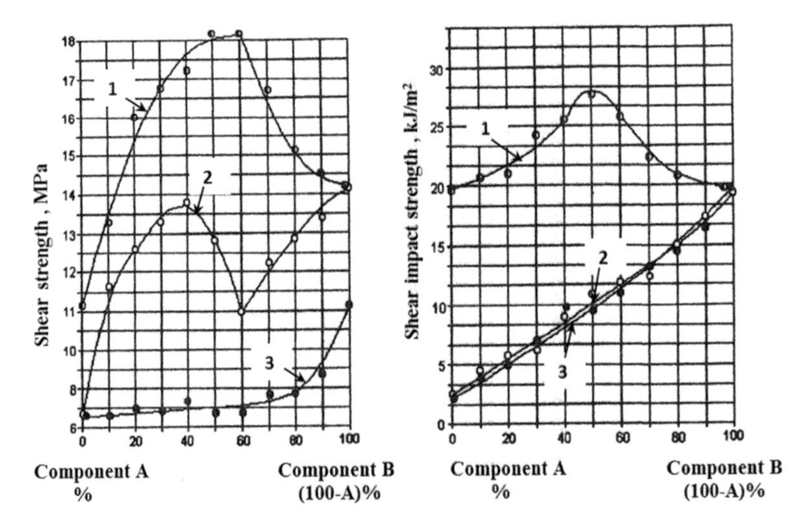

Figure 1.7 Strength characteristics of the adhesives based on two-component hardeners comprising the *Dendrepox™ Polyamide IB-100.*

Curve no.	Component A	Component B
1	Versamide 140	Dendrepox polyamide IB 100
2	Triethylenetetramine	Versamide 140
3	Triethylenetetramine	Dendrepox polyamide IB 100

Note: Curing at room temperature for 20–24 h and postcuring at 80°C for 3 h. The molar correlation of epoxy resin with mixed hardener was 120% to stoichiometric ratio.

1.1.7 Applications of IPN Technology

The IPNs have many applications, both proposed and in practice. Some commercial materials are shown in Table 1.3 [1, 3, 35].

Although most applications involve bulk polymeric materials, ion exchange resins make a particularly interesting application. IPN ion exchange resins usually have anionically and cationically charged networks within the same suspension particle, where as other types have charged networks on different particles. An outstanding example are artificial teeth. According to [3, 36] the material is substantially a homo-IPN of poly(methyl methacrylate).

One especially interesting are the nanostructured polymer compositions HNIPU, RubCon, SPC, and LEM (Table 1.3) in context of their highest properties and exciting prospective of application. These materials were developed and introduced into the industry

Table 1.3 IPN commercial materials

Composition	Application
SEBS[1]-polyester	Automotive parts
Silicone rubber-PU[2]	Gears or medical
PU-polyester-styrene	Sheet molding compounds
PP[3]-EP[4]	Automotive parts
Rubber-PE	
Rubber-PP	Tough plastic
EPDM[5]-PP	• Auto bumper parts, wire and cable
	• Tires, hoses, belts and gaskets
	• Outdoor weathering
	• Tubing, liners
	• Paintable automotive parts
Anionic-cationic	Ion exchange resins
Acrylic-urethane- polystyrene	Sheeting molding compounds
Acrylic-based	Artificial teeth
Vinyl-phenolic	Damping compounds
HNIPU[6]	• Industrial flooring
	• Paints
	• Coatings
	• Artificial leather
	• Rigid and flexible foams
	• Adhesives

(Contd.)

Table 1.3 (*Continued*)

Composition	Application
RubCon[7]	• Industrial flooring • Galvanic and electrolysis baths • Supports and foundations • Underground structures • High-speed railroad ties
SPC[8]	• Storage tanks for hot and cold acids • Lining of an equipment, reservoirs and building structures • Vessel heads • Vaults and diaphragms • Pickling baths for metallurgical plants
LEM[9]	• Rubber sheet linings • Pickle pipeline and tanks • Liquid complex fertilizers • Coating of bottom of an automobile • Protection from corrosive salts • Line mixing tanks • Refrigerant bottom boxes
Dendrepox^{TM} Polyamide	Highly effective hardener for epoxy compositions

Source: [1, 3, 35]. Reprinted from Ref. [1] with permission from *Engineering Journal of Don*.
[1] Styrene ethylbuthylene styrene. [2] Polyurethane. [3] Polypropylene. [4] Ethylene propylene.
[5] Ethylene propylene diene monomer. [6] Hybrid nonisocyanate polyurethane.
[7] Polymer concrete based on vulcanized polybutadiene matrix.
[8] Polymer concrete based on organic-silicate matrix. [9] Liquid ebonite mixtures.

by Nanotech Industries, Inc., and Polymate Ltd.-INRC (USA and Israel). The companies received "2015 Presidential Green Chemistry Challenge Award." From the official statement about getting the award: "*As a recipient of this prestigious award, you are distinguished at the national level as in innovator in green chemistry.*"

1.2 Nanostructured Liquid Ebonite Composition for Protective Coatings

1.2.1 Introduction

Rubber covering (rubberizing) finds wide application in chemical and other industries where it is used to provide corrosion protection for apparatus, equipment, and pipelines. However, the use of sheet rubber and traditional liquid rubberizing compounds, based on polychloroprene (Neoprene), polysulfide (Tiokol), polyurethane, and other cauotchoucs, is not possible, to ensure longevity, reliability, and effective protection. It demands adhesive substrate and fails to afford safety because of organic solvents present in the compounds. LEM are protective lining products whose function is superior to the current rubber products that are used in the marketplace. LEM is a cost-effective and functional solution for rubber-coating applications.

The novel liquid ebonite mixtures (LEM) for rubberizing based on linear low-molecular polybutadiene containing nanosize black carbon fillers allow to get rid of the deficiencies intrinsic to conventional rubber sheet and liquid rubberizing compounds.

With value-added design and engineering flexibility, LEM has the ability to effectively cover hard-to-reach places, successfully coating the complex surfaces of such areas as the mesh of sieves otherwise considered impossible to coat with conventional rubber sheets. With increased chemical resistance, LEM maintains material integrity by assuring long-term durability and effective corrosion protection. In addition to high performance properties, material production of LEM meets environmental safety standards, further elevating LEM as an advanced product that outperforms conventional rubber coatings and coverings. The cost–benefit factors based on anticorrosion reliability and application functionality will satisfy the demands of the high-performance coating industry.

Depending on the purpose and the type of rubber, used as a base, the LEM may be one- or double-pot composition. The one-pot LEM is low-viscosity thrixotropic composition, solvent-free, and hence safe in handling; their shelf life by 20–30°C is practically unlimited. The double-pot LEM are high-viscosity compositions intended to make thick-layer (up to 2.5 mm) coverings.

LEM is based on oligobutadienes without functional groups which are liquid polymers with a hydrocarbon structure and a high degree of non-saturation. The oligobutadienes are characterized by low values in molecular weight, the contrary to similar structured high-molecular elastomers. Oligomers of this type are mostly used as plastifying rubber mixtures based on high-molecular rubbers [38] and as a film-forming base of lacquer or paint coatings, for the complete or partial replacement of plant raw material [39].

We explored the possibility of preparing elastic and hard ebonite coatings, utilizing the properties of oligobutadienes without ending functional groups [40–43].

1.2.2 Structure and Properties of Oligobutadienes

Oligobutadienes without functional groups are similar in a structure and microstructure of its links to high-molecular butadiene rubbers (polybutadienes) and contain links of three different types:

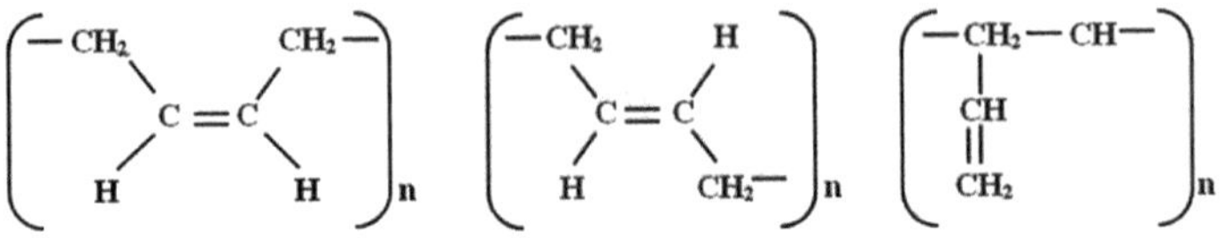

1,4 cys   1,4 trans   1,2

Depending on the conditions of polymerization and the kind of catalyst we can synthesize the oligobutadienes with various microstructure and molecular weight. The effect of this kind of catalyst is shown in Table 1.4 [37, 44].

Table 1.4 Dependence of liquid oligobutadiene's microstructure on kind of catalyst

Process of polymerization	Catalyst	Structure of polymer (%)		
		1,4-cis	1,4-trans	1,2
	Sodium, potassium	30–40	10–30	30–50
	Lithium		40	20
Anionic	Sodium, potassium + ether	—	15	85
	Lithium-butyl + ether		8	92
Kationic	Boron fluorides	≤5	70–90	10–25
Radical	Potassium persulfate or cumene hydroperoxide	≤40	30–60	10–30

Source: [37, 44]

The study of the rheological properties of more viscous polybutadiene oligomer (NMPB[a]) and less viscous (PBN[b]) shows the decisive

[a]Loss-molecular polybutadiene rubber, Russian Standard (TU) 38.103290-75.
[b]Loss-molecular polymer, Russian Standard (TU) TU 38.103641-87.

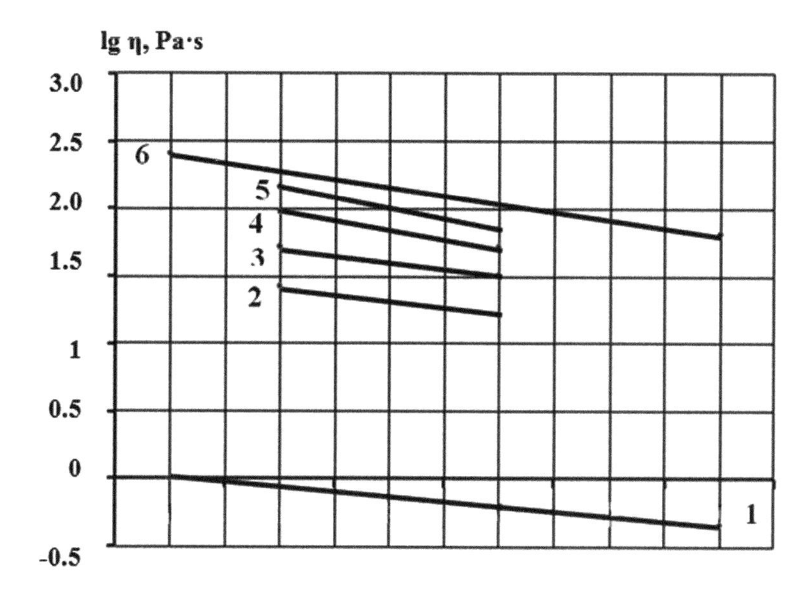

Figure 1.8 Dependence of viscosity η of NMPB and PBN mixtures (% mass) on temperature T: 1, 0/100; 2, 30/70; 3, 40/60; 4, 50/50'; 5, 75/25; and 6, 100/0 [37, 44].

influence of the latter, which can be explained by the role of low-viscosity polybutadiene oligomer as an intramolecular plasticizer. At the same time dependence of the viscosity of mixtures NMPB and PBN on temperature points to the influence of high-viscosity oligobutadiene (Fig. 1.8) [37, 44].

Thixotropic structuring reflects the ability of rubberized composition to form a physical and chemical structure of the relatively thick coating layer applied to the protected surface. The process of thixotropic recovery of the destroyed structure in the state of rest is characterized by an increase of strength in time. Modification compositions based on oligobutadiene by carbon black or surface-active substances such as lecithin, increase the thixotropic properties of the filled composition. Loading-unloading curves of compositions based on SKDNN[a] filled with a modified carbon black show a hysteresis loop; the loop area is far of excess of the hysteresis

[a] Loss-molecular cis-polybutadiene rubber, Russian Standard (TU) 38.103515-82.

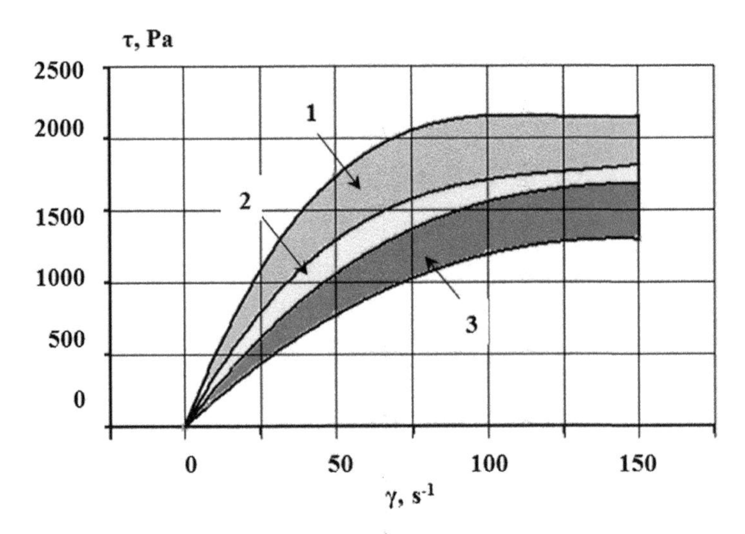

Figure 1.9 Hysteresis deformation loop of the composition based on SKDNN with no modification and modified by carbon black as a filling agent. The time of the deformation 140 s^{-1} is 20 min. 1, Carbon black modified by n-hexane; 2, carbon black modified by lecithin; and 3, unmodified carbon black [37, 44].

loop for the composition, modified lecithin, and even more so for non-modified composition (Fig. 1.9) [37, 44].

1.2.3 Vulcanization

The vulcanization of low-molecular polymers of butadiene is traditionally carried out using the same vulcanizing systems, which are employed in the vulcanization of analogous high-molecular butadiene rubbers, i.e., by sulfur with accelerators, organic peroxides, parachinonedioxime combined with lead, or manganese dioxides, among others. Rubber-like vulcanized products prepared with such systems are characterized as having insufficient mechanical durability and elasticity, caused by the high number of defects in the rare vulcanization net due to the low length of molecular chains of the oligomer (Fig. 1.10).

The study examined the vulcanization processes leading to formation of rubberizing ebonite coatings on samples of oligob-

Figure 1.10 Solidified sample of LEM.

utadienes of various molecular parameters and microstructures (Table 1.5) [37, 41].

Vulcanization was carried out by high-frequency current using sulfur with accelerators (bithiocarbamates, thiazoles, and thiurames) in the temperature interval of 100–170°C to form ebonites. The ebonites were evaluated according to the value of durability characteristics against the following: strain, firmness, content of bonded sulfur in the vulcanite, relative density of the vulcanite net, and the value of swelling in physically and chemically aggressive

Table 1.5 Principal characteristics of studied oligomers

Samples	Molecular weight Mn	Microstructure percentage of links			Viscosity (Pa·s)	Non-saturation (against iodine-number)
		1,4-cis	1,4-trans	1,2		
1	740	25	19	56	1.4	343.3
2	2,140	40	45	15	1.0	401.7
3	1,750	24	34	42	1.3	414.5
4	2,350	35	44	21	1.1	422.3
5	2,130	75.5	23	1.5	1.2	440.4

Source: [37, 41]

Table 1.6 Dependence of quantity of bound sulfur in the vulcanizates based on oligobutadiene (OB)

Quantity of introduced sulfur		Quantity of bound sulfur (% mass) at 150°C Time of vulcanization (hours)							
Mass parts per 100 mass parts of OB	**% mass**	**1**	**2**	**3**	**4**	**5**	**6**	**8**	**10**
20	16.7	37.4	67.7	—	69.2	—	66.0	68.1	
30	23.0	47.8	71.2		72.2		75.7	77.4	73.5
40	28.5	50.8	76.0	72.5	85.1	75.4		—	
50	33.3	47.8	79.0	54.7	99.8	98.8	99.7		—
70	41.2	61.1	78.0	63.5	63.4		—		

Source: [37, 44]

media and, following the elaboration of protective coatings their adhesion to carbon steel. The technical findings are discussed throughout.

In general terms vulcanization reaction of 1,4-cis-oligobutadiene at 150°C can be described by the following equation [44]:

$$\ln(S_{bb} - S_{cb}) = -K\tau + \ln S_{bb},$$

where S_{bb} is the amount of sulfur introduced into the oligobutadiene, % by mass weight; S_{cb} the amount of sulfur bonded with oligobutadiene in the course of vulcanization, % by mass weight; K the constant of speed of vulcanization; and $\tau$the duration of vulcanization time, hours. The amount of bounded sulfur during the vulcanization depends on the process time presented in Table 1.6 [37, 44].

Organic accelerators such as thiuram, dithiocarbamate, thiazole, guanidine, etc., introduce in ebonite composition for acceleration of sulfur vulcanization process (Table 1.7) [37, 44].

For the comparative estimation of the ability of oligomers with a different structure to vulcanize, studies were carried out on model mixtures with sulfur content 23% of the mass, at a temperature of 150°C. After sodium-initiated polymerization, the oligobutadiene sample 1 displayed low values of the molecular weight and non-saturation, and a high content of vinyl links. Results characterize the sample as very slow vulcanization, low durability of ebonites, a low content of bonded sulfur and considerable swelling in hexane.

Table 1.7 Main accelerators of vulcanization

Chemical name	Chemical formulae	Technical name
Zinc diethyldithiocarbamate	S ‖ $[(C_2H_5)_2N-C-S-]_2Zn$	Ethyl zimate®
Zinc dimethyldithiocarbamate	S ‖ $[(CH_3)_2N-C-S-]_2Zn$	Methyl zimate®
Tetramethylthiuram disulfide	S ‖ $[(C_2H_5)_2N-C-S-]_2$	Thiuram
Diphenylguanidine	$[(C_6H_5)NH-]_2C=NH$	Guanide
2-Mercaptobenzothiazole	N ⁄ ⫽ C_5H_4 C-SH ⧹ ⁄ S	Kaptax

Source: [37, 44]

Oligobutadienes from the lithium-initiated polymerization (samples 2–4, Table 1.5) contain mainly linear monomeric links. They are less saturated and therefore vulcanize more actively. Like oligobutadienes from sodium polymerization, oligomers of this type have a larger induction period before the start of vulcanization.

Out of the samples tested, sample 5—the vulcanization of oligobutadiene synthesized on nickel catalyst—was the most effective with the least period of induction. The structure of this sample is mostly linear with predomination 1, 4-cis links, characterized by a high-degree of non-saturation. Vulcanizates prepared from this oligobutadiene are very durable, contain a high concentration of bonded sulfur (17%) and display nearly no swelling in hexane, which means a high number of crosslinks in the oligomer. The microstructures of the monomeric links, confirmed by the roughly equal incline of the curves in Fig. 1.11 [37, 41] for all oligobutadiene samples, were shown to have little influence on the vulcanization rate in its active period (Table 1.8) [37, 41].

The increased dose of sulfur in the composition treated by vulcanization does not change the general influence character of the microstructure of the oligobutadiene links on its vulcanization.

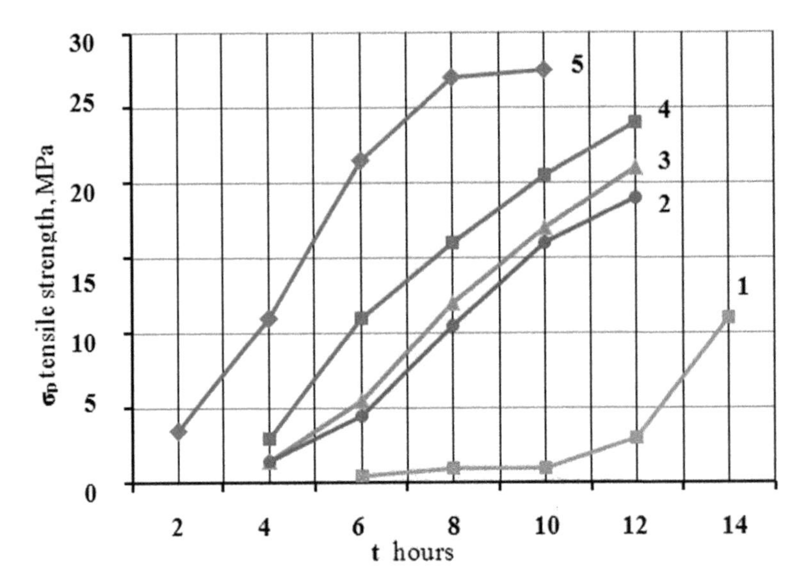

Figure 1.11 Tensile strength of sulfur vulcanizates based on various oligobutadienes (Table 1.2) depending on the duration time of the vulcanization at 150°C where the initial dose of sulfur in the composition is 23% of the mass [37, 41].

Vulcanization of 1,4-cis oligobutadiene is much more active than that of 1,2 oligobutadiene with a sulfur content of 23% or 33% of the composition mass.

The durability dependence, bonded sulfur content in the vulcanizate, and the relative density of the vulcanization net of the duration time of vulcanization of compositions based on high-active 1,4-cis oligobutadiene are shown in Fig. 1.12 [37, 41]. If the content of sulfur

Table 1.8 Oligobutadiene influence on the properties of vulcanizates prepared from a composition with 23% of its mass, sulfur

Parameter	**Sample number according to Table 1.5**				
	1	**2**	**3**	**4**	**5**
Tensile strength (MPa)	1.5	8.4	10.5	16.8	25.5
Swelling in hexane at 20°C during 10 h (%)	5.7	1.3	4.9	4.0	0.0
Content of bonded sulfur in the vulcanite, mass (%)	11.0	12.2	12.1	12.2	17.0

Source: [37, 44]

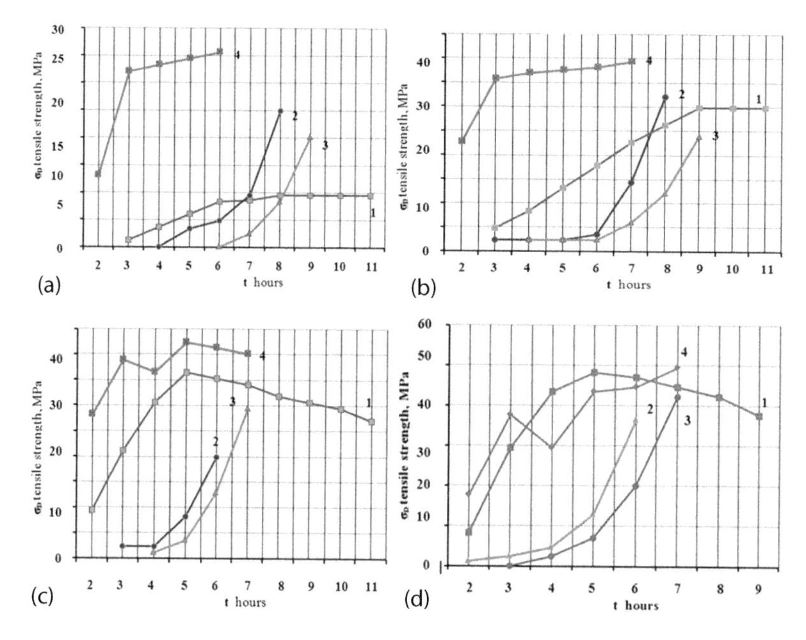

Figure 1.12 Correlation of the tensile strength in MPa (1), relative density of the vulcanizate net from the equilibrium swelling in hexane at 20°C (2) and 50°C (3), and the content of bonded sulfur in vulcanizate (4) with sulfur's initial dose—(a) 16.7%, (b) 23.0%, (c) 28.6%, and (d) 33.3%—versus the vulcanization duration time at 150°C [37, 41].

is less than 16.7% of the mass, the rate of vulcanization reduces and the prepared vulcanizate display low durability (Fig. 1.5a). The content of bonded sulfur in the vulcanite prepared from such a mixture seems high enough just after 2 h of vulcanization when the durability of the vulcanite and the relative density of the net are still very low.

Similar results have been obtained when the dose of sulfur in the composition increased to 23% of the mass that corresponds to 30% of the mass from the amount of oligobutadiene (Fig. 1.12b). The same results were obtained by increasing the dose of sulfur to 28.6% and 33.3% of the mass (Figs. 1.12c,d).

The increase in the duration time of vulcanization does not lead to a significant change in the amount of bonded sulfur interacting with oligobutadiene during the first 2 h; however, the durability of the vulcanite and the density of the vulcanization net seriously rise. Moreover, with a high enough dose of sulfur (28.6 and 33.3%), the increase in vulcanite durability of and the density of the vulcanization net in the period between two and 4 h of the vulcanization, may be accompanied by the reduction of the content of bonded sulfur after several moments.

The obtained experimental data prove that in the first stage of vulcanization, sulfur joins oligobutadiene to form rare polysulfide bonds which do not assure enough density of the vulcanization net and the vulcanite durability. In the second stage of vulcanization the polysulfide bonds are broken up on account of their insufficient durability, and transformed into bonds with a lower content of sulfides while the released sulfur forms new horizontal low-sulfide bonds that raise the density of vulcanization net. The results indicate that the process mechanism of the vulcanization of low-molecular polybutadiene by sulfur with the forming of ebonite, occurs in the following two stages [41]:

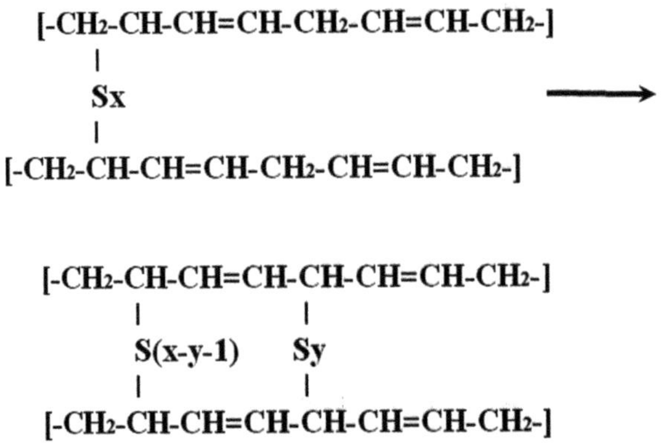

[-CH₂-CH-CH=CH-CH₂-CH=CH-CH₂-]
|
Sx ⟶
|
[-CH₂-CH-CH=CH-CH₂-CH=CH-CH₂-]

[-CH₂-CH-CH=CH-CH-CH=CH-CH₂-]
| |
S(x-y-1) Sy
| |
[-CH₂-CH-CH=CH-CH-CH=CH-CH₂-]

The stage may be accompanied by partial inner-molecular capture of sulfur by oligobutadiene, the traditional explanation of the thermoplasticity of ebonite [41]:

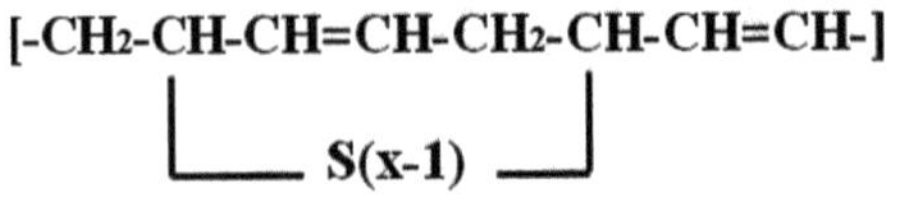

[-CH₂-CH-CH=CH-CH₂-CH-CH=CH-]
└── S(x-1) ──┘

The amount of sulfur within the mixture (a), as well as the amount of bonds in the vulcanization sulfur (x) is increased. Within the first 2 h, or the first stage of vulcanization, when rare polysulfide bonds are formed the amount of bonded sulfur a is proportional to the amount of sulfur in the composition b. For mixtures of 1,4-cis oligobutadiene with a sulfur content $b = 16.6$–28.6% of the composition mass, vulcanization process at 150°C is illustrated by the correlation [41]:

$$a = K \cdot b,$$

where K is the tangent of the angle of the incline of the graph against the abscissa axe ($K = 0.853$), and b corresponds to the inactive part of sulfur ($b = 2.56$).

The values of the constants of reaction rate for the vulcanization of the mixtures for this oligobutadiene, is found from experimental data, on the assumption that the initial capture of sulfur by oligobutadiene (during the first 2 h) satisfies the first order equation [41]:

$$\ln(a - x) - K/\ln a$$

The observed correlation provides evidence that in the initial stage, the formation process of vulcanization at a constant temperature is not limited by the occurrence of active centers in the oligomer. Furthermore, the vulcanization rate depends only on the concentration of sulfur in the composition if its dose varies from 16.7% to 33.3% of the mass. When the amount of sulfur in the composition is increased to 41.2% of the mass, its capture may be limited by the presence (or absence) of active centers in the molecular chain of the oligobutadiene.

1.2.4 Strength and Hardness of Vulcanizate

As a result of sulfuric vulcanization of the high-viscosity grade 1.2-oligobutadiene SKDSN[a] high-strength vulcanizate was obtained, without degradation signs due to the high thermal stability of the vinyl links. (Fig. 1.13) [37.44].

---

[a]Synthetic rubber, Russian Standard (TU) 38.103331-84.

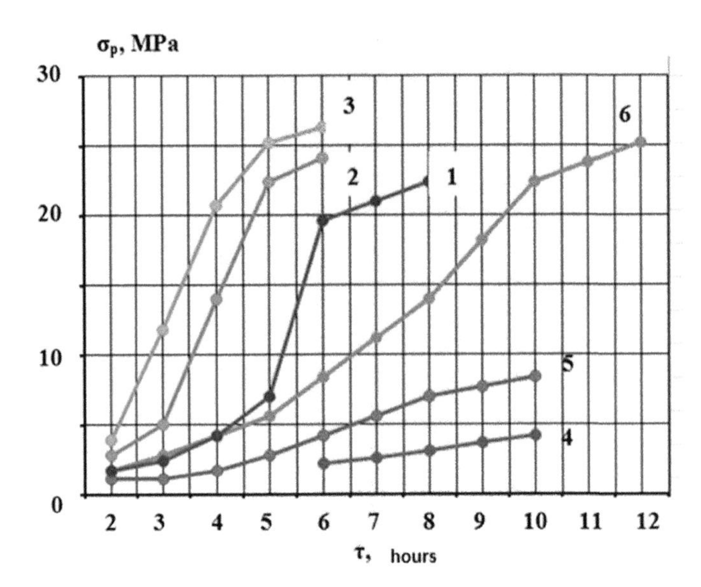

Figure 1.13 Dependence of strength of the vulcanizates (σ_p) based on oligobutadienes SKDSN (curves 1–3) and PBN (curves 4–6) on time of vulcanization (τ) at 150°C. Sulfur quantity (% mass): curves 1 and 4, 23%; curves 2 and 5, 28.6%; curves 3 and 6, 33.3% [37, 44].

The temperature of vulcanization significantly influences the strength of the vulcanites and adhesive properties of ebonite coatings. At 100°C, without effective accelerators, the sulfur vulcanization of the studied oligobutadienes cannot be carried out. Using ultraaccelerators like zinc biethylthiocarbamate or tetramethyltiurambisulfide, enables the vulcanization of oligobutadiene to form ebonite coatings with enough durability and chemical stability coatings at 100°C within 20–25 h of vulcanization.

Raising the temperature of vulcanization from 125 to 175°C leads to a near linear increase of the tensile strength of vulcanites prepared from compositions on the base of 1,4-cis oligobutadiene with sulfur without accelerators for the vulcanization duration of 6–8 h (Table 1.9) [37, 41].

The correlation between the temperature and the adhesion value of the coatings, prepared from the same compositions, is characterized by obvious extreme values, the maximum being at 150°C.

Table 1.9 Durability and adhesion of vulcanites and coatings prepared from compositions with a sulfur content of 23.3% of the mass

Vulcanization duration (hours)	Characteristics at the vulcanization temperature T (°C)					
	Tensile strength (MPa)			Adhesion at breaking (MPa)		
	125	150	175	125	150	175
2	1.2	4.8	20.2		2.3	1.2
4	1.3	10.2	29.6	<0.1	3.1	0.8
6	2.4	18.8	38.4		4.8	1.2
8	3.5	25.2	51.3		8.0	2.0

Source: [37, 41]

Table 1.10 [37, 44] shows the increase in hardness of vulcanizates based on oligobutadiene samples NMPB with a viscosity 29–311 Pa·s at a low-temperature vulcanization (20–25°C) in the presence of 1,4 benzoquinone dioxime and manganese dioxide. Rubbery vulcanizates get quite stable properties after 5–10 days of outdoor exposition.

Table 1.10 Hardness of the rubberlike vulcanizates based on oligobutadiene NMPB with 1,4 benzoquinone dioxime and manganese dioxide

Viscosity at 25°C Pa·s	Hardness (Shor A)					
	Cure time in air (days)					
	1	2	4	6	8	10
29	0	12		44	53	60
87	15	25	30	43	60	62
122	17	18		40	47	60
140	7	23	43	53	58	
165	15	27	42	56	60	
311	12	30	29	35	57	

Source: [37, 44]

Comparative properties of vulcanizates obtained at room temperature based on NMPB with use of the following vulcanizing systems are presented in Table 1.11 [37, 44]:

- 1,4 benzoquinone dioxime (1,4BD) with manganese dioxide (MD)
- sulfur with zinc diethyldithiocarbamate (ZDDC)

Table 1.11 Properties of vulcanizates and covering from compositions based on oligobutadiene NMPB produced at temperature 20–25°C

	Vulcanization agents and their dosage (mass parts per 100 mass parts of NMPB)		
Property	**1,4BD(5) +MD(15)**	**Sulfur (30) +ZDDC (15)**	**BP (4)**
Tensile strength (MPa)	0.2–0.4	-0.2–0.5	4.8
Eelongation per unit length	80–120	130–140	70–90
Rersidual elongation	10–25	8–10	3–5
Hardness (Shore A)	58–62	32–35	70–82
Change of mass at 20°C in hexane (%)	-23.9	-29.7	-21

Source: [37, 44]

- benzoyl peroxide (BP)

By increasing the dosage of sulfur (50 mass parts per 100 mass parts of oligobutadiene) and postvulcanization at 120–150°C can achieve a significant increase in the strength of vulcanizate (Table 1.12). The viscosity of the oligobutadiene has a significant impact on the strength of the vulcanizate obtained by the two-step process (Table 1.13) [37, 44].

1.2.5 Performance Characteristics of Vulcanizates

To investigate the principal application properties of ebonite vulcanizers and coatings, the study examined prepared samples based on the most active vulcanizable by sulfur 1.4-cis oligobutadiene with the optimal receipt and vulcanization regime that allowed a gain in the upper density of the vulcanization net characteristics in durability and adhesion. System properties are as follows.

1.2.5.1 Thermomechanical properties

Thermomechanical curves of ebonite vulcanizates in coordinates "relative lengthening vs. temperature" are monotonically rising curves (Fig. 1.14) which do not have a zone of glass transition, that proves the thermoplasticity of these polymeric materials. Using carbon black as a filler, the thermodeformation ability of ebonite

Table 1.12 Strength of vulcanizates based on the mixtures of oligobutadienes NMPB and SKDNN during the after-vulcanization process at 120–150°C

No.	NMPB/SKDNN (mass parts)	Dosage of the agents of sulfur vulcanization (mass parts per 100 mass parts of the mixture)		Strength of vulcanizate (MPa)	
		Sulfur	**Accelerator**	**150°C, 6 h**	**120°C, 20 h**
1	95/5	30	—	9.6	7.3
2	90/10	50	—	26.1	6
3			Kaptax (2)	30.5	24.6
4	95/5		Kaptax (2)	34.0	28.9
5			Kaptax (3)	35.5	29.0
6	90/10		Kaptax + Thiuram (2+2)	35.7	29.7
7	95/5		Kaptax + Methyl Zimat®	35.3	26.9

Source: [37, 44]

Table 1.13 Tensile strength of the vulcanizates based on NMPB

Viscosity of NMPB (Pa·s)	Strength (MPa) during time of vulcanization (hours)				
	2	4	6	8	10
27	10.0	19.8	21.4	23.2	28.4
208	6.5	12.3	15.0	17.3	17.4
465	3.5	4.8	9.2	12.5	16.8

Source: [37, 44]

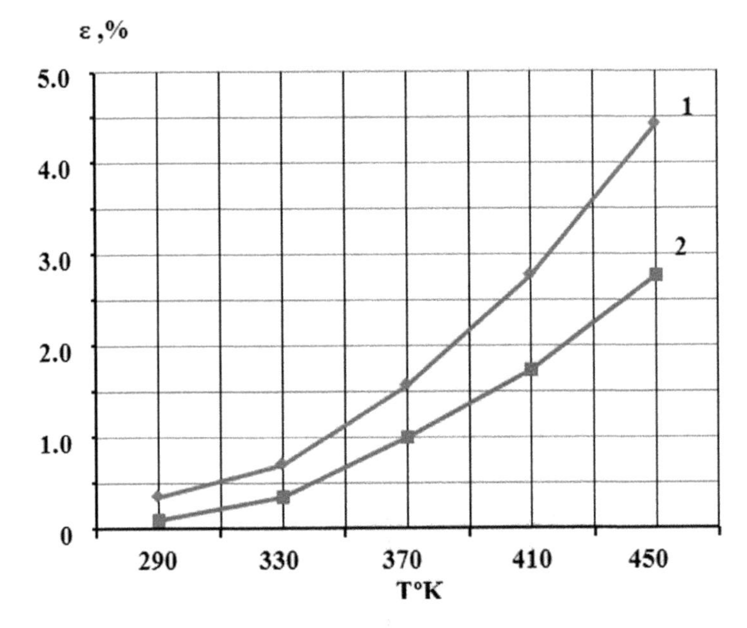

Figure 1.14 Thermomechanic curves for nonfilled (1) and filled (2) ebonite vulcanizates [37, 41].

vulcanite slightly decreases in comparison with a non-filled ebonite vulcanite [41].

1.2.5.2 Sorption and diffusion properties

Where water is considered the most widespread component of aggressive media acting, the study examined its effect on ebonite vulcanizates and coatings prepared from compositions based on oligobutadienes.

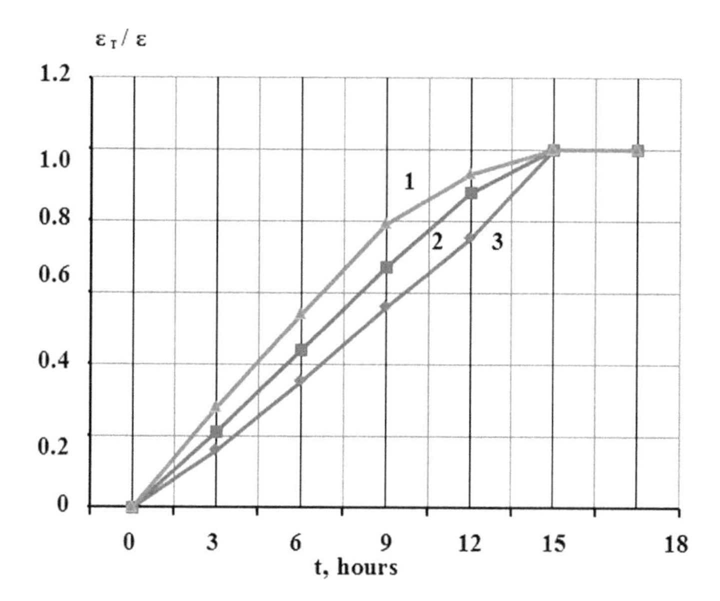

Figure 1.15 Correlation between the relative increasing of the linear deformation of ebonite and duration time of its staying in water at 80°C: 1, without filler; 2 and 3, with active carbon PM-15, 9, and 20 mass %, respectively [37, 41].

The kinetic curves of water sorption by ebonite vulcanizates have a long-lasting saturation plate. Involving the inert filler (technical carbon) reduces the amount of water sorbed in equilibrium. The initial parts of deformation curves of water sorption are straight lines (Fig. 1.15) representing the Fick-like diffusion of water into ebonite vulcanizates. The study of the correlation of water-sorption of ebonite vulcanizates, and temperature between 20 and 80°C, revealed that the values of the diffusion coefficient well satisfy the Arrhenius equation [41].

One can see that values of the activation energy during the process of water diffusion in ebonite vulcanizates do not significantly depend on the presence of filler in the composition, which are 49–53 kJ/mole only.

The weakening of the adhesive bond "metal-ebonite coating," is observed during absorption of water after a certain period of delay time; its numerical values are given in Table 1.14 [37, 44].

Table 1.14 Delay time of water penetration into ebonite covering

Composition	Time delay (hours) for thickness of the covering (mm)		
	0.2	1.0	2.0
No fillers	38.5	963	3861
PM (20% mass parts)	464	11583	46389

Source: [37, 44]

1.2.5.3 Chemical resistance

The density of the vulcanization net and the structure of vulcanization bonds are the main influence on the chemical stability of ebonite vulcanizates, which is determined by the type of vulcanization system selected and the conditions of carrying out vulcanization. Accelerators present in the vulcanizate increase chemical stability by reducing the amount of sulfides in the horizontal bonds as their density rises.

The chemical stability of the reagents also influences the chemical stability of ebonite vulcanizates. The comparison of the correlation between the rising of the mass of the sorbed liquid and the duration time of ebonites exposed in mineral acid solutions provides supporting evidence to conclude that their initial parts for water sorption are straight lines. The value of the coefficient of diffusion in acid solutions is equal to that of water and does not depend on their chemical composition. Nevertheless, after gaining saturation, water solutions of nitric and sulfuric acids exhibit a reduction of mass increase, meaning that chemical changes cause the removal of products for chemical reactions. The study found that nitric acid has a stronger reaction with ebonite vulcanizates, which can be attributed to its oxidizing ability. Solutions of hydrochloric and phosphoric acids cause positive swelling of ebonite vulcanizates [41].

Chemical resistance of the vulcanizates obtained at 150°C with use various types of oligobutadienes as a vulcanized basis is little different from each other (Table 1.15) [37, 44].

NPBM-oligobutadiene vulcanizates produced with use of mixed vulcanizing system consisting of (mass parts per 100 mass parts of NPBM)—1,4 benzoquinone dioxime (5 mass parts); Manganese

Table 1.15 Chemical resistance samples of vulcanizates based on oligobutadienes modified by thermoelastoplasts (700 h testing at 20°C)

Bonding base			Change of mass of the samples in media (%)		
Thermoelastoplasts	**Oligobutadiene**	**Hardness (Shore A)**	**50% H_2SO_4**	**35% HCL**	**20% NaOH**
Styrene-Butadiene Thermoplastic Elastomer DST-30R	Polyoil-110	96–100	0.4	1.2	0.4
			0.1	0.9	0.2
	Polyoil-130	94–98	0.3	2.1	1.5
—	SKDSN	89–92	0.5	4.0	2.3
Oil-extended thermoelastoplast TRS 75/90	Polyoil-110	81–85	0.8	29*	8*
	Polyoil-130	92–96	1.0*	13*	2.3*
		97–100	0.2	8.2*	1.7*
—	SKDSN	90–91	0.5	7.9	2.3
		97–100	0.3	4.5	4.1
Thermoelastoplast TRS 75/90, no oil	Polyoil-110		0.7	3.5	0.4
	Polyoil-130		0.8	3.3	0.6
—	SKDSN	95	0.4	3.7	
		96–100	0.2	2.6	0.5

Source: [37, 44]
*Increase of the mass is continued.

Table 1.16 Chemical resistance samples of vulcanizates based on NPDM oligobutadiene with application of mixed vulcanization system

Aggressive medium 20°C	Time of after-vulcanization at 150°C (hours)	Change of vucanizate mass during exposition (days)			
		15	30	45	60
20% NaOH	2	0.8	1.1	1.2	
	6	0.3	0.4	0.45	0.48
	8	0.2		0.3	
	10				
20% H_2SO_4	2	0.43	0.62	0.71	0.79
	6	0.35	0.45	0.56	0.60
	8	0.25	0.38	0.42	0.43
	10	-0.07	0.08	-0.08	
20% HCl	2	0.4	0.58	0.61	0.62
	4	0.38	0.5	0.60	
	6	0.65	0.42	0.43	
	8	0.18	0.27	0.32	
	10	0.1	0.16	0.17	0.19

Source: [37, 44]

dioxide (15 mass parts); Sulfur (30 mass parts); Technical carbonate (15 mass parts)—show satisfactory chemical resistance at a vulcanization temperature of 150°C (Table 1.16) [37, 44].

The high chemical resistance of NPDM vulcanizates is provided at 100°C (Table 1.17), in case of application of the vulcanization accelerator Ethyl Zimate® [37, 44].

The structure of the composition (mass parts per 100 mass parts of NPBM)is as follows:

- Ethyl Zimate® (15 mass parts)
- Sulfur 30 (mass parts)

Chemical resistance of the ebonite vulcanizates based on the various oligobutadienes modified by divinyl-styrene thermoelastoplast (48% of mass) is shown in Table 1.18 [37, 44].

Increasing of chemical resistance of LEM-coverings is achieved by adding oxide-contained additives with are interacting with diffusing aggressive media and formatting new phases of high-strength crystallohydrate complex-nonorganic glue-cement (Fig. 1.16) [37, 40].

Table 1.17 Chemical resistance samples of vulcanizates based on NPDM oligobutadiene with application Ethyl Zimate®

Aggressive medium 20°C	Time of after-vulcanization at 100/150°C (hours)	Change of vucanizate mass during exposition (days)			
		15	30	45	60
20% NaOH	2/2	−4/0.4	−6/0.6	−7/0.65	−8/0.66
	5/4	0.9/0.4	1.3/0.45	1.6/0.55	1.9/0.54
	10/6	0.5/0.18	1/0.3	1.1/0.4	1.2/0.45
	15/8	0.5/0.1	1/0.19	1.1/0.21	1.2/0.22
	20/10	0.4/0.1	0.6/0.19	0.9/0.21	0.9/0.22
20% H_2SO_4	2/2	−1/0.1	−1.6/0.15	−2/0.18	−2.2/0.2
	5/4	0.2/0.09	0.25/0.1	0.23/0.17	
	10/6	0.2/0.08	0.24/0.09	0.22/0.11	0.26/0.18
	15/8	0.21/0.09	0.25/0.13	0.24/0.13	0.25/0.15
	20/10	0.2/0.08	0.26/0.09	0.25/0.17	0.25/0.16
20% HCl	2/2	3.8/0.8	6/0.45	7.9/0.62	9/2
	5/4	0.04/0.35	0.05/0.41	0.05/0.56	0.05/0.6
	10/6	0.04/0.3	0.05/0.38	0.05/0.55	0.05/0.5
	15/8	0.2/0.27	0.4/0.36	0.6/0.43	0.6/0.47
	20/10	0.2/0.25	0.4/0.32	0.6/0.4	0.6/0.42

Source: [37, 44]

Table 1.18 Chemical resistance samples of vulcanizates based on same NPDM oligobutadienes modified by divinyl-styrene thermoelastoplast during exposition in 35% sulfur acid at 20°C (vulcanization at 150°C)

Kind of oligobutadiene	Change vulcanizate mass during eexposition (days)					
	2	4	10	14	35	261
Polyoil-110	0.46	0.47	0.49	0.60	1.0	2.2
Polyoil-110	1.0	1.24	1.48	1.76	2.5	6.0
SKDSN	1.1	1.76	2.24	3.25	4.3	9.2

Source: [37, 44]

1.2.5.4 Adhesion stability of ebonite coatings

The effectiveness and strength of ebonite coatings are determined by their adhesion properties, not only at the time of application but over time after exposure to various factors such as aggressive medium, temperature, and electrical field, among others.

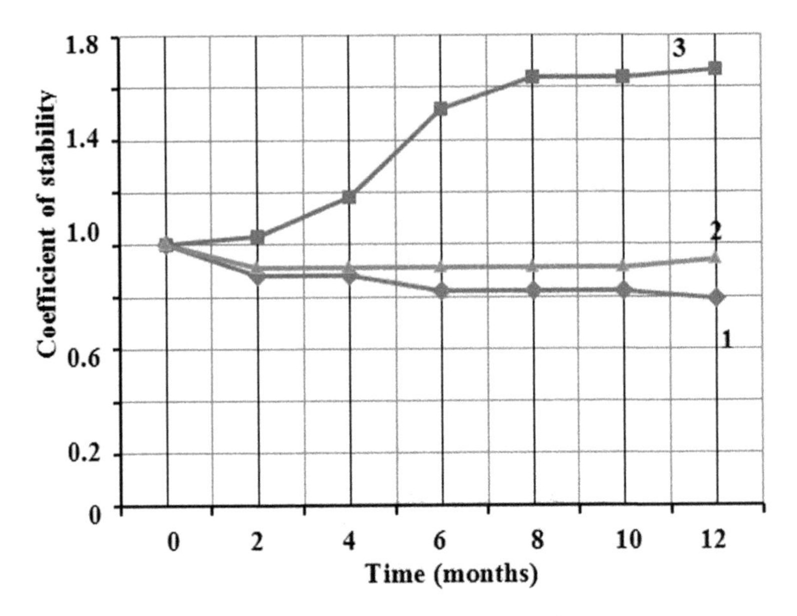

Figure 1.16 Dependence of chemical resistance of LEM coverings on oxide-contained additives: 1, without adding, 2, 2% of additives, 3, 7% of additives [37, 40].

The adhesion of ebonite coatings is significantly influenced by internal resting tensions following the thermal vulcanization and causing compression by 3.2–4.5%. The optimal regime of vulcanization results in the initial value of adhesion to most steels at approximately 8–12 MPa (for breaking). With water sorption, the weakening of the adhesion bond "metal-coating" is observed as a result of "retard" due to a long duration time of water penetration to the interphase surface.

When the speed of water transportation on the frontier "metal-ebonite coating" was measured, it was found that in such a case, water transportation is not through the capillary, but rather following the diffusive law, or the value of the diffusion coefficient, which is calculated from the water penetration constant. This parameter shows that the kinetics of the change of the adhesion durability of the ebonite coating in contact with water is determined by its diffusive properties, and one of the ways of stabilization of adhesion consists in the decrease of its diffusive properties. One way of stabilizing adhesion is by decreasing its diffusive permeability. This approach was considered as follows.

When water penetrates to the interphase surface of the ebonite coating, the adhesion decreases but is then restored after some time. For example, if the adhesion is decreased fast enough (within a few hours) the adhesion will be restored after a few weeks, or months. Testing found that once samples coated with the ebonite coating were completely dry, the adhesion fully returned to its original strength.

Thermoadhesiogrammes of adhesives prepared by the method "heating-cooling" exhibit an extreme temperature adhesion dependence. For temperatures below the critical range (140–144°C—about the melting point of sulfur), the curves display hysteresis, whereas for higher temperatures the curves appear close. The calculated value of energy activation for the process of adhesive bond breaking, when the coating is removed, is 97 kJ/mole [41].

The thermoadhesiogrammes of the joint "metal-ebonite coating" obtained in heating-cooling mode, show the extreme nature of adhesion depending on the temperature (Fig. 1.17). Below the critical temperature, strongly pronounced hysteresis is manifested, whereas above the critical temperature, the curves are almost identical [37, 44].

In many cases ebonite coating are intended to protect the inner surface of cylindrical apparatus. In this connection we consider the system in the form of a hollow circular cylinder with an ebonite coating on the inner surface taking into account the characteristic properties of the polymer at the interface with the metal (Fig. 1.18) [37, 44].

The calculation results of the internal stresses distribution in a circular sample ($R = 11$ cm) with an ebonite coating are presented in Table 1.19 [37, 44].

It should be noted that the layer of ebonite coating is compressed. Because of this, at defective places of the coating a phenomenon resembling the local buckling may be seen, resulting in the delamination of the structure.

The adhesion strength of the "metal-ebonite coating" system does not significantly change under electric field below the ultimate tension. The decreasing in adhesion takes place only in the zone

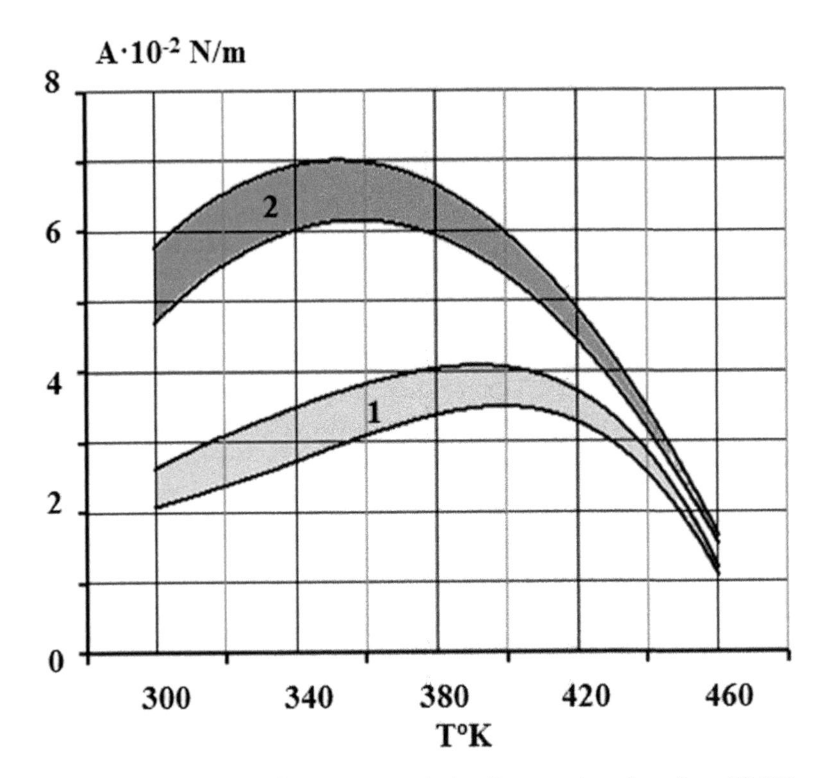

Figure 1.17 Thermoadhesiogrammes ebonite coverings based on NMPB (1) and mixture NMPB with SKDNN in the ratio 1:1 (2) [37, 44].

corresponding to the critical temperature. In this situation, there is a threshold of the field tension; its value depends on the sign of the potential applied to the coating and the coated material.

Table 1.19 Subsurface stresses into ring metal simples with inner ebonite covering

δ(mm)	Temperature (°C)	Subsurface stresses (10^5 Pa)		
		σ_R	σ_θ	σ_Z
1	70	-1.51	-84.5	
2	48	-1.10	-58.2	-59.4
3	28	-1.90	-32.4	-34.0

Source: [37, 44]

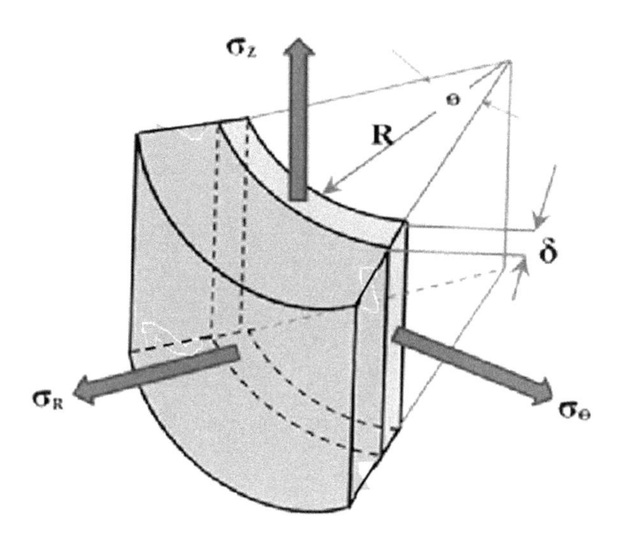

Figure 1.18 Ring sample of metal cylinder with inner ebonite covering [37, 44].

1.2.5.5 Electrochemical protection properties of ebonite coating

The results of electrochemical studies are in good agreement with measurements of adhesive strength. Coatings with high adhesion are characterized by the high protective properties. Increase of polybutadiene's viscosity in the system "metal-ebonite coating" tends to decrease the current of passivation resulting in increase of anticorrosion effect. Addition of functional hydroxyl- and carboxyl-containing groups in NMPB-based ebonite composition polymers improves both the adhesion and protective properties of the coatings.

Oligomers with oxygen-containing functional groups can be considered as surfactants capable to chemisorption interaction with the active centers of a metal surface. In response to the chemisorption of surfactant-oligomers, electric fields which arise in this area produce passive regions (Fig. 1.19) [37, 44].

The kind of vulcanizing system and the degree of ebonite coating vulcanization affect its electrochemical protective properties. Figure 1.20 [37, 44] demonstrates the dependence of the frequency dispersion coefficient K_0 of the anodic polarization (Fig. 1.19)

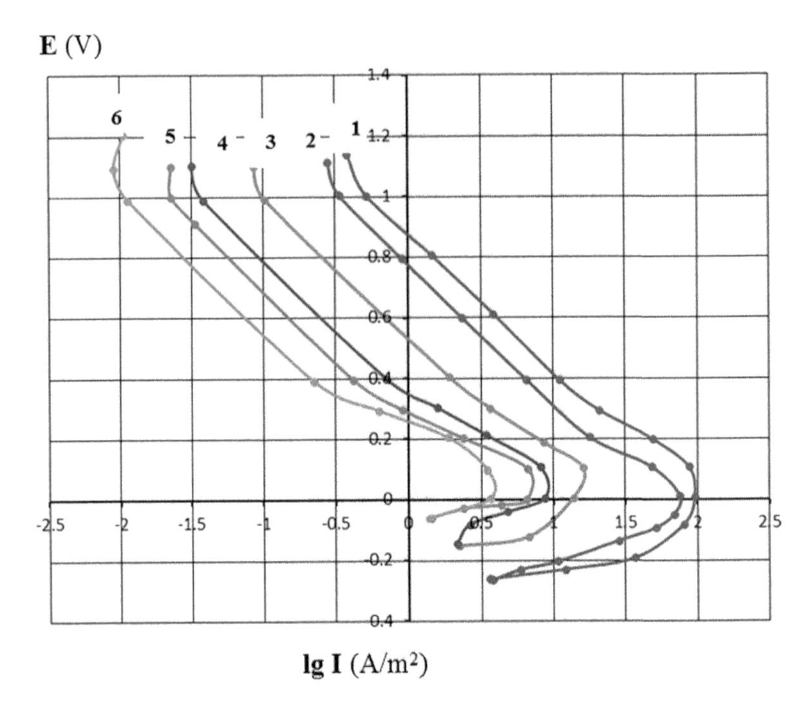

Figure 1.19 Curves of anode polarization of the low-alloyed steel with ebonite coating based on NPBM oligodiene. 1, No modificators; 2–6, with modificators containing following groups: 2, 1.17% OH; 3, 2.8% COOH; 4, 2.3% COOH; 5, 1.4% OH; 6, 1.5% OH in 0.1M H_2SO_4 [37, 44].

on exposure time ebonite coating based on oligobutadiene NMPB with thickness ±250 ±20 mkm vulcanized at temperature 150°C (coating no. 1)[a] and 100°C (coating no. 2)[b] in a few corrosive environments.

One can see that designer useful life ($K_0 = 1$) of the first coating is about 180 days exposition in solutions of hydrochloric and sulfuric acids whereas for second coating we have less than 130 days in 30% solution of sulfuric acid but more 200 days in 30% solution of hydrochloric acid. Therefore, at the deeper

---

[a]Structure of composition (mass parts): NMPB, 100; sulfur, 30; Kaptax, 3; carbon black, 20.

[b]Structure of composition (mass parts): NMPB, 100; sulfur, 30; Ethyl Zimate®, 103, carbon black, 20.

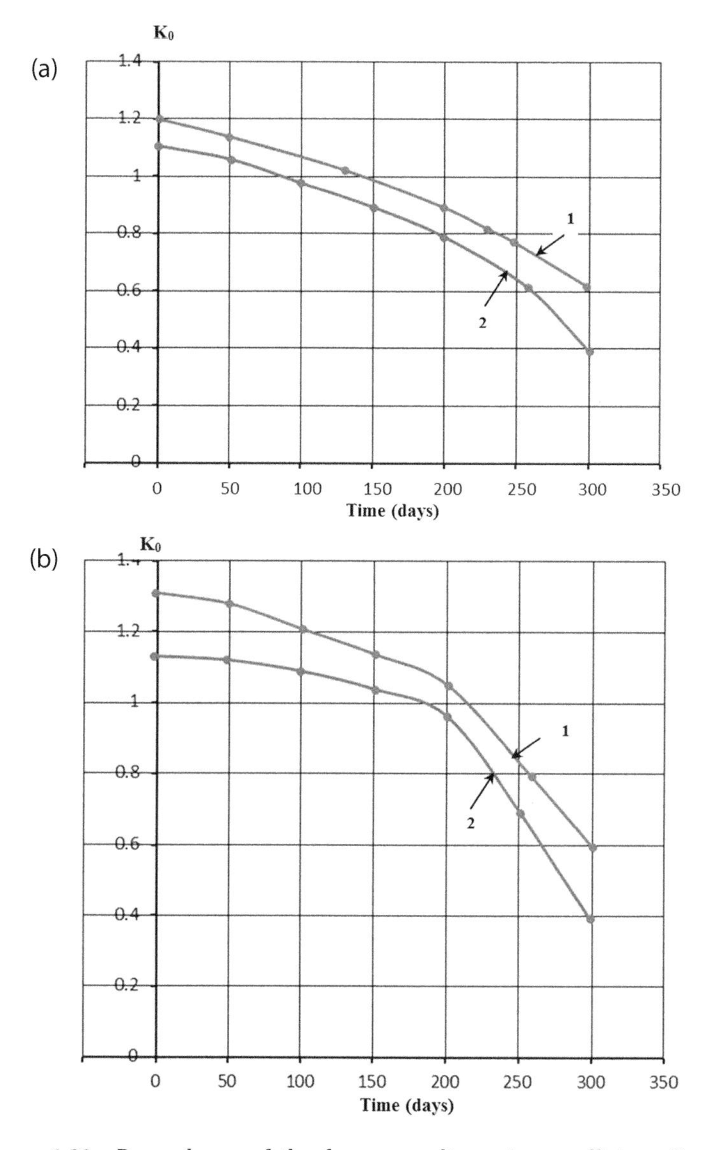

Figure 1.20 Dependence of the frequency dispersion coefficient K_0 on anodic polarization (Fig. 1.12) on exposure time ebonite coating in (a) 30% solution of sulfur acid; (b) 30% solution of hydrochloric acid. 1, sulfur vulcanizated coating at 150°C; 2, sulfur and Ethyl Zimate® vulcanizated coating at 100°C [37, 44].

vulcanization (150°C) dependence of kind of corrosive environment on the diffusive permeability of vulcanizate is less noticeable than for the low-temperature curing at a temperature of 100°C.

1.2.6 Technological Process of Manufacture and Properties of Ebonite Mixtures

The principal application characteristics of ebonite coatings, prepared from liquid non-solution compositions based on 1,4-cis oligobutadiene of the hydrocarbon structure are presented in Table 1.20 [37, 41].

The schematic diagram of the manufacturing of the liquid ebonite mixtures is shown in Fig. 1.21 [37, 44].

1.2.6.1 One-pot composition

One-pot ebonite compositions do not contain solvents and are produced on a basis of the low-viscosity oligobutadienes; therefore, they are safe during storage and use. A feature of these compositions is that all the ingredients are in one heterogeneous system.

As the bond matrix of the ebonite compositions ES-100_T and GES-1, oligobutadiene SKDNN was used (Table 1.21) [37, 44].

1.2.6.2 Double-pot composition

Double-pot compositions consist of an elastomer binder (basis) and the vulcanizer (paste). Basis of the double-pot compositions GES-2, GES-3 and GES-04M is butadiene and nitrile butadiene oligomer. Low-viscosity oligobutadiene is used as vulcanizer paste. Technological and operation properties of the double-pot compositions and coatings are presented in Table 1.22 [37, 44].

1.2.6.3 Examples of prepared LEM composition [43]

The components formulated into each sample and the amounts used are shown in Table 1.23.

Liquid ebonite compositions 1–4 were used to coat a cleaned steel plate. Each composition formed a coating layer with a thickness of 0.7 to 1.2 mm. The coating was then vulcanized at a temperature of 125°C for 15 h by using only dry hot air at a pressure of 3.5 atm.

Table 1.20 Properties of ebonite coatings prepared from oligobutadiene

No.	Parameter	Units	Value
	Liquid composition		
1	Color: yellow-gray, white, black		
2	Viscosity	Pa· s	80 to 280
3	Method of coating: brushing, spraying, dipping		
4	Condition of vulcanization (without pressure)		
	• heat carrier	°C	Hot dry air
	• temperature		100 to 150
5	LEM consumption per 1 mm of coating thickness	Kg/m^2	0.9 to 1.0
	Coating after vulcanization		
1	Volumetric shrinkage	%	3.2–4.5
2	Tensile strength	MPa	20–32
3	Shore hardness ("D")		55–60
4	Impact strength	N·m	>50
5	Adhesion strength to steel substrate by tear test	MPa	20–32
6	Strength of the contact with the metal:		
	• with Steel-3, at breaking	MPa	8–10
	• with aluminum, at stripping away N/m	175–400	
7	Thermal expansion at heating to 130°C	%	1.5–2
8	Thermal stability of the adhesive coating	°C	120–130
9	Diffusion coefficient	sq.m/s	(0.7–4.8)E-14
10	Cyclic stability (up to exfoliation) temperature cycle from –20°C to 100°C	90	
11	Electrical stability "heating-cooling" regime [(+20) – (+100) – (–20)°C]	MV/m	10–27
12	Chemical stability of coatings (maximal concentration):		
	a. In acids:		
	• nitric		10 (to 20°C)
	• sulfuric		70 (to 20°C)
			50 (to 60°C)
	• hydrochloric		35 (to 20°C)
	• fluoric		40 (to 20°C)
	• acetic		50 (to 20°C)
	• phosphoric %		80 (to 20°C)

(*Contd.*)

Table 1.20 *(Continued)*

No.	Parameter	Units	Value
	b. In salt solutions:		
	• Al/Zn sulfate		
	• Na/Ca chloride		20 (to 60°C)
	• K/Na dichromate		10 (to 60°C)
	c. In alkalis:		
	• Na/K hydroxide		40 (to 20°C)
	d. In organic media:		
	• acetone		
	• benzene		to 20°C
	• ethanol, butanol		
	• oils		to 60°C
13	Estimated delay of exploitation of coatings (2 mm thickness) in 30%—sulfuric and hydrochloric acids, at 20°C	years	4.1–5.3

Source: [37, 41]

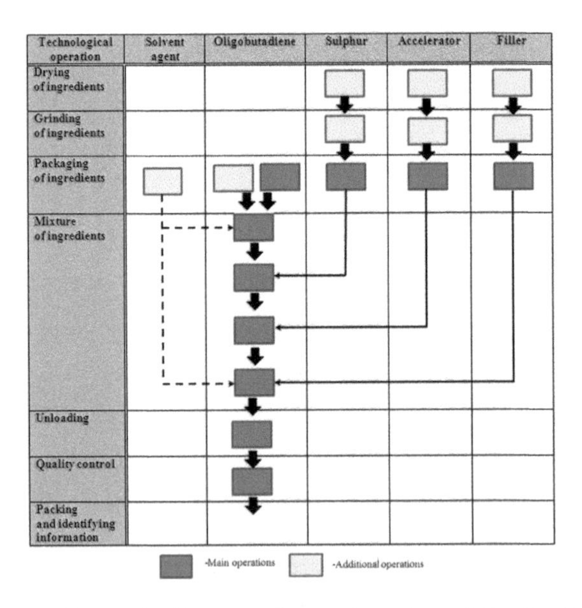

Figure 1.21 Schematic diagram of manufacturing of the liquid ebonite mixtures [37, 44].

Table 1.21 Composition and properties of rubber ebonite mixtures and coatings

Factor	Composition	
	GES-1	$ES\text{-}100_T$
Formulation of the mixture, mass parts		
SKDNN	100	
Sulfur	30	
Carbonate	20	
Thiuram	—	5
Zinc oxide	—	
Technological properties		
Surface appearance	Homogeneous viscosity liquids, black color	
Viscosity (75% solution in toluene at 20°C) (s)	50	45
Temperature of vulcanization (°C)	150	100
Time of vulcanization (hours):		
• Intermediate coat	2	5
• Last coat	8	25
Material consumption 0.4 at application of one layer (kg/m^2)	0.38	
Physical-mechanical properties		
Rupture resistance	26.0	12.5
Adhesion strength with steel substrate (MPa) (no less)	8.0	8.2

Source: [37, 44]

Heating to the vulcanization temperature and cooling-down from the vulcanization temperature were conducted at a rate of not more than 1°C per minute. These coated steel samples were used to determine tensile adhesion of each coating to the steel substrate according to the method prescribed by ASTM D 897.

For the purposes of tensile testing, the above procedure was modified by substituting a polytetrafluoro-ethylene (PTFE) sheet for the steel so that a free film of each of the 4 liquid ebonite compositions could be obtained by peeling the PTFE from the vulcanized film. The tensile properties, i.e., tensile strength and elongation at break, of each free vulcanized film were determined according to the method prescribed by ASTM D 638. Such samples were also used to determine the coefficient of chemical resistance of each coating by the procedure in which tensile strength was used

Table 1.22 Properties of two-pot ebonite compositions and coverings

	Technological properties			Operation properties	
Composition	**Method of application**	**Thickness of one-layer coating (mm)**	**Mode of vulcanization (°C · hour)**	**Rupture resistance (MPa)**	**Adhesion strength with steel substrate (MPa)**
GES-2	By brush	0.4–0.7	150° 4 ÷ 5	26	9.8
GES-3	By applicator	0.8–1.5	150° 3÷ 4	28	10.4
GES-4M	By brush or by applicator	0.3–0.5	150° 4.5÷ 5	35–42	10.0

Source: [37, 44]

Table 1.23 Preparing of LEM composition

	Compositions (in parts by weight per 100 parts of liquid rubber)			
	Composition no.			
Composition	**1**	**2**	**3**	**4**
Linear polybutadiene—SKDN-N (Efremov-Kautschuk GmbH) (82% cis-1,4, 25,000 mol. wt)	100	100	—	—
Linear polybutadiene —POLYOIL® 110 (Degussai (75% cis-1,4, 2,000 mol. wt.)	—	—	100	—
Linear polybutadiene —POLYOIL® 130 (Degussai (78% cis-1,4, 5,000 mol. wt.	—	—	—	100
Epoxy-terminated butadiene-nitrile rubber KR-207* (Kukdo) (tetrafunctional compound; viscosity 2,500 mPa·s, 25°C)	2	4.5	3	4
Sulfur —Code 104 Rubbermaker's (Hatwick Chem. Corp.)	38	33	32	30
Active filler				
Technical carbon black – FURNEX N-754 (Columbian Chemicals)	21	—	—	20
Silica white—SILENE 732D (PPG Industries)	—	7	9	—
Titanium dioxide—UNITANE (Kemira AY)	—	11	10	—
Accelerator				
THIURAM ME (Arrow Polychem, Inc.)	4.5	4.2	4.7	—
CAPTAX (R.T. Vanderbilt Co.)	0.2	0.3	0.1	2.0
Diphenylguanidine—DPG (Monsanto)	2.0	2.2	1.8	2.0
Zinc oxide—MO Brand	16.0	15.0	15.5	—

*Epoxy-terminated butadiene-nitrile rubber KR-207 is represented by the following structural formula:

```
      C-O-C-C-C-O-CO-[(C-C-C=C-C)x-(C-C)Y]-CO-O-C-C-C-O-C
      I                              I                I
C-C-C-C-C-O-C-C-C                   C≡N   C-C-C-O-C-C-C-C
      I        \ /                         I           I
      C-O-C-C   O                          O  C-C-C-O-C
           \ /                                   \ /
            O                                     O
```

Table 1.24 Properties of vulcanized liquid ebonite coatings

Characteristics	Composition no.			
	1	2	3	4
Physical-mechanical				
Tensile Strength, MPa	25.7	25.4	25.4	25.7
Elongation at Break (%)	5.9	6.1	5.7	6.2
Adhesion Strength, MPa	11.5	11.3	11.8	10.7
Chemical resistance K_{CR} after 360 days at 6°C				
35% Aqueous HCl	0.86	0.88	0.88	0.87
50% Aqueous H_2SO_4	0.94	0.93	0.94	0.96
35% Aqueous H_3PO_4	0.91	0.91	0.92	0.94

to determine K_{CR}. The properties of the four coatings samples are shown in Table 1.24.

1.2.7 Applications of Ebonite Coatings

With superior chemical resistance and functionality, LEM can replace the conventional, even-surface rubber sheet linings found in a multitude of products and industries and is commercially available in thicknesses of 2.5 to 4.0 mm. It can be easily applied using such simple coating techniques as brushing, rolling, spraying, flooding or dipping.

Before application, a vulcanizated paste is introduced into the composition to give preliminary curing to the LEM-layer in air at 20–25°C, which results in the formation of a rubber-like vulcanizate.

Unlike the traditional multilayered rubber sheets or liquid rubberizing compounds, LEM is most efficient in protecting the intricate shared and perforated parts of pumps, fans, centrifuge rotors, small diameter pipes and outlets, shut-off and control valves, stirrers and many other complex parts. As a breakthrough product, LEM offers a flexibility not found in its traditional predecessors (Fig. 1.22) [37, 45].

The recommended areas of application are given in Table 1.25.

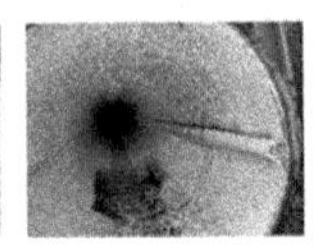

Figure 1.22 Examples of application of rubber covering [37, 45].

Table 1.25 The recommended area of LEM application

Area of industry	Protective equipment	Aggressive medium	Working temperature (°C)	Full time of testing in the working conditions
Chemical industry				
• Linings made from LEM will protect the intricate surfaces of filtered centrifuges, ventilators, and air ducts cleaning systems from blended acids solutions, vapors of HCl, HF, salt solutions and oxides of nitrogen, chlorine and bromine.				
• Parts coated with LEM are operating in oxalic acid working conditions up to 95–100°C.				
• LEM will effectively protect vacuum pumps from vapors of HCl and scrubbers from $NiCl_2$.				
• Sheet-steel cylinders of gas collectors coated with LEM will find protection from H_2S and vapors of HF				
	Filtered centrifuges	Master solutions		18
Chemical reagents	Ventilators	Vapors of HCl, HF and oxides of nitrogen	30	24
Chemical reagents	Ventilators and air ducts	Vapors of HCl, salt solutions, oxides of nitrogen and chlorine-bromine	21	
	Drying effects	Oxalic acid	95–100	16
	Vacuum pump	Vapors of HCl	30	24
	Scrubber	$NiCl_2$		12

(*Contd.*)

Table 1.25 *(Continued)*

Area of industry	Protective equipment	Aggressive medium	Working temperature (°C)	Full time of testing in the working conditions
Cryolite	Sheet-steel cylinders of gas collectors	H_2S and vapors of HF	48–54	18
Fertilizers	Picke pipeline	Solution of KOH and NaOH	80–90	24
Synthetic rubber	Mixing tank	H_2SO_4	80	16
Refrigerant	Bottom boxes	Vapors of HF	60–70	11
Agricultural industry	Tanks	Solution of KOH and NaOH, liquid complex fertilizers	20–30	24
Automobile industry	Bottom of cars	Chloride salts, sea water, big abrasive attack	(−40) – (+45)	
Marine	Variety of applications	Corrosive salts		

1.2.8 Conclusion

The above-mentioned results were obtained from conducting experimental studies on vulcanization processes for oligobutadiene. It was found to be the most effective bonding material for non-solution compositions to ebonite coatings which display excellent properties in durability and adhesion. Ebonite coatings may be applied to prepared surfaces by conventional methods used in lacquer-paint coating technology which decreases significantly the time needed for rubberizing of complex-profile or perforated techniques.

1.3 Nonisocyanate Polyurethanes Based on Cyclic Carbonates and Nanostructured Composites for Protective Coatings

1.3.1 Introduction

Polyurethanes (PUs) are products of the addition polymerization reaction between di- (or poly-) isocyanates and di- (or poly-) ols. PUs are among the most used polymers in many modern applications—foams, coatings, sealants, elastomers, thermoplastics, adhesives, fibers, and so on [21].

The polyurethane (PU) market today amounts to about 5% of the total polymer market and the worldwide consumption of PU has increased steadily. The demand in PUs has continued to increase and it will attain in 2016 a production of 18 million tons (~US$66.4 billion).

However, the use of isocyanates in the manufacturing process can render PU production extremely toxic and dangerous [22]. In fact, these compounds are harmful for human and environment. People exposed to isocyanates can develop a range of short-term health problems. More seriously, isocyanate exposure can lead to long-term asthma and dermatitis if individuals become sensitized. Sensitization is a condition in which the breathing or skin conditions can return with increasing severity on further exposures to the original sensitizing agent or to similar substances, even at very low exposures [22]. MDI (methylene diphenyl isocyanate) and TDI (toluene diisocyanate), the most widely used isocyanates in PU industry, are classified as CMR (Carcinogen, Mutagen and Reprotoxic) [46].

At the same time, conventional PUs have an inherent weakness depending on their molecular composition. Van der Waals forces mainly sets physically and mechanically properties of PU. The strength of these bonds is significantly lower in energy, however [21]. Therefore, PU unsatisfactorily stands up to the dynamic load, especially at elevated temperature.

Non-isocyanate sources for PU production have been sought for a long time.

There are several reactions, essentially based on transcarbamoylation, attracted researchers interest due to the possibility to form polyurethanes without isocyanate monomers in the formulations, which complies with the new environment and safety requirements. However, these reactions occur at high temperatures and accompanied by the formation of considerable amounts of by-products [47–49].

Non-isocyanate polyurethane (NIPU) based on the reaction of polycyclic carbonates and polyamines are known for more than 50 years. Recently, some reviews dedicated to the synthesis of cyclic carbonates (CC) and NIPU have been presented [35, 50–54].

NIPU networks are obtained by the reaction between polycyclic carbonate oligomers and aliphatic or cycloaliphatic polyamines with primary amino groups [1, 55–58]. This forms a crosslinked polymer with *p*-hydroxyurethane groups of different structure—polyhydroxyurethane polymer. Since NIPU is obtained without using isocyanates, the process of synthesis is relatively safe for both humans and the environment in comparison to the production of the conventional polyurethanes. The model scheme of the two options β-hydroxyurethane fragments of polymer chains formed in the case of bifunctional starting materials are shown in Fig. 1.23 [59].

Moreover, NIPU is not sensitive to moisture in the surrounding environment. Hydroxyl groups formed at the *p*-carbon atom of the urethane moiety also increase adhesion properties. Plurality of intra- and intermolecular hydrogen bonds [60] as well as the absence of unstable biuret and allophanate units [61] seems to be

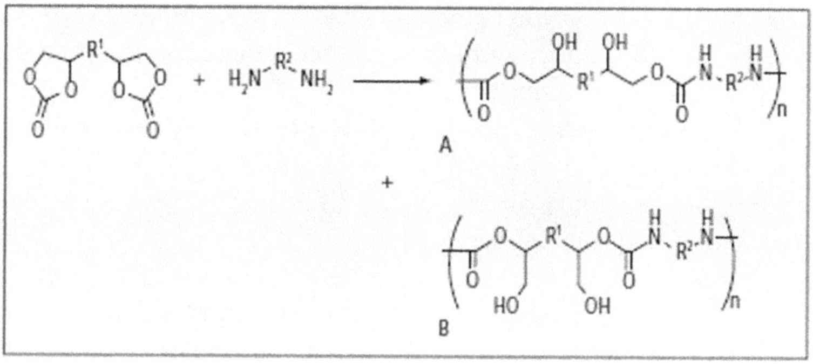

Figure 1.23 β-hydroxyurethane moieties of nonisocyanate polyurethanes: A, with secondary hydroxyl groups; B, with primary hydroxyl groups [59].

responsible for increased thermal stability and chemical resistance to non-polar solvents

The mechanism of the reaction of cyclic carbonates with amines, providing nonisocyanate urethanes, was studied by means of quantum chemical calculation in terms of DFT by the PBE/TZ2P method using as examples the reactions of ethylene carbonate and propylene carbonate with methylamine [62, 63]. Structural investigations have revealed four cyclic isomers with intramolecular hydrogen bond and six open conformers. This agrees well with previous results of IR and NMR spectroscopic investigations [55, 60]. The reaction can proceed through the one- or multistage path involving one or two amine molecules. The second amine molecule plays the role of the catalyst of the process, resulting in a substantial decrease in the activation energy of the reaction.

1.3.2 State of the Art in NIPU: Brief Description of the Latest Discoveries and Developments

A significant problem of the NIPU technologies is lack of commercially available multifunctional cyclic carbonates. Recent work in the field of new methods for preparing of cyclic carbonates is dedicated primarily to the development of new catalytic systems and the synthesis of mono-functional compounds [64, 65]. Similar catalyst systems are used also for the copolymerization of epoxides and CO_2

and ring-opening polymerization of cyclic carbonates [66, 67], and one or the other direction of the reaction depends on the process conditions. For this work the synthesis of mono- and polycyclic carbonates for research purposes were provided by the company Specific Polymers, France [68].

Bernard [69] proposes a method for preparing polyhydroxyurethanes, which comprises reaction of

- at least *one compound* having a cyclic carbonate functional group and at least one hydroxyl functional group;
- at least *one compound* having at least one linear carbonate functional group;
- at least *one compound* having at least one primary or secondary amine functional group.

The subject of this invention is to position predominantly aqueous formulations on the base of polyfunctionalized polyhydroxyurethane intermediates, as well as creating a method for preparing formulations for use especially in coatings, adhesives, and others. However, the described process is very complicated and time-consuming, requires large amounts of organic solvents, and the formulations for the practical application in most cases require the use of isocyanate-containing components.

Moeller et al. [70] also describe the bonding agent system that contains a *component (A)* carrying at least two cyclic carbonate groups and a *component (B)* carrying at least two amine functional groups to prepare a two-component NIPU adhesive. In this case, too, component (A) comprises the reaction product of hydroxyl group-containing cyclic carbonate with an isocyanate group-containing polyurethane prepolymer. Currently in the coatings industry cyclic carbonate raw materials are often suggested for the use in hybrid epoxy-hydroxyurethane compositions [71]. Use of such systems assumes preliminary production of adducts of the cyclic carbonates and amines. These adducts (also named as amino-urethanes) contain amine, urethane, and hydroxy groups and serve as hardeners for various oligomer compositions. Such compositions, named as hybrid NIPU (HNIPU), are well-known in the art. Mainly they relate to waterborne epoxy compositions.

Muller-Frischinger [72] describes a curable composition comprising

(a) Mixture of an epoxy resin, a cyclic carbonate,
(b) Curing agent—hybrid hardener, whereby said hardener is a blend of
    - an aminic compound, and
    - a dicyclopentadiene-phenol based Novolac.

Later Muller-Frischinger et al. [73] disclosed a curable composition comprising an epoxy resin and a hybrid hardener; wherein said hybrid hardener is a blend of adduct of amines or amidoamines and monocyclic carbonates (in particular) and a polyphenol Novolac. Such compositions are useful for rapid setting and protective coatings and adhesives in application fields like civil engineering, marine, architectural, and maintenance. Also, research by Huntsman Co. [74] proposed filled compositions on the base of NIPU or HNIPU and nanoclays. In this work, cyclic carbonates, synthesized by Polymate Ltd.-INRC, were used as raw materials.

Klopsch et al. [75] disclose the use of new cyclic carbonates with unsaturated bonds as reactive diluents in epoxy resin compositions. It has been found that addition of just small amounts of the new compounds to epoxy resins results in a significant increase in the reactivity of the epoxy resin composition, evident from a lower gel time following addition of a hardener. However, enhancing of other important properties has not been achieved.

Mecfel-Marczewski et al. [76] proposed substituted cyclic carbonates—2-oxo-1,3-dioxolane-4-carboxylic acid and derivatives thereof. It is assumed that these compounds will be widely used in the oligomer technology. The problem of obtaining NIPU materials based on renewable raw materials is given considerable attention in the research centers of the USA, Europe, and China [71].

Recently, a number of European academic centers have begun to actively develop NIPU from the direction of plant-based raw materials. Thus, researchers at the Institute of Macromolecular Chemistry at the University of Freiburg have investigated soy- and linseed oil-based polyurethanes prepared by curing carbonated soybean (CSBO) and linseed (CLSO) oils with different di-amines

[77]. Later they have reported on a very versatile new route to linear as well as crosslinked terpene-based non-isocyanate poly(hydroxyurethanes) (NIPU) and pre-polymers derived from the novel cyclic limonene dicarbonate (CL) [78]. The catalytic carbonation of epoxidized limonene with CO_2 was monitored in the presence of both homogeneous tetrabutylammonium bromide (TBAB) and heterogeneous silica supported 4-pyrrolidinopyridinium iodide (SiO_2-(I)) catalysts. The systematic variation of catalyst type, CO_2 pressure and temperature enabled quantitative carbonation in bulk and incorporation of 34.4 wt% CO_2 into CL. In contrast to conventional plant oil-based cyclic carbonates, such terpene-based cyclic carbonates afford much higher CO_2 fixation and do not contain ester groups. The absence of ester groups is essential to prevent side reaction with amines: ester groups react with the amine curing agent to afford amides and low molecular weight polyol byproducts which can cause undesirable emissions and plastification of NIPU. Novel linear NIPU and prepolymers were obtained by means of CL advancement with diamines such as 1,4-butane diamine (BDA), 1,6-hexamethylene diamine (HMDA), 1,12-dodecane diamine (DADO), and isophorone diamine (IPDA).

Cramail et al. [79] reported polyaddition of diamines with vegetable-based bis-carbonates to prepare new polyurethanes. The intermediate material (epoxidized compounds) were obtained in two steps by a transesterification starting from monoalkyl esters of unsaturated acids and diols, and subsequent epoxidation. Then the bis-carbonates were prepared from difunctional epoxides and supercritical CO_2 in the presence of ionic liquids and TBAB as the catalyst. Received cyclic carbonates were considered as polyhydroxyurethane precursors and further self-polycondensed with ethylene diamine and isophorone diamine to form polyurethanes. For preparation of bis-carbonates with terminal cyclic carbonate groups were used in a metathesis reaction with Hoveyda's catalyst.

The article [80] presents a new bio-based non-isocyanate urethane by the reaction of a cyclic carbonate synthesized from a modified linseed oil and an alkylated phenolic polyamine (phenalkamine) from cashew nut shell liquid. The incorporation of functional cyclic carbonate groups to the triglyceride units of the oil was done by reacting epoxidized linseed oil with CO_2 in the presence of a catalyst. Structural changes and changes in molar mass during the carbonation reaction were characterized. Also, the aminolysis reaction of the cyclic carbonate with phenalkamine was monitored, as well as the viscoelastic properties of the system and the time of gelation.

A novel bio-based, isocyanate-free poly-(amide urethane) derived from soy dimer acids is described in [81]. Three steps are involved in this one-pot synthesis: first, dimer fatty acids are condensed with ethylene diamine to produce amine-terminated oligomers intermediates. These intermediates are then reacted in a second step with ethylene carbonate to yield hydroxyl-terminated di-urethanes, which then undergo a transurethane polycondensation at 150°C for 9 h under vacuum to produce high-molecular-weight polymers. Although the polymers are produced at a high temperature, above properties still do not allow them to find a practical application.

There is continuous development being made in the direction of silicon-contained and nanostructured hydroxyurethane compounds. Hoşgör et al. [82] synthesized a novel carbonate-modified bis(4-glycidyloxy phenyl) phenyl phosphine oxide (CBGPPO) for preparing nonisocyanate polyurethane/silica nanocomposites. Spherical silica particles were prepared and modified with cyclic carbonate functional silane-coupling agent to improve the compatibility of silica particles and organic phase. The phosphine oxide-based and cyclic carbonate-modified epoxy resins and silica particles were used to prepare hybrid coatings using diamine as a curing agent. No damage was observed in the impact strength of the coatings. Incorporation of silica and CBGPPO into formulations increased modulus and hardness of the coating making the material more brittle. It was also observed, that the thermal stability of hybrid coatings enhanced with the addition of silica and CBGPPO.

A novel bis-urethane organosilane precursor has been developed via NIPU route in sol-gel processing conditions and employed as an organic precursor of organic-inorganic hybrid (OIH) coating systems. Coating formulations with variable proportions of this organic component were prepared and applied on aluminum substrate. These coatings were evaluated for mechanical, chemical properties and corrosion resistance and showed some improvements [83].

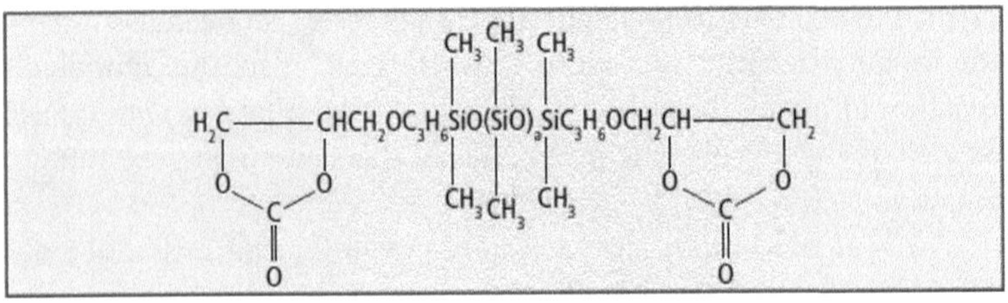

Figure 1.24 Example of five-membered cyclic carbonate polysiloxane compound [59].

Hanada et al. [84] disclosed a polysiloxane-modified polyhydroxy polyurethane resin being derived from a reaction between a five-membered cyclic carbonate polysiloxane compound and an amine compound, its production process and a resin composition (Fig. 1.24) [59].

New materials can be used for thermal recording medium, imitation leather, thermoplastic polyolefin resin skin material, weather strip material, and weather strip.

Development of the scientific direction "NIPU based on the reaction of polycyclic carbonates and polyamines" demonstrates a chronological list of patents and applications issued in the field (Table 1.26 [85]). The list also describes the interest of leading companies to commercialize the developments. (Hereinafter patents from Table 1.26 are numbered as P1, P2, etc.).

Table 1.26 List of patents and applications in the field NIPU based on the reaction of polycyclic carbonates and polyamines

No.	Patent no./ priority (year)	Applicant(s)/ inventor(s)	Brief description
1	US 2802022/ 1954	Groszos et al./ (American Cyanamid)	A method of preparing a polyurethane which comprises reacting together urea and mono- or poly-β-hydroxyurethane on the base of mono- or polyamine and alkylene carbonate

Table 1.26 (*Continued*)

No.	Patent no./ priority (year)	Applicant(s)/ inventor(s)	Brief description
2	US 2935494/ 1957	Whelan, Jr. et al./ (Union Carbide—today Dow Chemical)	A linear polyurethane product prepared by the reaction of erythritol dicarbonate and a polyfunctional amine
3	US 3072613/ 1957	Whelan, Jr. et al./ (Union Carbide—today Dow Chemical)	A resinous polyurethane product prepared by the reaction of bis-CC and a polyfunctional amine
4	US 3084140/ 1957	Gurgiolo et al./ (Dow Chemical)	A polyhydroxy polyurethane resin produced by reacting an aliphatic polyamine and a dicyclic carbonate and subsequent crosslinking the resin by reacting with a poly-functional reactant selected from the group consisting of aldehydes, dicarboxylic acid chlorides, thionyl chloride, and sulfuryl chloride and insolubilization of the resin
5	US 3305527/ 1964	Whelan, Jr., et al./ (Union Carbide—today Dow Chemical)	A resinous polyurethane product prepared by the reaction of bis-CC and a polyfunctional amine
6	SU 413824/ 1969	Gurgiolo et al./ (Dow Chemical)	A polyhydroxy polyurethane resin produced by reacting an aliphatic polyamine and a dicyclic carbonate and subsequent crosslinking the resin by reacting with a poly-functional reactant selected from the group consisting of aldehydes, dicarboxylic acid chlorides, thionyl chloride and sulfuryl chloride and insolubilization of the resin

(*Contd.*)

Table 1.26 (*Continued*)

No.	Patent no./ priority (year)	Applicant(s)/ inventor(s)	Brief description
7	SU 359255/ 1971	Whelan, Jr. et al./ (Union Carbide—today Dow Chemical)	A resinous polyurethane product prepared by the reaction of bis-CC and a polyfunctional amine
8	US 3929731/ 1973	Gurgiolo et al./ (Dow Chemical)	A polyhydroxy polyurethane resin produced by reacting an aliphatic polyamine and a dicyclic carbonate and subsequent crosslinking the resin by reacting with a poly-functional reactant selected from the group consisting of aldehydes, dicarboxylic acid chlorides, thionyl chloride and sulfuryl chloride and insolubilization of the resin
9	GB 1495555/ 1974	Whelan, Jr. et al./ (Union Carbide—today Dow Chemical)	A resinous polyurethane product prepared by the reaction of bis-CC and a polyfunctional amine
10	SU 529197/ 1975	Gurgiolo et al./ (Dow Chemical)	A polyhydroxy polyurethane resin produced by reacting an aliphatic polyamine and a dicyclic carbonate and subsequent crosslinking the resin by reacting with a poly-functional reactant selected from the group consisting of aldehydes, dicarboxylic acid chlorides, thionyl chloride and sulfuryl chloride and insolubilization of the resin
11	SU 563396/ 1976	Figovsky et al.	Polymer-concrete mix comprising trihydroxyurethane on the base of polyoxypropylene triamine and propylene carbonate, divinyl ether of ethylene glycol, catalyst and mineral fillers

Table 1.26 (*Continued*)

No.	Patent no./ priority (year)	Applicant(s)/ inventor(s)	Brief description
12	SU 630275/ 1976	Figovsky et al.	Polymer composition comprising polyoxyalkylenetriol tricyclic carbonate and adduct amine with epoxy resin
13	SU 668337/ 1977	Movsisyan, Figovsky et al.	Adhesion sublayer for polyethyleneterephtalate films comprising polyurethane acetals obtained by reacting nonisocyanate urethane-containing glycols with divinyl ethers of glycols
14	SU 671318/ 1977	Stroganov, Figovsky	Preparing of epoxy-cyclic carbonate resins
15	SU 704032/ 1976	Figovsky et al.	Composition of epoxy resin, polyoxypropylenetriol tricyclic carbonate, amine and fillers
16	US 4122068/ 1977	Meyer (Huntsman)	Resistance to thermal shock of anhydride cured epoxy resins is enhanced by addition of polyether dihydroxyalkyl carbamate additives on the base of polyoxyalkylene diamines (M_W 2000–3000) and monocyclic carbonates
17	SU 707258/ 1978	Stroganov, Figovsky et al	Preparing of epoxy-cyclic carbonate resins
18	SU 798126/ 1978	Mikheev et al.	Use of PU-glycols for synthesis of polyesters
19	SU 903340/ 1978	Kutsenok, Figovsky et al.	Polymer concrete on the base of hybrid composition: epoxy resin-dicyclic carbonate-amine
20	SU 908769/ 1978	Figovsky et al.	Polymer concrete on the base of hybrid composition: epoxy-cyclic carbonate resin and amine

(*Contd.*)

Table 1.26 (*Continued*)

No.	Patent no./ priority (year)	Applicant(s)/ inventor(s)	Brief description
21	RU 970856/ 1980	Sysoev, Mikheev et al.	NIPU on the base of aromatic polycyclic carbonates
22	CA 1247288/ 1982	Hesse (Hoechst)	Epoxy resin–based compositions comprise hydroxyurethane-modified amine hardeners
23	SU 1110783/ 1983	Stroganov, Figovsky et al.	Cyclohexyl cyclic carbonate and corresponding NIPU
24	SU 1126569/ 1983	Stroganov, Figovsky et al.	Aryl cyclic carbonates as adhesion additives in epoxy base compositions
25	SU 1240766/ 1983	Mikheev et al.	Water-soluble oligomeric urethane from amino-epoxy adduct and monocyclic carbonate and self-curable at 200°C; coatings on the base of thereof
26	US 4484994/ 1984	Jacobs III et al. (American Cyanamid)	Hydroxyalkyl urethane-containing resin having at least one tertiary amine and at least two hydroxyalkyl urethane groups per molecule; the polymer is obtained by reacting an epoxy resin with one or more amines having at least one secondary amine group and at least one hydroxyalkyl urethane group or a precursor thereof
27	US 4528363/ 1984	Tominaga (Kansai Paint)	Amino-(hydroxyurethanes) and epoxy or acrylic compounds as source blocks for heat-curable resin preparation
28	US 4758632/ 1985	Parekh et al. (American Cyanamid)	A self-crosslinkable acrylic polymer contains at least two hydroxyalkyl carbamate groups per molecule

Table 1.26 *(Continued)*

No.	Patent no./ priority (year)	Applicant(s)/ inventor(s)	Brief description
29	SU 1587899/ 1988	Figovsky et al.	Flooring on the base of hybrid composition: aromatic epoxy resin, aliphatic epoxy resin, polyoxypropylenetriol tricyclic carbonate and Mannich base amine
30	SU 1754747/ 1990	Stroganov, Figovsky et al.	Flooring on the base of hybrid composition: aromatic epoxy resin, aliphatic epoxy resin, dicyclic carbonate, aminophenol and polyoxypropylenediamine
31	SU 1754748/ 1990	Figovsky et al.	Flooring on the base of hybrid composition: aromatic epoxy resin, polyoxypropylenetriol triglycidyl ether, polyoxypropylenetriol tricyclic carbonate and aminophenol
32	US 5235007/ 1991	Alexander et al. (Huntsman)	A curing agent mixture for epoxy resin compositions comprises a di-primary amine (M_W 60–400) and an amine-carbamate which is the reaction product of an excess of the di-primary amine and a monocyclic carbonate
33	US 5175231/ 1992	Rappoport et al.	A method for preparing a urethane by reacting a compound containing a plurality of cyclocarbonate groups with a diamine in which the two amine groups have different reactivities with cyclocarbonate so as to form a urethane oligomer with amine end groups; the urethane oligomer can then be reacted in different ways to form a polyurethane

(Contd.)

Table 1.26 (*Continued*)

No.	Patent no./ priority (year)	Applicant(s)/ inventor(s)	Brief description
34	US 5340889/ 1993	Crawford et al. (Huntsman)	Liquid hydroxyurethane products having cyclocarbonate end groups are prepared by reacting a molar excess of a bis-carbonate of a bis-glycidyl ether of neopentyl glycol or 1,4-cyclohexanedimethanol with a polyoxyalkylenediamine; these products are useful for the preparation of PU, PU polyols, polyester PU polyols, and polycarbonate PU polyols
35	US 5677006, US 5707741, US 5855961, US 5935710/ 1993	Hoenel et al. (Hoechst)	Waterborne coating compositions, including one or more resins having amino-reactive groups, one or more polyamine curing agents, and one or more aminourethanes; the aminourethanes can be reaction products of oligomeric or polymeric compounds containing terminal cyclic carbonate groups, and amines
36	US 6120905/ 1998	Figovsky et al.	A hybrid non-isocyanate polyurethane network polymer formed by crosslinking at least one cyclocarbonate oligomer and at least one amine oligomer; the cyclocarbonate oligomer contains a plurality of terminal cyclocarbonate groups; at least one cyclocarbonate oligomer further comprises from about 4% to about 12% by weight of terminal epoxy groups

Table 1.26 *(Continued)*

No.	Patent no./ priority (year)	Applicant(s)/ inventor(s)	Brief description
37	US 6407198/ 1999	Figovsky et al. (Polymate)	Chemically resistant materials with high mechanical properties are provided by using polycyclocarbonates of special structure; the polycyclocarbonates are prepared by the reaction of oligocyclocarbonates containing ended epoxy groups with primary aromatic diamine
38	EP 1070733/ 1999	Figovsky et al. (Polymate)	Method of synthesis of polyaminofunctional hydroxyurethane oligomers and hybrid polymers formed
39	US 7045577/ 2002	Wilkes et al.	A preparation of novel carbonated vegetable oils (such as carbonated soybean oil) by reacting carbon dioxide with an epoxidized vegetable oil is described; the carbonated vegetable oils advantageously may be used for producing non-isocyanate polyurethane materials
40	US 7232877/ 2002	Figovsky et al.	A star epoxy compounds and their preparation and use in making star cyclocarbonates, star hydroxyurethane oligomers, and star NIPU and HNIPU are disclosed; acrylic epoxy compounds, acrylic cyclocarbonates, acrylic hydroxyurethane oligomers, acrylic NIPU and HNIPU and their methods of preparation also are described

(Contd.)

Table 1.26 (*Continued*)

No.	Patent no./ priority (year)	Applicant(s)/ inventor(s)	Brief description
41	US 7288595/ 2002	Swarup et al. (PPG Industries)	A reaction product having polyether carbamate groups formed from polyoxyalkylene amine, and cyclic carbonate, in equivalent ratios ranging from (1:0.5) to (1:1.5) is provided; further provided is a process for preparing the aforementioned reaction product
42	US 8143346 and 8450413/ 2003	Diakoumakos et al. (Huntsman)	A fast curable NIPU and HNIPU polymeric nanocompositions are derived upon crosslinking a mixture comprising natural or modified nanoclay with either a monomer(s) /oligomer(s) bearing cyclocarbonate group(s) or a mixture of the latter with an epoxy resin, with a hardener bearing amino groups
43	US 6960619/ 2003	Figovsky et al.	UV-curable liquid acrylic-based compositions (for sealing applications), which include products of reaction of non-isocyanate urethane diols with methacrylic or acrylic anhydride
44	US 7842773/ 2006	Herzig (Wacker)	Organosilicon compounds containing non-isocyanate urethane groups are prepared by reacting aminofunctional organosilicon compounds with CC, and, in a second stage, the first stage reaction products are optionally condensed with silanes bearing groups capable of condensation, to give higher molecular weight organosilicon compounds containing urethane groups

Table 1.26 *(Continued)*

No.	Patent no./ priority (year)	Applicant(s)/ inventor(s)	Brief description
45	US 8017719/ 2007	Bernard et al. (Rhodia)	Method for preparing polyhydroxy-urethanes from amino compounds and compounds carrying cyclic carbonate functions
46	US Application 12/315580, 2008	Birukov et al. (Polymate, NTI)	A liquid crosslinkable oligomer composition that contains a hydroxyurethane-amine adduct and a liquid-reacting oligomer is proposed; the hydroxyurethane-amine adduct is a product of an epoxy-amine adduct reacting with a compound having one or more terminal cyclocarbonate groups
47	US 7820779/ 2009	Birukov et al. (Polymate, NTI)	A nanostructured hybrid liquid oligomer composition including epoxy-functional component, cyclic carbonate component, amine-functional component, and, optionally, acrylate (methacrylate) functional component, wherein at least one epoxy, amine, or acrylate (methacrylate) component contains alkoxysilane units
48	US 7989553/ 2009	Birukov et al. (Polymate, NTI)	There are disclosed three-dimensional epoxy-amine polymer networks modified by a hydroxyalkyl urethane, which is obtained as a result of a reaction between a primary amine (one equivalent of the primary amine groups) and a monocyclic carbonate (one equivalent of the cyclic carbonate groups); such hydroxyalkyl urethane modifier is not bound chemically to the main polymer network

(Contd.)

Table 1.26 (*Continued*)

No.	Patent no./ priority (year)	Applicant(s)/ inventor(s)	Brief description
49	US 9115111/ 2009	Cramail et al. (C.N.R.S., France)	Synthesis of terminal bicarbonate precursors from plant base raw materials
50	US 8853322/ 2009	Mecfel-Marczevsky et al. (BASF)	Water-dispersible, cyclocarbonate-functionalized vinyl copolymer binder and an amine curing agent; in this binder the emulsifier groups are incorporated in the polymer chain, gives stable aqueous dispersions having a solids content of up to a 30% by weight
51	US 8703648, US 8975420, US 8951933/ 2009	Hanada et al.	A new polysiloxane-modified polyhydroxy polyurethane resin derived from a reaction between a 5-membered cyclic carbonate compound and an amine modified polysiloxane compound is disclosed
52	US 8741988/ 2010	Klopsch et al. (BASF)	Unsaturated monocyclic carbonates as reactive diluents for epoxy resin compositions
53	US 8742137/ 2010	Mecfel-Marczevsky et al. (BASF)	2-oxo-1,3-dioxolane-4-carboxylic acid esters transesterified with a polyol having a valency of 2 to 5 in the presence of an enzymatic catalyst or an acidic cation exchanger and further cured with polyamines
54	US Application 14/0030526/ 2011	Uruno et al.	Polyhydroxyurethane microparticles are spherical polymer microparticles having particle sizes of 0.1 pm to 300 pm and have been derived from cyclic carbonates and amines

Table 1.26 (*Continued*)

No.	Patent no./ priority (year)	Applicant(s)/ inventor(s)	Brief description
55	US Application 20140191156/ 2011	Marks et al. (Dow Global Technologies)	A novel cyclic carbonate monomer comprising a reaction product of at least one divinylarene dioxide and carbon dioxide; the poly(hydroxyurethane) compositions made from Divinylbenzene Dicarbonate and polyamines forms a reactive intermediate that can be used for making, for example, a poly(hydroxyurethane) foam product having an approximate volume expansion of 10
56	US 8653174/ 2011	Anderson et al. (Dow Global Technologies	Alternative synthesis of NIPU at ambient temperatures from the reaction of polyaldehydes with carbamate functional polymers using an acid catalyst
57	US Application 20140378648/ 2011	Soules et al. (University of Montpellier, France)	A method for preparing a compound comprising a hydroxyurethane unit or a y-hydroxy-urethane unit, comprising reacting a compound comprising a cyclocarbonate reactive unit with a compound B comprising an amino reactive unit (–NH_2) in the presence of a catalyst, said method being characterized in that said catalyst comprises an organometallic complex and a cocatalyst selected from the group of Lewis bases, or salts of tetra-alkyl ammonium
58	US 8586653, US 8877837/ 2011	Yu et al. (BASF)	HNIPU on the base of epoxy resins and cyclic carbonates with unsaturated bonds using mixtures of amino hardeners and catalysts

(*Contd.*)

Table 1.26 (*Continued*)

No.	Patent no./ priority (year)	Applicant(s)/ inventor(s)	Brief description
59	US 9079871/ 2011	Mezger et al. (BASF)	Preparation of tricycle carbonates by reaction of trioxyalkyl amine with (2-oxo-1,3-dioxolan-4-yl)methyl chloroformate in an aqueous/organic two-phase system in the presence of an auxiliary base and of a phase transfer catalyst
60	US 9150709/ 2011	Klopsch et al. (BASF)	HNIPU on the base of epoxy resins, cyclic carbonates with an electron- withdrawing organic group and amino hardeners
61	US 9062136/ 2011	Porta Garcia et al. (BASF)	Homo- and copolymers obtained by homo- or copolymerization of alkylidene-1,3-dioxolan-2-one monomers and to the use thereof as a component in 2K binder compositions by crosslinking with amine hardeners
62	WO 2012171659 2011	Bahr et al. (University of Freiburg, Germany)	Terpene or terpenoid derivatives comprising at least two cyclic carbonate groups and isocyanate-free polyurethanes obtainable form the reaction of monomers with at least two cyclic carbonate groups with amine monomers comprising at least two amino groups
63	US 9102829/ 2012	Birukov et al. (Polymate, NTI)	A method of obtaining hybrid polyhydroxyurethane compositions based on unsaturated fatty acid triglycerides and crosslinked at ambient temperatures is disclosed

Table 1.26 (*Continued*)

No.	Patent no./ priority (year)	Applicant(s)/ inventor(s)	Brief description
64	US 9193862/ 2012	Gehringer et al. (BASF)	HNIPU on the base of epoxy resins, amine hardeners (0.3–0.9 equiv.), catalyst and monocyclic carbonate for fiber-reinforced cured compositions
65	US Application 20150024138/ 2013	Figovsky et al. (Polymate, NTI)	Method for forming a sprayable non-isocyanate foam composition
66	CA 2876736, 2014	Figovsky et al. (Polymate, NTI)	Radiation-curable bio-based flooring compositions with nonreactive additives (bio-based hydroxyurethanes and silane-based hydroxyurethanes)
67	US Application 20150353683, 2014	Birukov et al. (Polymate, NTI)	A new hybrid epoxy-amine hydroxyurethane network polymers with lengthy epoxy-amine chains and pendulous hydroxyurethane units
68	US Application 20150247004/ 2014	Lombardo et al. (Dow Global Technologies)	A method of forming non-isocyanate based polyurethane includes providing a cyclic carbonate, an amine, and a cooperative catalyst system that has a Lewis acid and a Lewis base
69	WO 015107113/ 2014	Darroman et al. (University of Montpellier, France)	The invention relates to a composition that includes cork or a cork-based material and a binder which includes at least one polyhydroxyurethane prepared using raw materials that are not highly toxic, preferably from biomass

Source: [85]

1.3.3 Recent Achievements in the Field of NIPU

1.3.3.1 Polyhydroxyurethanes

Usually polyhydroxyurethane polymers (PHUs) have poor water resistance due to the plurality of hydroxyl groups. It is, however, possible to prepare water-resistant materials in some formulations. For example, on the base of acrylic epoxy oligomers, cyclocarbonate acrylic polymers with high water and weather stabilities were prepared. A paint was developed with curing temperature 110°C, in 2–3 h. Unfortunately we need to use solvents for this composition [55].

Research on the synthesis of aliphatic multifunctional cyclic carbonates from corresponding epoxies and carbon dioxide and NIPU based on them was carried out. Some compositions of polyfunctional carbonates synthesized in the laboratory were examined, namely: trimethylolpropane tricyclocarbonate (TMPTCC) and chlorine containing aliphatic tricyclocarbonates (on the base of chlorine containing aliphatic epoxy resins Oxilin) and various diamines:

- 2-methylpentamethylene diamine (MPMD)—Dytek A, Invista Co.;
- meta-xylenediamine (MXDA)—Mitsubishi Gas Chem. Co.;
- polyetheramine Jeffamine EDR-148—Huntsman Co.;
- diethylenetriamine (DETA)—D.E.H. 20, Dow Chemical Co.

The properties of these materials are shown in Table 1.27 [59].

One can see that some results are significantly higher than previously achieved levels and offer good prospects for their practical use.

Table 1.27 Properties of the polyhydroxy urethanes

Cyclocarbonate	Amine	Tensile strength (MPa)	Elongation (%)	Water absorption (%)
TCCTMP	MPDM	33–47	3.3–3.8	n/a
CC Oxilin 5	MXDA	18	4	n/a
CC Oxilin 6B	EDR-148	0.8	10	4.0
CC Oxilin 6	DETA	1.6	16	n/a

Source: [59]

PHUs produced by the reaction between dicyclocarbonate and diamine groups are often presented as possible candidates to substitute for classical polyurethanes based on isocyanate precursors. In the literature, the synthesis of this class of polymers is often performed according to arbitrary conditions of time and temperature without any scientific justification. As such, the real potential of PHUs is probably not fully known. Numerous contradictions in published results seem to support this hypothesis. Authors of the work [86] propose two methodologies based on dynamic rheometry to determine optimized conditions for the synthesis of PHUs. The case of a PHU formed by the reaction between 1,10-diaminodecane and a dicyclocarbonate bearing a central aromatic group is described more precisely.

The first approach consists of conducting various rheological experiments (kinetics, thermomechanical analyses) in situ on the reaction mixture.

The second one retains the same technique to qualify the viscoelastic properties of PHUs synthesized according to various conditions. In this latter case, all samples show thermomechanical behavior of amorphous thermoplastic polymers. But discrepancies are observed with regard to the value of the glass transition temperature and the existence or not of a rubbery zone.

The comparison of these data with size exclusion chromatography results shows that these differences are direct consequences of the polymer molecular weight that can be predicted using macromolecular theory. The properties of the PHUs obtained after optimization of the polymerization reaction were compared with literature data in order to complete the evaluation of the efficiency of the rheological methodology.

1.3.3.2 Hybrid non-isocyanate polyurethanes

Mechanical and physical-chemical characteristics of NIPU are inferior to conventional PU, so almost immediately after the first reports of NIPU started developing of hybrid compositions (HNIPU), namely subsequent crosslinking the resin by reacting with a polyfunctional reactant (P4); combining NIPU with epoxy resins and amine hardeners (P5, P6).

HNIPU is a modified polyurethane with lower permeability, increased chemical resistance properties and material synthesis that does not use the toxic isocyanate at any stage of production. An intermolecular hydrogen bond is formed during HNIPU synthesis that improves hydrolic stability well above that of conventional polyurethanes. Materials that contain intermolecular hydrogen bonds display chemical resistance 1.5 to 2 times greater than materials of similar structure without such bonds.

The synthesis of HNIPU is safe and easy and hardens at ambient temperature without using toxic components in the process. Due to their superior structure and excellent resistance to degradation, HNIPUs are ideal for numerous application, including crack-resistant composite materials, chemically resistant coatings, industrial flooring sealants, and glues. Their outstanding properties are beneficial to many different industries (Table 1.28).

Recently, BASF Corporation (USA) has paid serious attention to such areas as the synthesis of HNIPU on the base of epoxy-amine compositions and CC with unsaturated bonds (P52, P58) or CC with an electron-withdrawing organic group (P60) or conventional alkylene carbonate (P64). Syntheses of poly-CC were disclosed:

Table 1.28 Properties of HNIPU compared to conventional polyurethane (PU)

Properties	Conventional PU	Hybrid nonisocyanate PU HNIPU
Solid %	76–100	100
Pot life	4–6 h	4–6 h
Thorough cure time	7 days at 18–22°C	4 days at 18–20°C
Film appearance	clear smooth	clear smooth
Pencil Hardness	H-2H	H-2H
Elasticity	1 mm	1 mm
Impact	40–50 kg/cm	50 kg/cm
Adhesion mark	1–2	1
Coefficient of chemical resistance		
• H_2SO_4 10% at 60°C	0.75–0.80	0.90–0.95
• NaOH 10% at 60°C	0.80–0.85	0.95–1.0
• H_2O	0.85–0.90	0.95–1.0

- on the base of mono-CC contained acid esters and a polyols (P53);
- by reaction of trioxyalkyl amine with (2-oxo-1,3- dioxolan-4-yl)methyl chloroformate in an aqueous/organic two-phase system (P59);
- by homo- or copolymerization of alkylidene-1,3- dioxolan-2-one monomers (P61).

Water-dispersible, cyclocarbonate-functionalized vinyl copolymer binder and an amine curing agent are described in P50.

Practical application of HNIPU on the basis of the epoxy-amine compositions and five-membered CC (1,3- dioxolan-2-ones) in coatings, sealants, adhesives, etc., were largely developed by O. Figovsky, V. Mikheev, V. Stroganov et al. in the 1970–1990s (P10–P15, P17–P21, P23–P25, P29–P31).

Polymate Ltd.-INRC (Israel)[a] proposed compositions based on oligomer systems which contain hydroxylamine adducts on the base of aliphatic mono- and polycyclic carbonates (Cycloate A) as hardeners [87]. This composition was used for 100% solid flooring coating with high abrasion resistance and mechanical properties. As a result practically used formulations on the basis of HNIPU, using the two formulations F1 and F2 was developed (Table 1.29) [59].

Recently Polymate Ltd. developed a new hybrid epoxy-amine hydroxyurethane network polymer with lengthy epoxy-amine chains and pendulous hydroxyurethane units [P68]. The cured linear hybrid epoxy-amine hydroxyurethane-grafted polymers by novel structure have a controlled number of crosslinks and combine increased flexibility with well-balanced physical-mechanical and physical-chemical properties of conventional epoxy-amine systems. In particular, new materials have tensile strength up to 12 MPa and elongation at break 70–275%. They may be used for various applications, for example, for manufacturing of synthetic/artificial leather, soft monolithic floorings, and flexible foam.

[a]Polymate Ltd.-INRC was awarded *2015 Presidential Green Chemistry Challenge Award, USA, 2015* (together with company NanoTech Industries, Inc. USA) "For developing a safer, plant based polyurethane for use on floor, furniture and in foam industries...."

Table 1.29 Composition and properties of HNIPU compounds

Composition	Parts by weight F1	F2
Hydroxyl-amine adduct "1" (on the base of Cycloate A)	50.0	—
Hydroxyl-amine adduct "2" (on the base of Cycloate A)	—	50.0
Epoxy resin D.E.R. 324 of Dow Chemical	45.0	40.0
Polycyclic carbonate Cycloate A	5.0	—
Reactive acrylic oligomers (mixture) of Sartomer	—	10
Titanium dioxide	5.0	5.0
Carbon black	—	0.1
Byk-A530 (surface active additive of Byk Co.)	2.0	—
Byk-320 (surface active additive of Byk Co.)	—	1.5
Properties	**Values**	
Mixed viscosity, 25°C (mPa·s)	1,450	970
Pot life, 25°C (min)	30–60	30–60
Tack free, 25°C (h)	4	6
After seven days room temperature, substrate—concrete		
60° film gloss	100–105	115–120
Hardness (Shore D)	70–80	70–80
Tensile strength (MPa)	50–60	60–70
Elongation at break (%)	5–7	3–4
Taber abrasion, 1,000 cycles/1,000 g, CS-17 wheel (mg)	27	29
Impact resistance (N.m)	20	20

Source: [59]

Following is the schematic structural formula of the novel polymer:

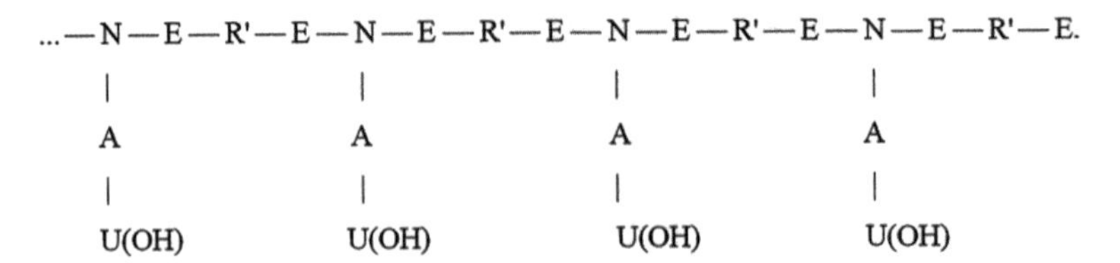

where E—R′—E is a residue of a diglycidyl ether, which reacted with amine hydrogens; E is a converted epoxy group, i.e., $-CH_2-CH(OH)-CH_2-O-$; N is a nitrogen atom; A is a residue of a di-primary amine; U(OH) is a hydroxyurethane group, i.e., $-R^1-NH-CO-O-CH(R^2)-CH(OH)-R^3$, and =N—A—U(OH) is a residue of aminohydroxyurethane with the number of free amine hydrogen atoms equal 2.

Following is a schematic structural formula of the novel polymer with the directions of the possible crosslinks (shown by arrows):

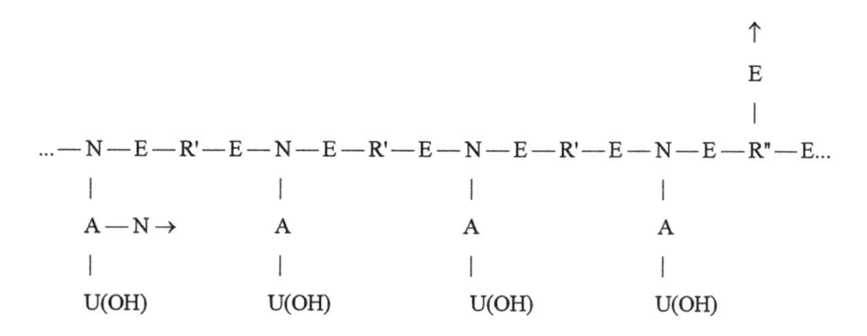

where $E-\overset{E}{\overset{|}{R''}}-E$ is a residue of the polyfunctional epoxy resin, other designations being the same as above. Polyamines with a number of free amine hydrogen atoms more than 2 also may be used for crosslinking.

1.3.3.3 Application HNIPU for flooring and paints

Industrial floors are one of basic elements of a building and simultaneously its most loaded part. The primary function of a floor is to act as a work surface for the manufacture, storage and/or movement of raw materials or finished goods. Floors experience the enormous loadings connected with abrasive and mechanical deterioration, thermal loadings, chemical and impact influences, etc.

The general requirements for industrial floors are as follows [71]:

- ***Wear resistance***

  Floors are exposed to very significant abrasion influence under the operating conditions. Abrasive deterioration is the most destructive factor attacking on floors.

- ***Mechanical load resistance***
  Where floors are exposed to the greatest mechanical deterioration, coverings with raised strength characteristics should be applied. Heavy load trolleys with small diameter wheels produce significant shear stresses in covering.
- ***Impact resistance***
  Falling of heavy articles on a floor should not result in damage of floor covering solidity and defects.
- ***Chemical resistance***
  The floor covering should protect the substrate from destroying influence of chemically aggressive media and to keep in this case its operational characteristics.
- ***Temperature resistance***
  Temperature difference is one of the main reasons of a floor covering destruction. In deciding on a polymeric covering it is necessary to consider the working temperature, probability of local increases and downturn of temperature of a floor.
- ***Impenetrability for liquids***
  Protection of an environment demands impenetrability of floor covering to action of aggressive media washing-up liquids, vapor, etc. It is especially important for the floors exposed to liquids.
- ***Crack resistance***
  Crack resistance of a floor covering increases its service life. Crack resistance is necessary for coverings in industrial refrigerators, open parking places, entrance stages, etc. Ability of a polymeric floor covering to overlap cracks considerably raises impact resistance of a covering.
- ***Resistance to sliding***
  Resistance to sliding is a traffic safety of pedestrians and transport. A floor covering should satisfy this requirement both in dry and in a damp condition.
- ***Fire safety***
  Modern floor coverings should prevent distribution of a flame in case of a fire. It is especially important for the floors in a zone of evacuation: lobbies, staircases and elevators halls, general corridors, foyer, etc.

- ***Effective sound absorption***
  Good sound proofing and sound absorption of floor coverings are very important requirements at construction, reconstruction and repair of buildings.
- ***UV radiation resistance***
  This requirement is important for spaces with the big glazing and for open-air surfaces. The floor should not fade on the sun and change its operational properties under influence of a UV radiation.
- ***Antistatic, current-carrying or dielectric properties***
  In premises with sensitive electronic equipment, in "pure rooms" of pharmaceutical plants, in spaces with high probability of formation of explosive dust concentration in air a floor covering should not accumulate static electricity charges.
- ***Attractive appearance and color palette***
  Competently selected color gamut exerts positive influence on the person, its physical condition, mood, labor productivity. Color is used for separation of various working zones from each other.
- ***Easiness of cleaning and service***
  Floor coverings should prevent from pollution penetration; thus general expenses for cleaning and service of premises will be reduced.
- ***Fast putting into operation***
  The easy of floor covering structure, small laboriousness, and short work cycle allow to put new buildings into service quickly, reduce time of reconstruction or repair. Fast repair is a controlling consideration of a continuity of technological processes.
- ***Placement of a floor on freshly placed or damp concrete***
  As a rule placement of a covering on the concrete basis is made later 28 days after the concrete works termination. Now there are polymeric floor coverings, allowing placement in 3–5 days after concrete laying.
- ***Hygienic properties of coverings***
  Pharmaceutical, cosmetic, food, chemical, and electronics industries have very high requirements of hygiene. These industries

require "pure premises" where floors should be absolutely dust-free and can be easily cleaned.

From the big variety of the polymeric materials applied for flooring, polyurethane coverings take a special place. Polyurethane floors have important advantages over all known coverings (concrete, linoleum, tile, etc.) on a number of parameters. Monolithic coverings on the polyurethane basis have high mechanical strength at compression and tension, wear resistance. They are elastic, have high chemical stability to aggressive environments action including acids, alkalis, solvents, oils, etc. During operation a floor covering on polyurethane basis endures high differences of temperatures and greater impact loadings.

The fields of application of monolithic floor coverings are quite extensive and highly diversified. These are industrial workshops and adjoining platforms, garage complexes and multistorey parking, car-care centers and car washes, warehouse and shopping centers, industrial refrigerators and freezing chambers, sports constructions, corridors, staircases, etc.

Durability and chemical resistance of monolithic floor covering depends on properties of a binder. A high build brush or roller applied polyurethane coating for concrete, granolithic, sand/cement and polyester type of a base course allows long-term stability of monolithic covering and high crack resistance. It has excellent resistance to attack from spillage of a wide range of contaminants. However, hydrolytic stability of existing polyurethanes is quite poor owing to a lot of pores. Other essential lack is process of manufacture of polyurethanes with use of toxic isocyanates.

Hybrid nonisocyanate polyurethanes are excellent example of material could be applied for monolithic floor coverings having high chemical-, crack-, wear-, and fire resistance and minimal absorption at toxic or radioactive contaminations.

Quantum-mechanical calculation, IR and NMR spectroscopic investigations have confirmed the stability of the resulting product with essentially low hydrolytic activity. The material does not have pores because the reaction of its formation is insensitive to the moisture of fillers or substrate surface. The chemical resistance of the obtained material containing intramolecular hydrogen bonds is

Table 1.30 Basic operational properties of the covering

Properties	Indices
Solid (%)	100
Pot life (doubling initial viscosity) (hours)	2–4
Curing time	4–7 days at 18–22°C
Film appearance	Clear smooth
Pencil hardness	>2H
Impact (Kg cm)	50
Coefficient of chemical resistance:	
• H_2SO_4 10% at 60°C	0.90–0.95
• NaOH, 10% at 60°C	0.95–1.0
• H_2O, at 60°C	0.95–0.90

Source: [71]

1.5–2 times greater than material of the same chemical structure but without such bonds.

Mechanical and operational properties of two-component HNIPU flooring illustrated in Table 1.30 [74]. Each test sample includes 150 weight parts (w.p.) of quartz filler per 100 w.p. of the binder.

Flooring and paint application developed by company Polymate Ltd.-INRC is shown in Table 1.31.

Table 1.32 and Fig. 1.25 [71] illustrate physical-mechanical, chemical, and operational properties and general view of the HNIPU flooring ECPU 5851W.

1.3.3.4 UV curable HNIPU floorings and coatings

The photochemistry involved in UV curable materials is very complicated and usually is tailored to the specific process with its method of application, UV source, pigments, and desired properties of the cured material. UV curable concrete floor coatings offer a durable, high-performance and eco-friendly opportunity for UNIPU

Table 1.31 Flooring and paint selection criteria

Trade name	Main properties and applications		
	Floorings		
ECPU 5851W	High chemical, wear and impact resistance including high-humidity media and special sanitary requirements	Indoor application	Curing at RT
ECPU 5851 W-FC			Fast curing at RT
ECPU 5851 LP			Extended pot life
ECPU 3968 K			Curing at low temperature
ECPU 4761 W			
ECPU 4761 S		Outdoor application,	UV resistant Ultra UV resistant
	Paints and varnishes		
ECPU 5851 W-P	High chemical resistance/high adhesion for substrates metals, concrete, ceramic, plastic and wood	Indoor application	Corrosion resistance
ECPU 5851 W-CP			
ECPU 4761 W-P		Outdoor application UV resistant	Varnish Ultra UV resistant
ECPU 4761 S			

(a)

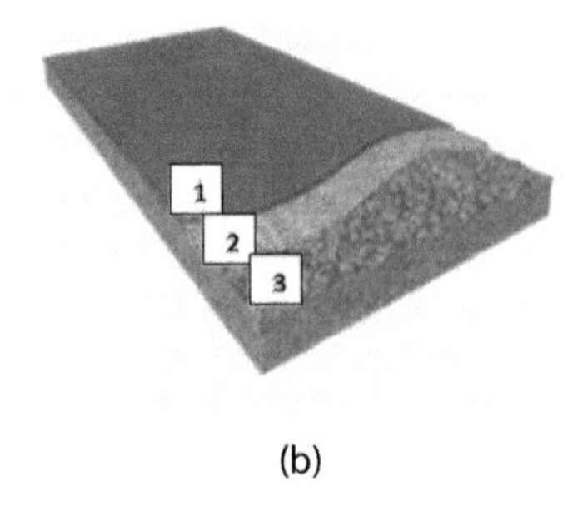

(b)

Figure 1.25 (a) Floor covering based on ECPU 5851W and (b) fragment of flooring: 1, substrate; 2, primer; 3, covering ECPU 5851W with a of thickness 0.5–3 mm [71].

chemistry. These thin-film coating systems cure instantly, thereby minimizing the downtime of any facility. Other benefits include excellent chemical resistance, easy cleanability, little odor, and the ability to coat in cold conditions.

The surface preparation and application of UV coatings is similar to that of traditional concrete floor coatings. UV coating systems

Table 1.32 Properties of the hybrid nonisocyanate polyurethane floor covering ECPU 5851W

Properties	Indices
Operational properties	
Ratio of components on weight (binder: hardener)	100:55
Standard colors	According to catalog of basic colors
Pot life	40–60 min at 25°C
Substrate	Concrete, cement cover, asphalt, wood, etc.
Primer	Depends on substrate
Thickness of the coating	0.5–3 mm
Application temperature	15–25°C
Curing time	Walk-on in 24 h at 25°C, full chemical load in 7 to 10 days
Physical-mechanical properties (after 10 days at 25°C)	
Viscosity of binder Brookfield RVDV II, (Spindle 29, velocity 100 rpm)	No more 3000 cps at 25°C
Rupture strength	3–4 kg/mm^2
Elongation at rupture	4–8%
Hardness (Shore D)	More than 72
Chemical properties (after 14 days at 25°C)	
Water Petrol Aviation petrol Oil NaCl (5%)	Resistant
Formaldehyde Sulfuric acid (20%) NaOH (20%)	Resistant during short time
Benzene Alcohols	No resistance

Source: [71]

include both clear and pigmented systems. The clear system consists of a primer and a topcoat, which is available in different finishes ranging from high gloss to matte. The topcoat finish can be further enhanced by broadcasting additives for decorative or performance purposes.

Single-coat systems are also available. The thin film thickness can range between 0.3 mm and 0.8 mm. Unlike conventional UV

Table 1.33 Properties of HNIPU UV-cured flooring compared to conventional UV-cured flooring

Properties	Standard	Conventional UV-cured flooring	HNIPU UV-cured flooring
Adhesion	ASTM D 3359-07, B	3B	5B
Pencil hardness	ASTM D 3363-05	3H	
Solvent resistance	ASTM D 5402-06	200+	4H
Gloss	ASTM D 523	84	90
Abrasion resistance, CS-17, 1,000 g, 1,000 cycles, mg	ASTM D 1044	150–200	100
Thickness applied, mm		0.065–0.1	03–0.8
Primer		Required	Not required for properly prepared substrates
Number of layers		2+	1

Source: [59]

curable coatings, formulations developed by us with higher elasticity while maintaining the basic strength characteristics. The use of these compounds improves the adhesion of the cured composition to concrete, allowing reduction in the number of coating layers to two, and for special coatings even to a single layer. The uniqueness of this compound is the possibility to apply one layer up to 0.8 mm thickness, at the rate of polymerization which allows the use of standard curing technology and standard equipment. The introduction of the new hydroxyurethane modifier (HUM) based on vegetable raw materials and adducts obtained on its basis allow improving hardness and wear resistance, while maintaining the other properties of the system (Table 1.33) [59].

The uniqueness of the developed formulation and the possibility of coating concrete sometimes without a primer, with a layer thickness of 0.3–0.8 mm, allows the covering of even open areas. Application is done by spraying, eliminating the negative effects of sunlight during the coating process and uses sunlight during the curing process, which reduces the total polymerization time even more.

1.3.3.5 Hydroxyurethane modifiers

All known polymer compositions with hydroxyalkyl urethane monomers require specific chemical reactions such as transetherification, transamination, or self-crosslinking). These reactions are carried out at elevated temperatures, in the presence of organic solvents, and/or in water-dispersion media, sometimes in the presence of catalysts [48]. Polymate Ltd.-INRC (Israel) proposed a novel concept of generating new multifunctional modifiers.

The HUM, which possesses a wide range of hydrogen bonds, is embedded in an epoxy polymer network without a direct chemical interaction. Patent P48 discloses novel "cold" cure epoxy-amine composition modified with a hydroxyalkyl urethane (HUM), which is obtained as a result of a reaction between a primary amine (C1) and a monocyclocarbonate (C2), wherein modifier (C) is represented by the following formula:

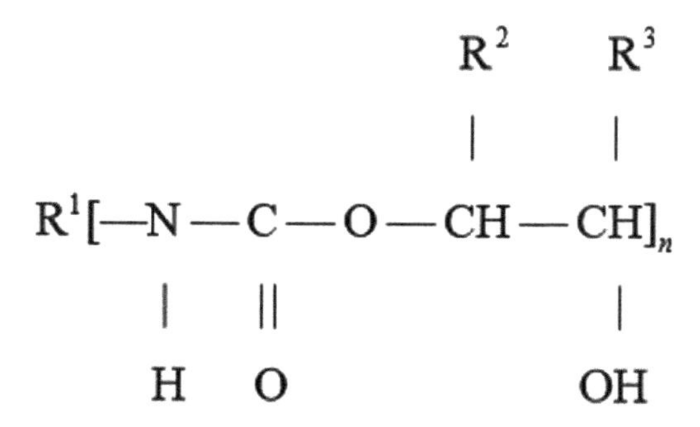

where R^1 is a residue of the primary amine, R^2 and R^3 are the same or different and are selected from the group consisting of H, alkyl, hydroxyalkyl, and n satisfies the following condition: $n > 2$. Diluents, pigments, and additives can be used. The new modifier primarily affects a significant acceleration of the curing process as well as an increase in abrasion resistance.

Doping with the HUM imparts to the cured composition superior coating performance characteristics (pot life/drying, strength-stress properties, bonding to a variety of substrates, appearance in a well-balanced state). Since the structure of the polymer network is not broken, other characteristics, such as weathering and chemical resistance, do not worsen. Use of the modifiers in different structures with varying amounts allows obtaining an epoxy-based coating with tunable hydrogen bonds. It has recently been confirmed by the positive effect of H-bonding moieties into epoxy-amine network. In the work [87] was shown, that the incorporation of acetamide into the covalent networks suggests preemptive healing behavior which helps to reduce the internal stresses arising during the curing of the coating. The fully covalent network and the hydrogen-bonded network did not show a pronounced difference in strength under dry conditions and maintain extreme good wet adhesion. Also, the introduction of hydrogen bonds on epoxyamine coatings impart a superior relaxation of the mechanical stresses and promotes flow, at temperatures just below T_g, which assists the self-healing of mechanical damage.

A nanostructured hybrid liquid oligomer composition including at least one epoxy-functional component (A), at least one cyclic carbonate component (B), at least one amine-functional component (C), and, optionally, at least one acrylate (methacrylate) functional component (D), wherein at least one epoxy, amine or acrylate (methacrylate) component contains alkoxysilane units was described in patent application [P47]. The composition is highly curable at low temperatures (approximately 10 to 30°C) with forming of nanostructure under the influence of atmospheric moisture and the forming of active, specific hydroxyl groups by reaction of cyclic carbonates with amine functionalities. According to the present invention, the cured composition has excellent strength-stress properties, adhesion to a variety of substrates, appearance, and resistance to weathering, abrasion, and solvents.

Hybrid epoxy-hydroxyurethane compositions crosslinked at ambient temperatures were obtained on the base of renewable raw materials [P63]. In the compositions were used HUMs on the base of carbonated-epoxidized soybean oil and monoamines, without the use of isocyanate intermediates. Compositions can apply to the preparation of curable polymeric foam and other materials (coatings, sealants, adhesives).

A non-isocyanate composite modifier was used in the acrylic radiation-curable composition [P66]. The additive comprises (a) a bio-based hydroxyurethane additive of formula (1.1):

$$R^1[-NH - COO - CR^2H - CR^3H(OH)]_2, \qquad (1.1)$$

where R^1 is a residue of the bio-based primary diamine, and R^2 and R^3 are the same or different and are selected from the group consisting of H, alkyl, and hydroxyalkyl; and b) a silane-based hydroxyurethane additive of formula (1.2):

$$(R^6)_{3-n}\,(OR^5)_n\,Si - R^4 - NH - COO - CR^2H - CR^3H(OH), \qquad (1.2)$$

where R^2 and R^3 are the same as stated above, R^4 is generally an aliphatic group having from 1 to 6 carbon atoms, R^5 and R^6, independently, are hydrocarbon radicals containing from 1 to 20 carbon atoms and selected from the group consisting of aliphatic, cycloaliphatic, and aromatic groups or combinations thereof, and n is equal to 1, 2, or 3.

Two advanced curing agents were elaborated: Uramine 5851 and Uramine 4761. Table 1.34 [59] illustrates comparative properties of Uramines and conventional amine hardeners. Comparison of Uramine 5851 with Vestamin TMD and Ancamin 2379 was carried out using epoxy resin D.E.R. 331. Comparison Uramine 4761 and Vestamin IPD was carried out using hydrogenated epoxy resin ST-3000

As shown in Table 1.34, uramines accelerate the curing of epoxy composites, improve the appearance of coatings, reduce abrasive wear. Uramines virtually have no effect on the strength and elasticity and also on the chemical resistance of the polymer. The dependence of curing characteristic and abrasion resistance from the content of HUM-01 is shown in Fig. 1.26 [59]. The following sections provide examples of successful applications of HUM into other oligomeric systems.

Summing up, we can say that Uramine 4761 and Uramine 5851 are hardeners for epoxy base compositions of novel type. Due to presence of hydroxyurethane modifiers, Uramines impart to cured compositions the best properties of epoxy and urethane coating materials. Created on the nonisocyanate base, Uramines do not have the toxicity of conventional urethanes.

Table 1.34 Comparative properties of Uramines and conventional amines

Technical data	Sample 1	Sample 2	Sample 3	Sample 4	Sample 5
Ratio: **A (DER-331): B (Curing agent)**	100:21	100:46	100:55	100:19	100:25
Pot life, min	45	20	10	240	60
Gel time (min)	65	30	20	360	120
Dry to touch (h)	>20	>2	1.5	8	3.5
Shore D, 25°C, 24 h	79	75	75	70	70
Shore D, full cure	82	80	81	80	78
Tensile strength (kg/mm^2)	6.4	5.8	6.6	5.5	4.7
Ultimate elongation (%)	3.1	4.4	4.8	4.4	5
Abrasion, Taber, loss of mass (mg/1000 cycles)	57	36	20	58	31
Weight gain at immersion (%)					
• in water (24 h @ 5°C)	0.1	0.3	0.2	0.2	0.3
• in 20% H_2SO_4	0.7	0.5	0.4	1.2	1.2
• in 20% NaOH	0.4	0.2	0.2	0.1	0.1
Appearance	Sticky 24 h, semi gloss	Gloss	Gloss	Matte surface	Gloss

Source: [59]
Sample 1: DER 331 + TMD; Sample 2: DER 331 + Ancamine 2379; Sample 3: DER 331 + Uramine 5851; Sample 4: ST-3000 + IPD; Sample 5: ST-3000 + Uramine 4761.

Uramines aimed for receiving a wide range of coating materials (clear or colored, filled floorings, paints, etc.) with needed properties depending on the base used, in particular,

- low VOC, 100% solid;
- water-based;
- high abrasion resistance (20–30 mg at 1000 cycles, 1000 g, CS-17, ASTM D4060);
- high weather resistance (at least 3000 h, UVA, ASTM G154);

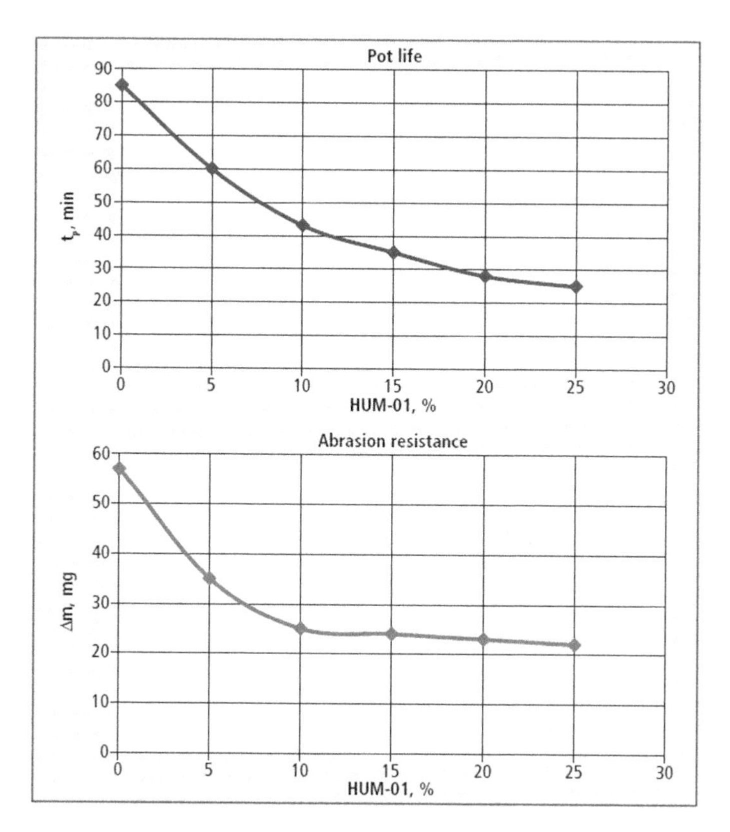

Figure 1.26 Influence of modifier HUM-01 (trimethyl-hexamethylene-diamine + propylene carbonate) on the properties of epoxy composition based on D.E.R. 331 [59].

- improved chemical resistance, flexibility, impact resistance, adhesion to substrates, etc.;
- fast curing (at 25°C.: gel time 20–60 min; thin film set time 1.5–2.0 h), but if needed gel time may be prolonged to 4 h;
- excellent appearance.

1.3.3.6 Hydroxyurethane compounds from renewable plant-based raw materials

In contrary to the polymerization of carbonate ring containing monomers, molecules containing multiple cyclic carbonate rings are also available by an addition of carbon dioxide to the appropriate multi-epoxy compounds. The largest class of such reagents is epoxidized vegetable oils or their derivatives like fatty acids or their dimers [35]. The reports on carbonation of epoxidized vegetable oils are in most cases based on similar procedures using comparable catalytic systems and reaction conditions (Table 1.35) [59].

Table 1.35 The reaction conditions of carbonation of epoxidized vegetable oils

No.	Exposited vegetable oil	Carbonation condition
1	Soybean Epoxol 7-4	TBAB* (2.5 mol% per epoxy groups) CO_2 (5.65 MPa), 22 h, 140°C
2	Soybean Shanxi Chemical Factory, China	TBAB and $SnCl_4 \cdot 5H_2O$ (2.5 mol% per epoxy groups) CO_2 (high pressure), up to 30 h, 140°C
3	Soybean (ESO, Vikoflex 7170) Arkema	TBAB (5.9 MPa), 46 h, 140°C
4	Vernonia oil Vertech, Inc	TBAB (5.9 MPa), 46 h, 140°C
5	Soybean, Flexol EPO, Dow Chemical	TBAB (1.25–5 mol% per epoxy groups) CO_2 (5.65 MPa), 110–180°C
6	Fatty acids dimers from sunflower oil	TBAB (3 wt%) CO_2 (5 MPa up to 18.5 MPa), 60–140°C
7	Soybean oil Paraplex G-62 from CP.Hall Co.	TBAB (5 mol% per epoxy groups) CO_2 (medium flow), 70 h, 140°C
8	Soybean oil from Cognis	TBAB (3 mol% per epoxy groups) CO_2 (1 MPa), 20 h, 140°C
9	Linseed oil (ELSO from HOBUM Oleo chemical	TBAB (3 mol% per epoxy groups) CO_2 (1 MPa), 20 h, 140°C
10	Soybean oil NOPCO Colombiana	TBAB (3–7 mol% per epoxy groups) CO_2 (continuous flow), 100–140°C microwaves, 40–70 h
11	Soybean oil VIKOFLEX 7170 from Atofina	TBAB or TBAOH**CO_2 (3.4 MPa), 100°C
12	Cottonseed Xinjjland Wulumuqi Xinsai Oil & fat Co., Ltd., China	TBAB (1.25–6 mol% per epoxy groups) CO_2 (1–3 MPa), 100–150°C
13	Soybean oil from Cognis Turkey	TBAB, CO_2 (0.45 MPa), 12 h, 110°C
14	Linseed oil (Dehysol B316) from Cognis GmbH	TBAB (5 mol% per epoxy groups) CO_2 (medium flow), 72 h, 110°C

Source: [59]
*TBAB - Tetrabutylammonium bromide.
**TBAOH - Tetrabutylammonium hydroxide.

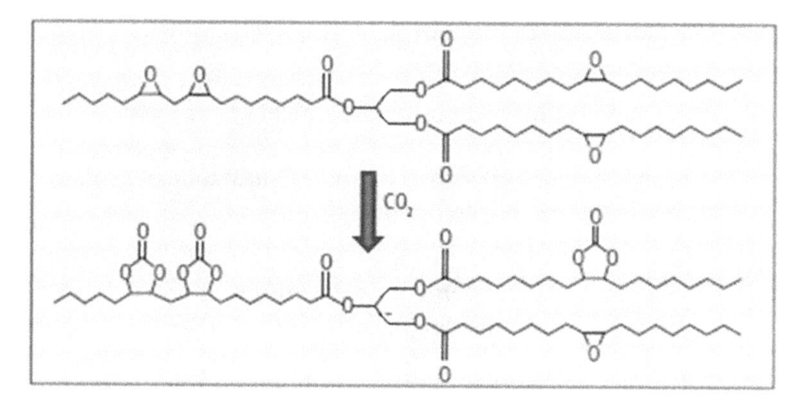

Figure 1.27 Obtaining carbonated epoxidized unsaturated fatty acid triglycerides [59].

Polymate Ltd.-INRC proposed a new method of producing a hybrid polyhydroxyurethane network comprising [31]

(a) reacting epoxidized unsaturated fatty acid triglycerides with carbon dioxide in the presence of a catalyst to obtain carbonated-epoxidized unsaturated fatty acid triglycerides, wherein conversion of oxyrane groups to 2-oxo-1, 3-dioxolane groups (cyclic carbonate groups) for said carbonated-epoxidized unsaturated fatty acid triglycerides ranges from 35% to 85% (Fig. 1.27) [59];
(b) mixing and reacting the carbonated-epoxidized unsaturated fatty acid triglycerides with a compound having an amine functionality comprising at least one primary amine group realized at stoichiometric or within nearly balanced stoichiometry;
(c) mixing and reacting the product of (b) with a compound having amine functionality comprising at least two primary amine groups realized at excess of an amine-functional compound;
(d) mixing the product of (c) with a compound having amino-reactive groups and selected from the group comprising

- a compound having epoxy functionality, and
- a mixture of the compound having epoxy functionality with carbonated-epoxidized unsaturated fatty acid triglycerides, a ratio of the sum of amino-reactive groups to the

sum of amine groups being stoichiometric or within nearly balanced stoichiometry;

(e) curing the resulting composition at ambient temperature.

The proposed method can significantly reduce time of synthesis and improve quality of the final products.

1.3.3.7 Silane-containing and nanostructured hydroxyurethane compounds

The concept of generating silica from alkoxysilanes by the sol-gel method within a macromolecular organic phase (in situ) is widely known in the art. The organic and inorganic components of these materials are present as co-continuous phases of a few nanometers in lateral dimensions. The new types of NIPUs based on cyclic carbonate-epoxy resin systems and aminoalkoxysilanes were studied [89]. The proposed dendro-aminosilane hardeners give the possibility for the introduction of siloxane fragments into the aromatic structure of BPA epoxy-amine and cyclocarbonate network polymers which improves the service properties of the network polymer. Additional hydrolysis of organosilane oligomers creates a secondary nanostructured network polymer.

Known in the art as hybrid organic-inorganic compositions include mixtures of epoxy resins, amine hardeners, functional silanes, and/or polysiloxanes and cure in the presence of water in an amount sufficient to bring about substantial hydrolytic polycondensation of the silane [71].

A novel nanostructured hybrid polymer composition was synthesized on the base of epoxy-functional components, cyclic carbonate components, amine-functional components, and acrylate (methacrylate) functional components, wherein at least one epoxy, amine, or acrylate (methacrylate) component contains alkoxysilane units [29]. The composition is highly curable at low temperatures (approximately 10–30°C) with generating nanostructure under the influence of the forming of active, specific hydroxyl groups by reaction of cyclic carbonates with amine functionalities.

These hydroxyurethane functionalities activate by hydrolytic polycondensation of alkoxysilanes by means of atmospheric

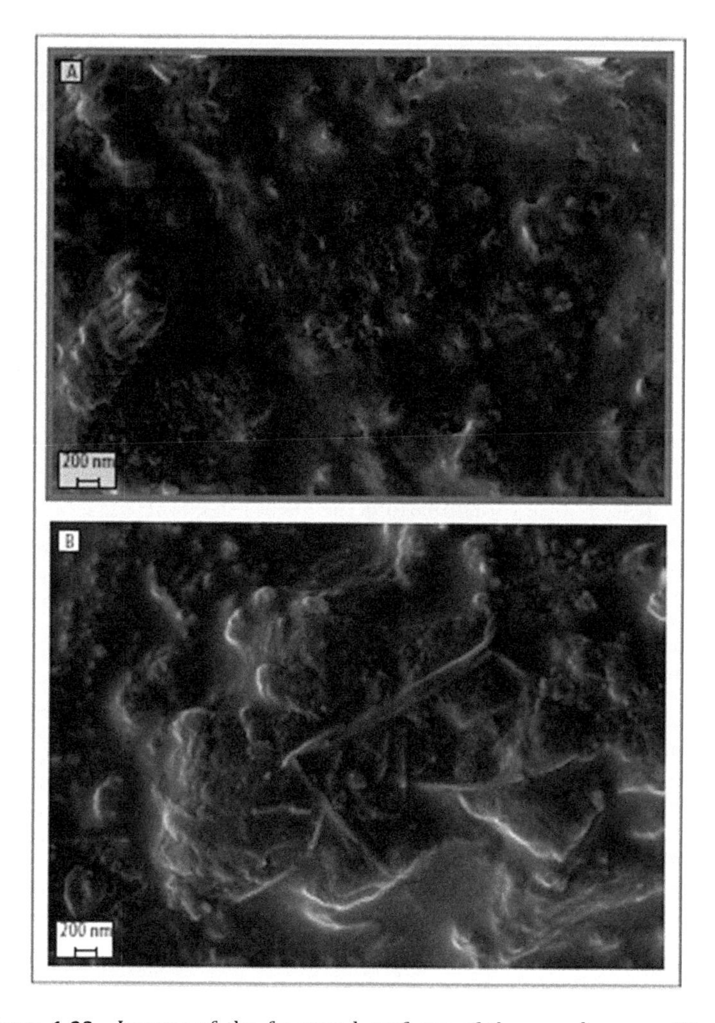

Figure 1.28 Images of the fractured surfaces of the cured compositions at 20,000× magnification (Auriga cross-beam): A, Neat epoxy-amine composition; B, epoxy-silane-amine composition [59].

moisture, thus producing an organic-inorganic nanostructure without a special procedure of water embedding or addition of nanofillers. The cured composition has excellent strength-stress properties, adhesion to a variety of substrates, appearance, and resistance to weathering, abrasion, and solvents (Fig. 1.28) [59].

1.3.3.8 Sprayable foam

The vast majority of methods for the application of sprayable polymer foams onto various substrates use air or airless spraying equipment. The main advantage of these methods is the rapid formation of a polymer structure to obtain a non-flowing foam on vertical surfaces. However, non-isocyanate resin foams require some other approach since they exhibit longer durations of gelation and solidification, which can lead to flow on vertical surfaces and a collapse of the foam.

Olang [90] disclosed hybrid spray foams that use a urethane reactant, a crosslinker, and an (optional) epoxy and/or acrylic resin, along with a blowing agent and rheology modifier to produce a quick-setting foam that remains in place until the foam forms and cures. In some other formulations the author used the NIPU adducts of cyclic carbonates and di- or poly-amines. Unfortunately, the use of rheology modifiers in practice increases the viscosity of the compositions and imparts to them for foams.

As known the foam-formation process in which a blowing agent forms cells in a resin during curing, depends on a number of factors. Most importantly are the rate of cure and the blowing gas generation rate, which must be properly matched. The aforementioned components that define a foamable nonisocyanate polymer composition form a relatively slow reacting system. At ambient temperature, the gel time of such compositions is not less than 5 min. On the other hand, premature application of the forming foam product onto a substrate must be avoided because as soon as the foam composition is applied, the foam rapidly expands, and this may cause the expanding foam to collapse as a result of inadequate strength of the walls surrounding the individual gas cells. In other words, synchronization of curing and foaming processes is a very important factor that is not provided by conventional methods of producing sprayable non-isocyanate polymer foams.

As noted above some non-isocyanate compositions related to hybrid systems on the basis of epoxy, hydroxyurethane, acrylic, cyclic carbonate, and amine raw materials in different combinations. Patent [91] discloses foamable, photopolymerizable liquid acrylic-based compositions for sealing applications, which include products of reaction of non-isocyanate urethane diols with methacrylic or acrylic anhydride. Patent [92] describes hybrid non-isocyanate foams and coatings on the basis of epoxies, acrylic epoxies, acrylic cyclocarbonates, acrylic hydroxyurethane oligomers, and bifunctional amines. However, all these compositions are used "in-place" (in situ) and are not suitable for spray applications.

Sprayable nonisocyanate polymer foam composition consisted of an amino-reactive component, a blowing agent, and additives. The components are separated into two parts, i.e., part (A) on the basis of an amino-reactive compound, and part (B) on the basis of an amino-containing compound.

Various mounting options provide foaming of the compositions with simultaneous application on horizontal, inclined, or vertical surfaces, or injecting the composition into voids.

Unlike conventional systems where mixing of the components occur in the mixing head of the gun, we have suggested the new method provides mixing of components and aging of the mixture.

For spray application, parts **A** and **B** are loaded under pressure into a foam-spraying apparatus (Fig. 1.29) [59]. The installation comprises containers (1 and 3) for the components **A** and **B** of the composition, respectively, and loading meters (2 and 4) which are intended for dosing the components **A** and **B** into the mixer (5) in proportions determined for the specific composition.

The reaction between the amine-containing component A and the basic component B is accompanied by the generation of heat, which is consumed simultaneously for the activation of a polymerization reaction and evaporation of the blowing agent (when a physical blowing agent is used). The polymer formation process occurs under the quasi-adiabatic conditions [93], i.e., without heat exchange with the environment, but with continuous movement of the reaction mass. In accordance with the above-described conditions, the temperature in the intermediate chamber (6) well correlates with the degree of chemical conversion of the reaction mass and, consequently, with the strength of the cell walls of the foam and their ability to retain the blowing agent.

Although in the description of this apparatus the constituents A and B are called "components." In fact, each such component may consist of several sub-components. For example, the first component

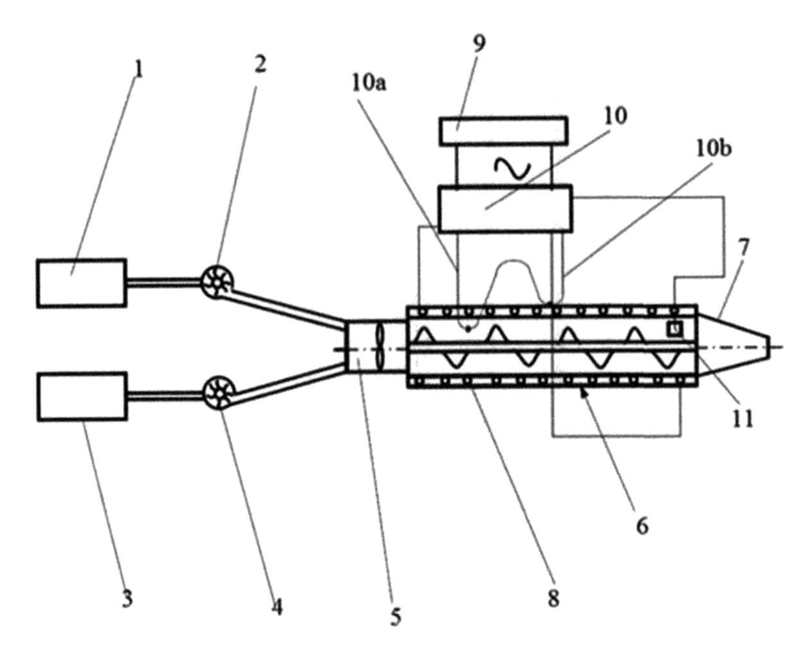

Figure 1.29 Schematic of the laboratory installation for preparing and spraying HNIPU: 1, Container for component **A**; 2, dispensing device for loading the component **A**; 3, container for component **B**; 4, metered loading device for component **B**; 5, mixer; 6, intermediate chamber; 7, discharge nozzle; 8, heater; 9, power supply; 10, temperature control unit; 10a and b, differential thermocouple; 11, temperature sensor [59].

A, which comprises an amino compound as an active ingredient, an integral component may also contain a foaming agent, surfactant, or other additives (accelerators, dyes, etc.). The second component B representing a base composition, i.e., an epoxy-cyclocarbonate component, may also contain various additives for adjusting the final properties of the foam.

With dosing meters, components **A** and **B** are delivered in predetermined proportions to the mixer (5) where the components are uniformly mixed and begin to react with each other. From the mixer (5), the reaction mixture is transferred to the intermediate chamber (6) through which the reaction mixture flows to a spraying device (7).

In the intermediate chamber the chemical process of polymer formation occurs under quasiadiabatic conditions, i.e., without heat exchange with the environment, and with continuous movement of the reaction mass. Under such conditions, the temperature is well correlated with the degree of chemical transformation and, thus, with the strength of the walls of the foam cells and their ability to retain the blowing agent.

The residence time of the reaction mixture in the chamber (6) is defined specifically for each composition and depends on the curing reaction rate and the nature of the blowing agent. The estimated residence time of the composition in the chamber (6) should provide a predetermined period needed for completing the curing reaction and forming foam directly on the insulated surface.

As a result of delay in the chamber the mixed composition reach of the state necessary for spraying. The residence time of the mixture in the chamber and the heating temperature are controlled based on the rate and the exothermic of the curing reaction as well as the boiling point of blowing agent.

Thus, the temperature of the composition in the intermediate chamber may serve as a parameter most suitable for optimal control of the foam formation process in spray application. During the heating of the reaction mixture, it becomes possible to reach the boiling point of the blowing agent, and the mixture acquires properties necessary for foaming the composition after spraying the latter onto the insulation surface. If conditions during residence of the reaction mixture in the intermediate chamber (6) are adjusted correctly, the foam applied onto the treated surface is obtained without defects and provides the required parameters.

To ensure quasi-adiabatic conditions, in addition to thermal insulation, the intermediate chamber (6) is provided with a heating system. Such an additional heating system may comprise a heater (8), for example, a resistive heater powered from a power source (9) connected with the heater (8) via a temperature control unit (10). A temperature control unit (10) includes a differential thermocouple and a temperature sensor for determining the temperature required to foam the mixture at the outlet of the intermediate chamber (6). One junction (10a) of the thermocouple assembly is located within the intermediate chamber, and the other junction (10b) is located on an insulated outer wall of the intermediate chamber (6). Turning on and off the heater (8) depends on the temperature difference between the two thermocouple junctions (10a, 10b).

It is important to note that in order to ensure continuity of the foam application process, operations associated with mixing, component supply, and mixture delivery must be strictly coordinated for continuity of the flow of the material from the loading device to the spray device (7).

In various embodiments, the installation may also include supply of compressed air, which may be required, for example, for purging the mixer (5), the intermediate chamber (6), and the exit nozzle (7) after the foam formation and application process are completed. The loading devices (2 and 4) may comprise, for example, metering pumps.

Polymate Ltd.-INRC has developed a basic formulation and technique for mixing and foaming insulating spray foam and a procedure for obtaining conventional foams has been updated in respect to HNIPU application.

Technical specifications of this newly developed composition correspond to the mean values of a similar urethane-based thermal insulation. The ultimate strength of the new material is about 1.5 times higher than that of the polyurethane foam, and the heat-insulating properties are at the same level as in polyurethane foam with open cell structures. The possibility of application by spraying was tested on pilot equipment.

The composition of the foam was studied with regard to the use of renewable raw materials, including new HUM. As a result, hard and elastic foams were obtained with properties not inferior to polyurethane foams (Table 1.36) [59].

A comparison of the properties of conventional polyurethane and HNIPU foam is given in Table 1.37 [59].

It can be seen from Table 1.37 that strength and thermal insulation properties of the newly developed foam are not inferior to conventional SPF of the same class.

Table 1.36 Rigid HNIPU foam

Properties	Standard	Rigid foam
Viscosity (Brookfield RVDV II, Spindle 29, 20 rpm) at 25°C (cP)	ASTM D2393	
• Base "A"		2800–3200
• Base "B"		3600–4100
• "A" + "B" (3–5 s after mixing)		≤ 3.700
Pot life at 25°C (77°F) (s)		8–10
VOC	ASTM D2369	Compliant
Gel time (s)		2–4
Dry to touch (s)		30–40
Curing for transportation (min)		15–20
Appearance of rigid foam		White
Compressive properties of rigid cellular plastics, 24 h (MPa)	ASTM D1621	0.2–0.4
Apparent density of rigid cellular plastics (kg/m^3)	ASTM D2369	30–40
Thermal transmission properties (h·ft^2°F/Btu.in)	C 518	4.5–5.0

Source: [59]

Table 1.37 Comparison physical properties of conventional insulation SPF and HNIPU foam

Characteristics	Conventional SPF	HNIPU
Apparent density (kg/m^3)	30–100	30–70
Breaking stress (MPa)	0.15–1.0	0.17–1.2
Thermal conductivity (W/m*K)	0.8–1.0	0.8–1.0
Number of closed cells (not less %)	>90	>60
Hygroscopic property (vol.%)	2.0–3.5	2.5–3.5
Possibility of practical use	Depends on the purpose of the premises	No limitations

Source: [59]

1.3.4 Conclusion

It is evident from the present review that considerable effort has been made during the last years to develop environmentally friendly methods of obtaining polymer materials contained urethane groups.

The most promising method is obtaining HNIPUs based on modification of different polymers by NIPU fragments. Hybrid coatings, comprised epoxy matrix modified by NIPU, are commercially available under the name Green Polyurethane™ as an isocyanate-free and phosgene-free alternative to conventional materials and represent the first successful application of HNIPUs in the industry. HNIPUs offer several advantages with respect to conventional polyurethanes. They are solvent-free, more resistant to chemical degradation, 20% more wear resistant, can be applied on wet substrates, and cured under cold conditions. They exhibit up to 30% higher adhesion than that of conventional polyurethane.

The authors believe that the most promising development ways of NIPU must include the following:

- creation of production of polyfunctional cyclocarbonates, development of optimal technology and equipment: carbonized vegetable oils and terpenes; carbonized aliphatic compounds, including chlorinated ones; and carbonized polyfunctional silicones
- development of waterborne HNIPU formulations
- development of NIPU formulations for sealants and adhesives
- development of production of amines modified with hydroxyurethane groups
- elaboration of non-amine RT curing agents for oligomer compositions
- development of self-extinguishing compositions of HNIPU
- development of silicone-based HNIPU
- development of NIPU and HNIPU-based foams
- development of formulations for UV-cured compositions

References

1. Figovsky O., Beilin D. (2015). Nanostructured composites based on interpenetrated polymer networks. Kinds, classification, properties, synthesis, application, *Engineering Journal of Don*, 3, www.ivdon.ru/ru/magazine/archive/n3y2015/3113.
2. Grigoryeva O. (2004). Reactive functionalization and compatibilization of components of interpenetrating polymer networks, *Journal Scientific Israel—Technological Advantages*, **6**(3–4).

3. Sperling L. H. (1994). Interpenetrating polymer networks: an overview, pp. 3–38, DOI: 10.1021/ba-1994-0239.ch001.
4. Shivashankar M., Mandal B. K. (2012). A review on interpenetrating polymer network, *International Journal of Pharmacy and Pharmaceutical Sciences*, **4**(Suppl 5), 1–7.
5. Myung D., Waters D., Wiseman M., Duhamel P.-E., Noolandi J., Ta C. N., Frank C. W. (2008). Progress in the development of interpenetrating polymer network hydrogels, *Polymers for Advanced Technologies*, **19**(6), 647–657.
6. Myung D., Ta C., Frank C., Won-Gun Koh, Noolandi J. (2007). Interpenetrating polymer network hydrogel corneal prosthesis, US Patent 0179605 A1.
7. Hermant I., Damyanidu M., Meyer G. C. (1983). *Polymer*, **24**, 1419.
8. Morin A., Djomo H., Meyer G. C. (1983). *Polymer Engineering and Science*, **23**, 394.
9. Djomo H., Widmaier J. M., Meyer G. C. (1983). *Polymer*, **24**, 1415.
10. Hermant I., Meyer G. C. (1984). *European Polymer Journal*, **20**, 85.
11. Jehl D., Widmaier J. M., Meyer G. C. (1983). *European Polymer Journal*, **19**, 597.
12. Jin S. R., Meyer G. C. (1986). *Polymer*, **27**, 592.
13. Jin S. R., Widmaier J. M., Meyer G. C. (1988). *Polymer*, **29**, 346.
14. Gillham J. K. (1986). *Encyclopedia of Polymer Science and Engineering*, 2nd ed., Wiley-Interscience: New York, Vol. 4, p. 519.
15. Allen G., Bowden M. J., Blundell D. J., Hutchinson F. G., Jeffs G. M., Vyvoda J. (1973). *Polymer*, **14**, 597.
16. Allen G., Bowden M. J., Blundell D. J., Jeffs G. M., Vyvoda J., White T. (1973). *Polymer*, **14**, 604.
17. Allen G., Bowden M. J., Lewis G., Blundell D. J., Jeffs G. M. (1974). *Polymer*, **15**, 13.
18. Allen G., Bowden M. J., Lewis G., Bludell D. J., Jeffs G. M., Vyvoda J. (1974). *Polymer*, **15**, 19.
19. Allen G., Bowden M. J., Todd S. M., Blundell D. J., Jeffs G. M., Davies W. E. (1974). *Polymer*, **15**, 28.
20. Blundell D. J., Longman G. W., Wignall G. D., Bowden M. J. (1974). *Polymer*, **15**, 33.
21. Thomson T. (2005). *Polyurethanes as Specialty Chemicals: Principles and Applications*, CRC Press: Boca Raton, 190 p.
22. Meier-Westhues U. (2007). *Polyurethanes: Coatings, Adhesives and Sealants*, Vincentz Network GmbH & Co KG: Hanover, 344 p.

23. Soviet Union patents: SU529197, 1976; SU563396, 1977; SU628125, 1978; SU630275, 1978; SU659588, 1979; SU671318, 1984; SU707258, 1984; SU903340, 1982; SU908769, 1982; SU1126569, 1984; SU1754747, 1992; SU1754748, 1992.
24. Potashnikova R., Leykin A., Figovsky O., Shapovalov L. (2014). Spraying of nonisocyanate polyutrethane insulative foams, *Engineering Journal of Don, 2 (in Russian).*
25. Figovsky O., Shapovalov L., Leykin A., Birukova O., Potashnikova R. (2014). Progress in elaboration of nonisocyanate polyurethanes based on cyclic carbonates, *Engineering Journal of Don*, 3, 2530 (in Russian).
26. Guan J., Song Y., Lin Y., Yin X., Zuo M., Zhao Y., Tao X., Zheng Q. (2011). Progress in study of non-isocyanate polyurethane, *Industrial and Engineering Chemistry Research*, **50**, 6517–6527.
27. Figovsky O. (2000). Hybrid nonisocyanate polyurethane network polymers and composites formed therefrom, US Patent 6,120,905 A.
28. Figovsky O., Shapovalov L. (2007). Preparation of oligomeric cyclocarbonates and their use in nonisocyanate or hybrid nonisocyanate polyurethanes, US Patent 7,232,877.
29. Birukov O., Beilin D., Figovsky O., Leykin A., Shapovalov L. (2010). Nanostuctured hybrid oligomer composition, US Patent 7,820,779.
30. Birukov O., Figovsky O., Leykin A., Shapovalov L. (2011). Epoxy-amine composition modified with hydroxyalkyl urethane, US Patent 7,989,553.
31. Birukov O., Figovsky O., Leykin A., Potashnikov R., Shapovalov L. (2012). Method of producing hybrid polyhydroxyurethane network on a base of carbonated-epoxidized unsaturated fatty acid triglycerides, US Patent Application 2012/0208967.
32. Figovsky O., Potashnikov R., Leykin A., Shapovalov L., Sivokon S. (2015). Method for forming a sprayable nonisocyanate foam composition, US Patent Application 2015/0024138.
33. Figovsky O., Leykin A., Potashnikov R., Shapovalov L., Birukov O. (2014). Radiation-curable biobased flooring compositions with nonreactive additives, US Patent Application 14/160,297.
34. Birukov O., Figovsky O., Leykin A., Shapovalov L. (2014). Hybrid epoxy-amine hydroxyurethane-grafted polymer, US Patent Application 14/296,478.
35. Moshinsky L. (2002). Dendrisynthesis of new highly branched polyaminopolyamides (PAMAM) based on the epoxy-amine reactions, *Journal Scientific Israel - Technological Advantages*, **4**(1–2).

36. Roemer F. D., Tateosian L. H. (1984). European Patent 0,014,515.
37. Figovsky O., Beilin D. (2015). Nanostructured liquid ebonite composition for protective coatings, *Journal Scientific Israel—Technological Advantages*, **17**(3).
38. Harper C. A. (ed.) (1996). *Handbook of Plastics, Elastomers and Composites*, New York, NY.
39. Weismantel G. E. (ed.) (1981). *Paint Handbook*, New York, NY.
40. Figovsky O., Blank N. (1995). Liquid ebonite mixtures for anticorrosive coverings, *NACE International; Italia Section/Associazione Italiana di Metallurgia by NACE International Italia Section*, Monza, pp. 93–596.
41. Pushkarev Y., Figovsky O. (1999). Protective vulcanizate coatings on the base of oligobutadiene, *Anti-Corrosion Methods and Materials*, **46**(4), 261–267.
42. Pushkarev Y., Figovsky O. (1999). Protective vulcanizate coatings on the base of oligobutadiene without functional groups, *Corrosion Reviews*, **14**(7), 33–46.
43. Figovsky O. (2011). Liquid solventless synthetic-rubber-based composition, US Patent 7,989,541 B2.
44. Pushskarev Y. (2012). Ebonite compositions and covering based on oligobutadienes, Burun Kniga, Kharkov, 172 p. (in Russian).
45. http://www.helyx.ru/antikorroziynaya-zaschita/gummirovanie/.
46. Merenyi S. (2012). REACH: regulation (EC) No 1907/2006: consolidated version (June 2012) with an introduction and future prospects regarding the area of Chemicals legislation, GRIN Verlag.
47. Deepa P., Jayakannan M. (2008). Solvent-free and nonisocyanate melt transurethane reaction for aliphatic polyurethanes and mechanistic aspects, *Journal of Polymer Science Part A: Polymer Chemistry*, **46**(7), 2445–2458.
48. Pan W. C., Lin C. H., Dai S. A. (2014). High-performance segmented polyurea by transesterification of diphenyl carbonates with aliphatic diamines, *Journal of Polymer Science Part A: Polymer Chemistry*, **52**(19), 2781–2790.
49. Li S. Q., Zhao J. B., Zhang Z. Y., Zhang J. Y., Yang W. T. (2015). Synthesis and characterization of aliphatic thermoplastic poly(ether urethane) elastomers through a non-isocyanate route, *Polymer*, **57**, 164–172.
50. Figovsky O., Shapovalov L., Leykin A., Birukova O., Potashnikova R. (2013). Recent advances in the development of non-isocyanate

polyurethanes based on cyclic carbonates, *PU Magazine*, **10**(4), 256–263.

51. Kathalewar M. S., Joshi P. B., Sabnis A. S., Malshe V. C. (2013). Non-isocyanate polyurethanes: from chemistry to application, *RSC Advances*, **3**(13), 4110–4129.
52. Besse V., Camara F., Voirin C., Auvergne R., Caillol S., Boutevin B. (2013). Synthesis and applications of unsaturated cyclocarbonates, *Polymer Chemistry*, **4**(17), 4545–4561.
53. Maisonneuve L., Lamarzelle O., Rix E., Grau E., Cramail H. (2015). Isocyanate-free routes to polyurethanes and poly(hydroxyl urethane)s, *Chemical Reviews*, **115**(22), 12407–12439.
54. Datta J., Wloch M. (2016). Progress in non-isocyanate polyurethanes synthesized from cyclic carbonate intermediates and di- or polyamines in the context of structure-properties relationship and from an environmental point of view, *Polymer Bulletin*, **73**(5), 1459–1496.
55. Figovsky O., Shapovalov L. (2006). Cyclocarbonate based polymers including non-isocyanate polyurethane adhesives and coatings, *Encyclopedia of Surface and Colloid Science*, V. 3, pp. 1633–1653. CRC Press, NY, Taylor & Francis.
56. Figovsky O., Beilin D. (2014). *Advanced Polymer Concretes and Compounds*, CRC Press, NY, Taylor & Francis.
57. Figovsky O., Shapovalov L., Leykin A., Birukova O., Potashnikiova R. (2013). Recent advances in the development of nonisocyanate polyurethanes based on cyclic carbonates, *PU Magazine*, **10**(4), 256–263.
58. Figovsky O., Shapovalov L., Leykin A., Birukova O., Potashnikova R. (2014). Progress in elaboration of nonisocyanate polyurethanes based on cyclic carbonates, *Engineering Journal of Don*, 3, 2530.
59. Figovsky O., Beilin D., Leykin A. (2015). Nanostructured composites based on interpenetrated polymer networks. Nonisocyanate polyurethanes based on cyclic carbonates and nanostructured composites Parts I, II, *Engineering Journal of Don*, 2; 3, 3131; 3132.
60. Figovsky O., Shapovalov L., Leykin A., Beilin D. (2012). Nanostructured hybrid nonisocyanate polyurethane coatings, *Proceeding of PPS Americas Conference*, Niagara Falls, Ontario, Canada, May 21–24, pp. 396–397.
61. Tomita H., Sanda F., Endo T. (2001). Structural analysis of polyhydroxyurethane obtained by polyaddition of bifunctional five-membered cyclic carbonate and diamine based on the model reaction, *Journal of Polymer Science Part A: Polymer Chemistry*, **39**, 851–859.

62. Zabalov M. V., Tiger R. P., Berlin A. A. (2011). Reaction of cyclocarbonates with amines as an alternative route to polyurethanes: a quantum-chemical study of reaction mechanism, *Doklady Chemistry*, **441**(2), 355–360.
63. Zabalov M. V., Tiger R. P., Berlin A. A. (2012). Mechanism of urethane formation from cyclocarbonates and amines: a quantum chemical study, *Russian Chemical Bulletin*, **61**(3), 518–527.
64. North M., Pasquale R., Young C. (2010). Synthesis of cyclic carbonates from epoxides and CO_2, *Green Chemistry*, **12**(9), 1514–1539.
65. Pescarmona P. P., Taherimehr M. (2012). Challenges in the catalytic synthesis of cyclic and polymeric carbonates from epoxides and CO_2, *Catalysis Science & Technology*, **2**(11), 2169–2187.
66. Kember M. R., Buchard A., Williams C. K. (2011). Catalysts for CO_2/epoxide copolymerization, *Chemical Communications*, **47**(1), 141–163.
67. Guillaume S. M., Carpentier J.-F. (2012). Recent advances in metallo/organocatalyzed immortal ring-opening polymerization of cyclic carbonates, *Catalysis Science & Technology*, **2**, 898–906.
68. www.specifi cpolymers.fr/medias/downloads/nipur.pdf.
69. Bernard J.-M. (2011). Method for preparing polyhydroxy-urethanes, US Patent 8,017,719; US Patent Application 2011/0288230.
70. Moeller T., Kinzelmann H.-G. (2012). Two component bonding agent, US Patent 8,118,968.
71. Leykin A., Beilin D., Birukova O., Figovsky O., Shapovalov L. (2009). Non-isocyanate polyurethanes based on cyclic carbonate: chemistry and application (review), *Journal Scientific Israel—Technological Advantages*, **11**(3–4), 160–190.
72. Muller-Frischinger I. (2011). Coating system, US Patent 8,003,737.
73. Muller-Frischinger I., Gianini M., Volle J. (2012). Coating system, US Patent 8,263,687.
74. Diakoumakos C. D., Kotzev D. L. (2012). Nanocomposites based on polyurethane or polyurethane-epoxy hybrid resins prepared avoiding isocyanates, US Patent 8,143,346; Non-isocyanate-based polyurethane and hybrid polyurethane-epoxy nanocomposite, polymer compositions, US Patent Application 2012/0149842.
75. Klopsch R., Lanver A., Kaffee A., Ebel K., Yu M. (2011). Use of cyclic carbonates in epoxy resin composition, US Patent Application 2011/0306702.

76. Mecfel-Marczewski J., Walther B., Mezger J., Kierat R., Staudhamer R. (2011). 2-oxo-1,3-dioxolane-4-carboxylic acid and derivatives thereof, their preparation and use, US Patent Application 2011/0313177.
77. Bähr M., Mülhaupt R. (2012). Linseed and soybean oil-based polyurethanes prepared via the non-isocyanate route and catalytic carbon dioxide conversion, *Green Chemistry*, **14**(2), 483–489.
78. Bähr M., Bitto A., Mülhaupt R. (2012). Cyclic limonene dicarbonate as new monomer for non-isocyanate oligo- and polyurethanes (NIPU) based upon terpenes, *Green Chemistry*, **14**(5), 1447–1454.
79. Cramail H., Boyer A., Cloutet E., Gadenne B., Alfos C. (2012). Bicarbonate precurors, method for preparing same and uses thereof, US Patent Application 2012/0259087.
80. Mahendran A. R., Aust N., Wuzella G., Müller U., Kandelbauer A. (2012). Bio-based non-isocyanate urethane derived from plant oil, *Journal of Polymers and the Environment*, **20**(4), 926–931.
81. Hablot E., Graiver D., Narayan R. (2012). Efficient synthesis of bio-based poly(amideurethane)s via non-isocyanate route, *PU Magazine International*, **9**(4), 255–257.
82. Hoşgör Z., Kayaman-Apohan N., Karataş S., Mencelocğlu Y., Güngör A. (2010). Preparation and characterization of phosphine oxide based polyurethane/silica nanocomposite via non-isocyanate route, *Progress in Organic Coatings*, **69**(4), 366–375.
83. Kathalewar M., Sabnis A. (2012). Novel bis-urethane bis-silane precursor prepared via non-isocyanate route for hybrid sol-gel coatings, *International Journal of Scientific and Engineering Research*, **3**(8), 1–4.
84. Hanada K., Kimura K., Takahashi K., Kawakami O., Uruno M. (2012). Five-membered cyclic carbonate polysiloxane compound. polysiloxane-modified polyhydroxy polyurethane resin, US Patent Applications 2012/0231184, 2012/0232289, and 2012/0237701.
85. Figovsky O., Leykin A., Shapovalov L. (2016). Non-isocyanate polyurethanes: yesterday, today and tomorrow, *International Scientific Journal for Alternative Energy and Ecology*, 03-04.
86. Benyahya S., Boutevin B., Caillol S., Lapinte V, Habas J.-P. (2012). Optimization of the synthesis of polyhydroxyurethanes using dynamic rheometry, *Polymer International*, **61**(6), 918–925.
87. Birukov O., Beilin D., Figovsky O., Leykin A., Shapovalov L. (2010). Liquid oligomer composition containing hydroxy-amine adducts and method of manufacturing thereof, US Patent Application 2010/0144966.

88. Villani M., Deshmukh Y. S., Camlibel C., Esteves A. C. C. (2016). Superior relaxation of stresses and self-healing behavior of epoxy-amine coatings, *RSC Advances*, **6**(1), 245–259.

89. Figovsky O., Shapovalov L., Buslov F. (2005). Ultraviolet and thermostable non-isocyanate poly-urethane coatings, *Surface Coatings International Part B: Coatings Transactions*, **88**(B1), 67–71.

90. Olang F. N. (2012). Hybrid polyurethane spray foams made with urethane prepolymers and rheology modifiers, US Patent Application 2012/0183694.

91. Figovsky O., Shapovalov L., Potashnikov, R., Tzaid Yu., Bordado J., Letnik D., De Schijuer A. (2005). Foamable photopolymerized composition, US Patent 6,960,619 B2.

92. Figovsky O., Shapovalov L. (2007). Preparation of oligomeric cyclocarbonates and their use in non-isocyanate or hybrid non-isocyanate polyurethanes, US Patent 7,232,877 B2.

93. Figovsky O., Potashnikov R., Leykin A., Shapovalov L., Sivokon S. (2015). Method for forming a sprayable non-isocyanate foam composition, US Patent 20150024138 A1.

Chapter 2

Nanocomposites Based on a Hybrid Organosilicate Matrix

2.1 Introduction

In this review, we discuss a particular group of nanocomposites—organic-hybrid composites. In practice, nanocomposite materials contain reinforcing elements with an extremely high specific surface area, immersed, for example, in a polymer matrix. In this case, the organic and inorganic components form independent phases, so the contact is achieved at the phase boundary.

Promising modern composite materials are those in which the organic and inorganic components interact at the molecular level. They were called "polymer hybrids" [1–4]; the concept of a "hybrid" was made in order to emphasize the nature of the molecular interaction between the components.

Hybrid materials are materials produced due to the interaction of components with different chemical properties. Most often it is the organic and inorganic substances that form a certain spatial structure. These structures differ from that of the initial reagents but often inherit certain motives and functions of the original structures.

Green Nanotechnology
Oleg Figovsky and Dmitry Beilin

ISBN 978-981-4774-10-9 (Hardcover), 978-1-315-22928-7 (eBook)
www.panstanford.com

A feature of the new composite materials is the fact that they have nanometer parameters of their structural elements. The size of at least one of the directions is not more than 100 nm. This is either nanometer distances between the lattices and the layers formed by polymer and inorganic ingredients or the nanometer size formed of particles, including particles containing metals [5].

The inorganic compounds—precursors—typically used are oxides of silicon, aluminum, titanium, zirconium, vanadium, molybdenum, clays, layered silicates and zeolites, phosphates, metal chalcogenides, iron oxychloride, graphite, various metals, etc. Carbochain and organometallic polymers (silicone polymers), are used as the polymer component.

From an environmental point of view are optimal drainless methods of obtaining composite materials, in particular, the sol-gel or spin-on-glass process. This method allows the exclusion of multiple washing steps because a compound that does not introduce impurities into the final product composition is used as a starting material [6].

A sol is a colloidal dispersion of solid particles in a liquid. A colloid is a suspension in which the dispersed phase is so small (1–1000 nm) that the gravitational forces may be neglected. Here are dominant short-range forces, such as van der Waals, and also Coulomb forces (attraction and repulsion between the surface charges). The inertia of the dispersed phase is small, so there is Brownian motion of the particles (Brownian diffusion), that is, random jumps caused by the kinetic energy imparted by the collision of the sol particles with each other and with the molecules of the dispersion medium. The important factor is that the dispersed particles are not molecules that are aggregates consisting of a plurality of molecules [4].

Colloidal gel formation occurs by a different mechanism. The particles of the dispersed phase (micelles) under the influence of attraction dispersion forces interact with each other to form a skeleton of the inorganic polymer.

The gel is obtained from a polymeric sol formed during polymerization of the monomers, and the polymers are in a sol. In this process, gradually from polymerizable branched oligomers is formed a gigantic cluster. When the cluster reaches a macroscopic size and spreads to the entire volume of the sol, it is said that there is a sol-gel transition. In this case, the gel will comprise, on the one hand, a continuous structural grid—a solid skeleton (core)—and on the other a continuous liquid phase.

Colloidal gel formation occurs by a different mechanism. The particles of the dispersed phase (micelles) under the influence of attraction dispersion forces interact with each other to form a skeleton of the inorganic polymer. The gel comprises continuous solid and fluid phases that are of colloidal size (1 to 1000 nm). These phases are continuous interpenetrating systems.

2.2 Sol-Gel Technology

The most studied of the sol-gel chemistry, certainly system based on silica that also received by the slow hydrolysis of the ester of silicic acid. In this case the formation appeared historic starting chemistry of sol-gel processes [7]. For the first time in 1845, Ebelmen used a transparent material of silica gel in the acidification of alkali metal silicates. The concept was known to scientists and chemists earlier, but no one gave the practical value of this process. At the first stage of the sol-gel process, pure silica is mainly used to form ceramic. In the early stages of the sol-gel process study, pure silicon dioxide is mostly used to form ceramic. However, it soon became clear that the process may also be used for the formation of other metal oxides [8]. Furthermore, it was shown that the mixture of several starting materials allows obtaining the materials of a more complex composition. However, in such complicated systems for achieving material homogeneity, it is necessary to know the properties and behavior of each individual component in the conditions of the synthesis implementation.

A summary of the sol-gel process is reflected in the scheme shown in Fig. 2.1 [33].

With respect to other methods for the synthesis of inorganic oxide materials, including nanoparticles [4, 9], the sol-gel process has a number of significant advantages [10]. In particular, these include the following:

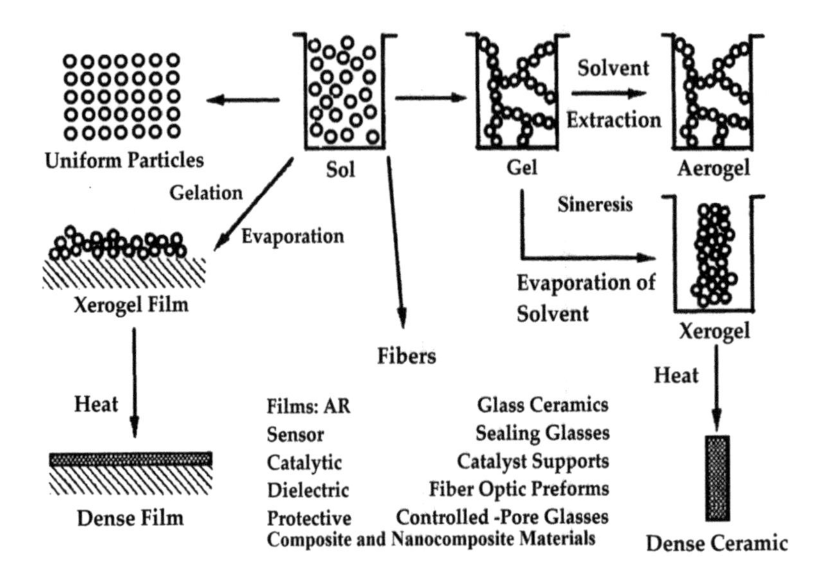

Figure 2.1 Overview of the sol-gel process [33].

- High purity in the starting material and the resulting product (especially in the case of alkoxides)
- The homogeneity of the distribution of components, including the small modifying additives
- The possibility of achieving homogeneity of the resulting compounds, which can go down to the molecular and ionic levels of the material structure
- The possibility of obtaining new crystalline and amorphous phases, materials with cations in unusual oxidation states, the synthesis of which through traditional methods is difficult or impossible
- Regulation of the rheological properties of sols and nanoparticle dispersions, which allows obtaining a wide range of products, ranging from coatings to monoliths

Typically, for the implementation sol-gel processes, two traditional approaches [11] are used, which, however, have a number of branches:

- Colloidal method: Hydrosols gelation occurring due to the association of particles in water suspension (e.g., through hydrogen bonds between groups belonging to different particles). A variation of this method is the direct deposition and polymerization of the hydrated oxides of chemical elements from solutions of their salts, such as soluble silicates;
- Alkoxide method: Hydrolytic polycondensation of the starting compounds in aqueous-organic media. The starting materials for this process may be alkoxides, nitrates, etc. Removal of the liquid phase from the obtained products structures is carried out either under atmospheric or under supercritical conditions. Recent years have seen the use of the nonhydrolytic method. This is an alternative way which consists of the interaction of a metal halide with oxygen donors—the metal alkoxide in an anhydrous medium.

2.2.1 Alkoxide Method of Sol-Gel Synthesis

There are alternative reaction schemes when forming the oxide material by precipitation, hydrothermal treatment, or a sol-gel process. The sol-gel process is the most interesting process due to high technology applications in such advanced areas as thin films in electronic or optical devices. It begins with molecular precursors, and the formation of the oxide grid occurs at rather low temperatures [12]. In contrast to classical solid-phase reactions, the material formation is usually carried out in solution. Thus, reactive reagents are dispersed at the molecular level, which provides a low diffusion length of reacting substances and thus high reaction rates under mild conditions. In addition, the molecular precursors show the advantage that they can be purified by conventional methods such as rectification and chromatography. Consequently, for the formation of materials are available very pure starting substances, which are very important in application areas such as electronics, optics, and biomedical devices.

One of the bases of nanotechnology is that the primary size of the initial structural elements formed in a sol-gel process is in the nanometer size range. There are several technologies where the sol-gel process is the most advanced state of the art, for example, wear-resistant or antireflective coatings. At present, this process is widely used in the production of nanoparticles.

The sol-gel process provides control of the structures of various length scales and thus enables the formation of hierarchically structured materials. The advantage of the sol-gel process with respect to the production of nanocomposite materials is the ability to control the mechanism and kinetics of the existing reaction steps. This allows the formation of the hierarchical materials, for example, to control the properties of materials ranging from the macroscopic to the molecular level. Moreover, because this process takes place under mild conditions, it is possible to make modifications of materials, not possible in the case of the classical high-temperature ceramic synthesis. For example, due to the low temperature and the presence of solvent, organic or biological components and groups may be included in the material structure. This makes it possible to carry out the formation of organomineral hybrid materials or nanocomposites that exhibit properties that are completely different from conventional materials [4].

Thus, the sol-gel process is more similar to the polymerization process leading to the formation of a three-dimensional ceramic structure, as in the case of formation of the polymer network. In this it differs from the classical high-temperature inorganic solid-phase process. Due to this similarity, the sol-gel process is ideally suited for the formation of nanocomposites that contain both inorganic and organic polymer structures.

The sol-gel process is a chemical reaction that starts from an ion or molecular compound and allows the formation of a three-dimensional polymeric network through the occurrence of bridging oxo-bonds between the ions (Fig. 2.2 [27]) and the release of water or other small molecules. Thus, this process is a polycondensation reaction that leads to a three-dimensional polymer network.

When applying the sol-gel process in an aqueous solution, a special kind of radical is formed in the first stage—an M–OH bond that is unstable and reacts with other types of radicals. This first step is hydrolysis. In the second step, the labile group M–OH condenses with other M–OH or M–OR (when the initial product of the sol-gel process was used alkoxides of elements) groups to form M–O–M

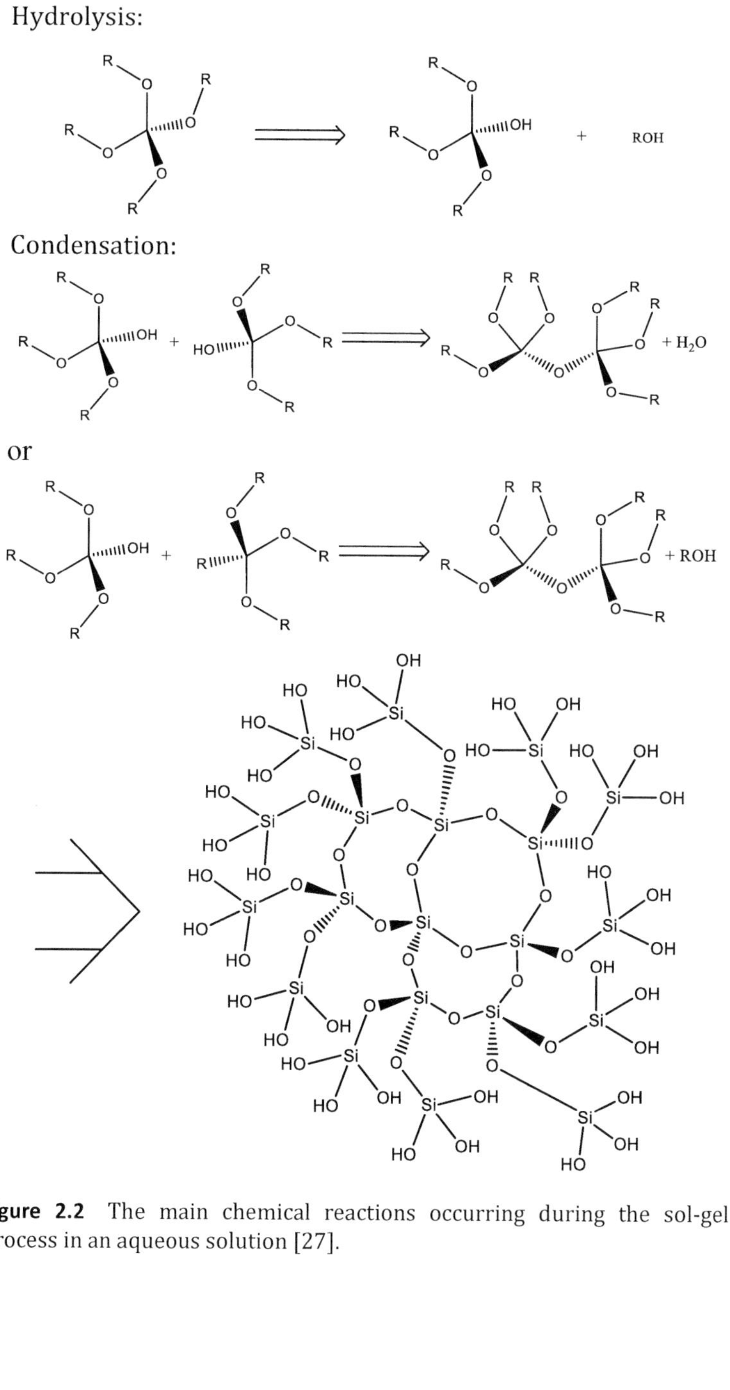

Hydrolysis:

Condensation:

or

Figure 2.2 The main chemical reactions occurring during the sol-gel process in an aqueous solution [27].

bonds and there is elimination of water or alcohol. Thus is formed a three-dimensional lattice. Typically, the obtained intermediate is not completely condensed in the process, as a consequence of steric and kinetic difficulties. Water or OH groups get included into its structure. Therefore, the products obtained can correctly be classified as hydrated oxides.

The progress of hydrolysis and condensation leads at first to the formation of solid particles that are suspended in a liquid, the so-called sol. Particles in the condensation stages contain at their surface active groups and, therefore, they are cross-linked to the gel. The gel is formed as a solid openwork net and framework that contains the liquid phase in the pores.

As a rule, the hydrolysis of silicon alkoxides is a pretty slow process. Thus, typically to accelerate the sol-gel processes, an acid or a base is used as a catalyst. The catalysts have a significant impact on the final structure of the resulting network. Furthermore, there is also a different reactivity, with no condensed or partially condensed intermediate particles of silicic acid, which leads to the formation of various silicate structures. This stage of network forming is statistical in nature. The resulting silicate structure is best described using fractal geometry.

The term "fractal" was introduced by Benoit Mandelbrot in 1975, and it became widely known with the release in 1977 of his book *The Fractal Geometry of Nature* [12]. The word "fractal" is used not only as a mathematical term. Fractal is an object that has at least one of the following properties:

- It has a nontrivial structure on all scales. This differs from regular geometric figures such as a circle, an ellipse, or a graph of a smooth function. If we consider a small fragment of a regular figure on a very large scale, it will be like a fragment of the line. For fractal, zoom in does not lead to the simplification of the structure, that is, at all scales, we can see the same complicated picture.
- It is self-similar or approximately self-similar.
- It has a fractional metric dimension or metric dimension that exceeds the topological dimension.

Typically, the acid as a catalyst leads to an extended structure similar to polymers, while the base leads to a structure consisting of separate interconnected particles. In the case of gels based on alkoxysilanes, the size, structure, and cross-linking of polymer chains depend on the ratio of Si–OR in Si–OH and the rates of hydrolysis and condensation.

For understanding mechanisms of these processes, the electronic structure of the silicon atom and silanol and siloxane bonds formed by the silicon atom should be considered. In general, there are two important differences between organic derivatives of elements of carbon subgroups from similar derivatives of boron subgroups. Thus, the elements of carbon subgroups have low polarity of the E–C bond (where E = Si, Ge, Sn, or Pb) and octet stable configuration at the central atom in the compounds of the binary type ER_4. An important issue that has a direct relation to the mechanism of the substitution reactions, the nature of multiple bonds, and the explanation of these elements is the presence of hypervalent compounds of the type $[SiX_6]^{2-}$.

Traditionally it is believed that 3d-orbitals of the silicon atom are involved in hybridization (sp^3d and sp^3d^2). However, more recent studies have shown that the d-orbitals of silicon are placed too high in energy and do not contribute significantly to the formation of bonds. An alternative to this idea is the concept of negative hyper-conjugation (Fig. 2.3 [27]). In the formation of "multiple" Si–O bonds in the silanols, antibonding orbitals σ^*(Si–C) can act as electron acceptors.

Increased acidity of silanol compounds compared with conventional alcohols explains the presence of a partial π-character of the chemical bond Si–O. Another important property of the elements, starting from silicon, is hypervalency. Hypervalency is an increase of the coordination number >4 for nontransition elements, or, in a general form, the violation of the octet rule. This phenomenon is not necessarily connected with the participation in the binding of nd-orbitals of the central atom. The structure of these compounds can be explained on the basis of three-center interactions, for example, three 4e3c–bonds $X(\sigma)$-E(p)-$X(\sigma)$, in octahedral complexes EX_6. The need to use d-orbitals as basic functions in the quantum

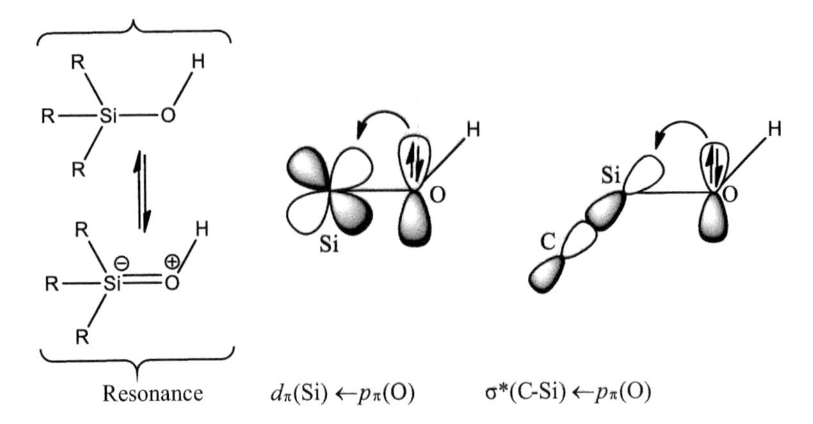

Figure 2.3 An example of negative hyperconjugation manifestation in a silanol structure [27].

chemical calculations, as hypervalent, and "ordinary" nontransition elements compounds is due to the fact that they allow taking into account the polarization of the electrons.

In the sol-gel process, an acid is catalyzed in the first stage and there is a rapid protonation of the alkoxyl group. This reaction is a nucleophilic substitution reaction. In it, the attack is carried out by a nucleophile—a reagent carrying a lone electron pair. The alkoxyl group is substituted with a water molecule by the reaction scheme with the mechanism S_N2. The S_N2 reaction mechanism or bimolecular nucleophilic substitution reaction takes place in one step without the formation of an intermediate. In this case, the nucleophilic attack and cleavage of the leaving group occurs simultaneously. The S_N2 reaction rate depends on both the concentration of the nucleophile and the concentration of the substrate:

$$r_a = k_a \times [Si(OR)_4] \times [H_3O^+] \quad (2.1)$$

$$r_b = k_b \times [Si(OR)_4] \times [OH^-] \quad (2.2)$$

Because of the reaction with a nucleophile attack, hydronium ions H_3O^+ or OH^- ions may occur on only one side. The result of the reaction is the inversion of the stereochemistry of the resulting product. This phenomenon may be useful in the preparation of biologically active nanocomposites, using the methods of stereoselective synthesis.

Stereoselective synthesis is also called chiral synthesis, asymmetric synthesis, or enantioselective synthesis. This is a chemical reaction in which stereoisomeric products are formed in unequal amounts. The methodology for the stereoselective synthesis plays a role in the pharmacology because different enantiomers and diastereomers of one molecule often have a different biological activity.

Thus, the acid-catalyzed hydrolytic reactions of nucleophilic substitution are more likely to occur at the end of the resulting oligomers, with the preferred formation of linear polymers (Fig. 2.4 [27]).

From the above, it becomes clear why there is no nucleophilic substitution at the silicon atom by a dissociative mechanism (D, S_N1), which was to take place through the formation of intermediate silyl-cations. Instead, associative mechanism participation (A, S_N2) is assumed. This assumption is confirmed by the dependence of the rate of substitution, the nature of the attacking nucleophile, slowing of the reaction in the presence of electron-donating substituents R, while also the often observed inversion of configuration at the silicon atom.

The direct use of the criterion of inversion of symmetry in silicon chemistry is not possible because of the possible increase in the coordination number of the central atom and the rearrangement of trigonal bipyramidal intermediates, through pseudorotation. Such frontal attack is possible due to the presence of free silicon atom d-orbitals and low-lying antibonding σ^* orbitals, which stabilize the coordination number 5. Often observed in nucleophilic substitution reactions, racemization can lead to an erroneous conclusion about the dissociative mechanism. Actually, however, the racemization is no evidence for the formation of intermediate silyl-cation as hypercoordinate intermediates can undergo rearrangements (pseudorotation), which may lead to loss of chirality in the resulting inversion or preservation of chirality from a statistical probability.

In an alkaline medium, polycondensation occurs much faster and the reactivity increases with the decreasing number of alkoxy groups associated with a silicon atom. The mechanism, in this

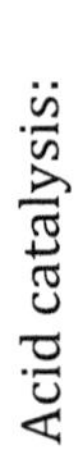

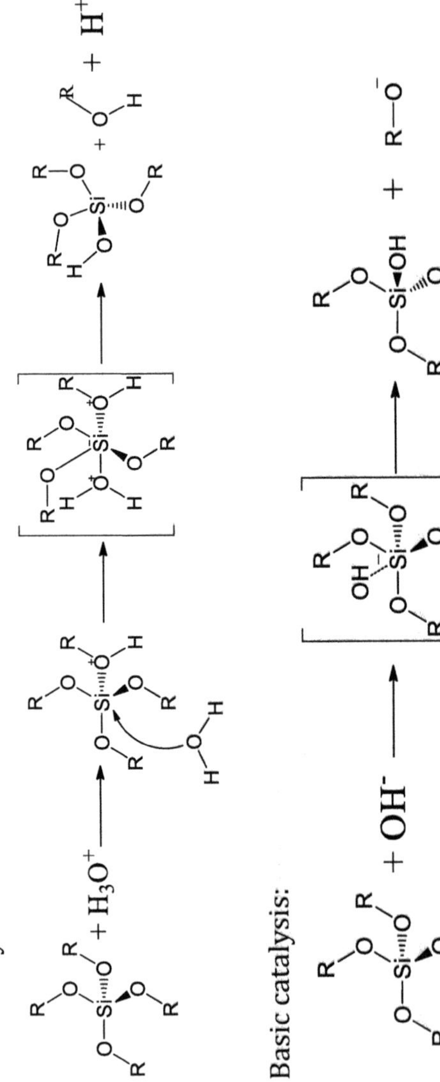

Figure 2.4 Mechanism of formation of silanol groups, depending on the catalyst that is used [27].

case, is based on the interactions of the nucleophilic hydroxyl anion with a silicon atom that belongs to the alkoxysilanes. The hydrolysis reaction occurs through the formation of a negatively charged intermediate product with a coordination number of 5. The condensation of silanol groups preferably occurs not at the chain ends, and the internal centers of oligomers, which leads to a highly branched, dense structure. Thus, the small spherical particles are formed.

Usually, simple mineral acids or metal hydroxides are used as catalysts, but fluoride ions F^- can also be used.

Sol-gel transition depends upon the concentration of the starting reactant, the amount of water, the catalyst, the temperature, and the pH. The final solid material has a plurality of surface OH groups, which can be stabilized by hydrogen bonds with the solvent and residual water. Furthermore, after completion of gelation, in a large amount in the obtained material are present residual alkoxide groups and free OH groups that are not participating in the condensation reaction. In the aging step, these groups react with each other to form additional quantities of water and alcohols. Furthermore, in the aging process is observed substance transfer of gel particles from the outer zone to the contact zone between the particles, thus increasing the size of the particles that form a gel. The aging time has great influence on textural properties of the material. A further condensation step leads to compaction of the material and compression of the gel. The aging can be accelerated by increasing the temperature. But it can lead to crack formation in pure gels.

For subsequent applications, the gel should be dried. Removing the liquid from a gel leads to a sharp compression of the gel structure; as a result, a product gets the most shrinkage as compared to the original form. Compression of the gel structure is known as syneresis. Shrinkage of the gel at syneresis can be up to 50–70% of its original size. During the syneresis occur two types of processes. First substance transfer occurs from the outside of the gel particles, which form a gel in the inner part of the zone of contact between them. Thus, in the contact area of the particles takes place the formation of bridges between particles, which are formed out of gel material. The second process is determined by the movement of the particles relative to each other, with a gradual decrease in the pore spaces in the gel. This process is also caused by the transfer of substances of the gel and certain fluidity of the gel material.

Upon receipt of a nanocomposite volatile components must be removed from the final material before its application in the respective products. This process is also important to obtain a high-quality material. When removing the liquid phase from the gel structure in the gel, pores have a free liquid surface, and thus the capillary forces arise, which tend to destroy the gel structure. If the resulting capillary forces exceed the strength of the gel structure, a phenomenon of decryptation—the destruction of the gel structure due to its cracking—takes place. In some cases, even the formation of a powdered material occurs. Therefore, the proper conduct of the operation of aging the gel and the proper carrying out of syneresis and drying processes provide high-quality products in the implementation of the sol-gel process in the preparation of nanocomposites.

Typically, as the feedstock for the implementation of the sol-gel process alkoxides of corresponding chemical elements are used. In the case of silicon, the most known alkoxides are the tetra-methoxysilane $Si(OCH_3)_4$ (TMOS) and tetra-ethoxysilane $Si(OCH_2CH_3)_4$ (TEOS). The TMOS hydrolysis rate is much higher compared to that of TEOS. Thus, as a result of the reaction, methanol is obtained. It is not always acceptable the alcohol in the sol-gel technology because of its toxicity. Both agents are liquid at standard conditions and may be purified by rectification. Typically, the substitution pattern and hence the organic radicals in the precursors have a large influence on the kinetics of the sol-gel process.

As shown above, the use of TMOS or TEOS as precursors in the sol-gel process with an average thermal treatment leads to a three-dimensional lattice of silica. However, the sol-gel process is well known for the production of hybrid materials that include organic functional groups that are attached to the inorganic lattice. This requires different starting materials that contain Si-OR groups and can be hydrolyzed and Si-C bonds that are stable to hydrolysis. The result of the use of such intermediates for the sol-gel reaction is the introduction of the organic group into the final material.

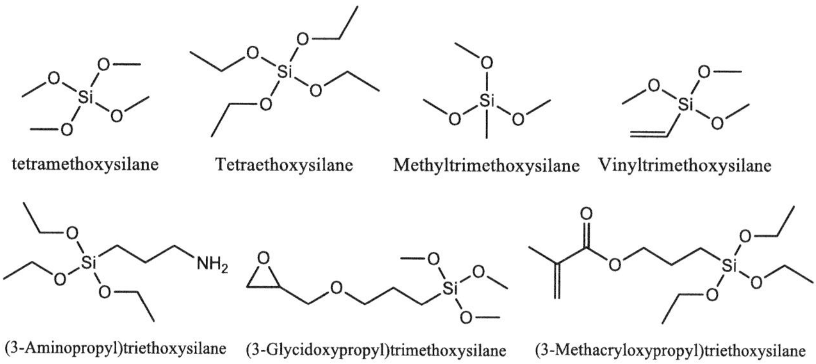

tetramethoxysilane Tetraethoxysilane Methyltrimethoxysilane Vinyltrimethoxysilane

(3-Aminopropyl)triethoxysilane (3-Glycidoxypropyl)trimethoxysilane (3-Methacryloxypropyl)triethoxysilane

Figure 2.5 Selection of commonly used alkoxysilane compounds [27].

The application of this methodology makes it easy to incorporate organic functional groups into the resulting inorganic network. The result is a final material that may bear certain organic functional groups. These groups can help to obtain materials with certain optical or electronic properties and modify the chemical reactivity and polarity of the silica lattice. Formation of the lattice may be possible only in a case where the precursor is used, having at least three possible locations for cross-linking. Both tetra-alkoxysilanes $Si(OR)_4$ and tri-alkoxysilanes $(RO)_3SiR^0$ possess this ability (Fig. 2.5 [27]).

Other alkoxides of type $(RO)_2SiR^0_2$ or $(RO)SiR^0_3$ can also react by hydrolysis and condensation, but the bis-alkoxides may form only chained molecules and mono-alkoxides form only dimers. If they are used in the ordinary sol-gel process, they only allow the modification of the inorganic network. For example, if mono-alkoxysilane is connected to the surface of the silica lattice, then the result will be a certain amount of functional groups attached to the surface of the inorganic substance. Although the mono-alkoxysilanes are not used in the ordinary sol-gel process, they can be used for surface modification of the inorganic component by surface reactions.

Molecules that contain more than one silicon alkoxide group, for example, a system containing two or more alkoxy groups—tri-alkoxide $(RO)_3Si\text{-}R^0\text{-}Si(OR)_3$—are also used in sol-gel processes [4]. These starting substances allow the insertion of the organic

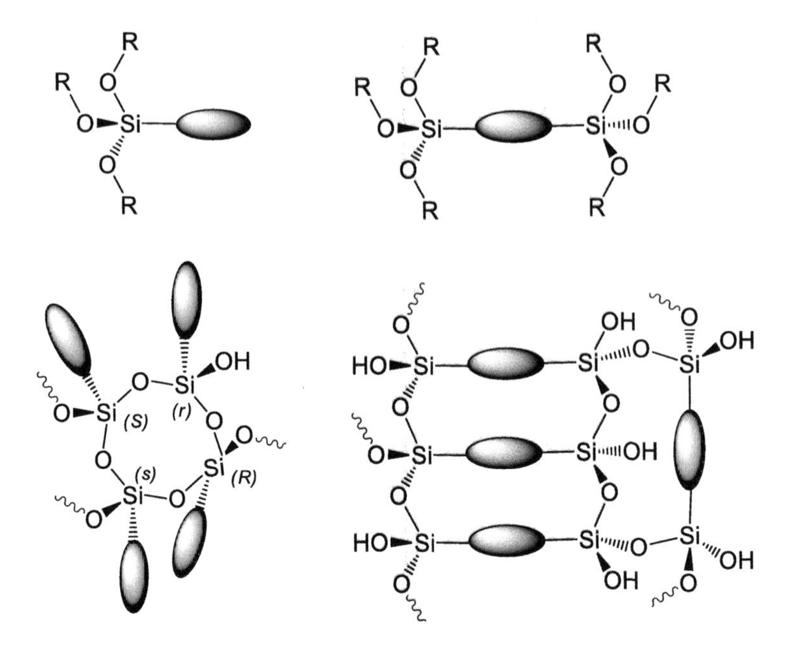

Figure 2.6 The difference between the organo-silanes of the type $(RO)_3SiR^0$ and $(RO)_3Si\text{-}R^0\text{-}Si(OR)_3$, in the reaction of formation of the siloxane-organic network [27].

functional group directly in the lattice of the solid material. This means that the organic functional groups are part of the lattice. Thus, molecules of the type $(RO)_3SiR^0$, with an attached functional group R^0, form the network (Fig. 2.6 [27]).

On the one hand, in the sol-gel processes, a mixture of tetra-alkoxysilanes and tri-alkoxysilanes is used to obtain a dense silica lattice by their hydrolysis and condensation. On the other hand, the introduction of tri-alkoxysilanes in the reaction mixture is used to include organic functional groups into the silicate lattice and for the formation of hybrid materials.

Important parameters for the optimization of the sol-gel process are ratios of the various components, for example, the ratio of water to alkoxide $C = H_2O/M(OR)_n$; the use of catalysts; and the nature of the alkoxide precursor. The kinetics of the process can change significantly depending on the type of the alkoxy-group used in the precursor.

Materials based on silicon oxide, obtained by the sol-gel process, are often porous, in connection with the fact that the final material is a gel. Its pores are filled with solvent, water, and alcohol, formed from the initial silicon alkoxide. Furthermore, the formation of a gel in the process of gelation of the sol does not mean that the hydrolysis and condensation reactions are stopped in the reaction vessel. The gelation point determines only the time point when there is a sharp increase in the viscosity of the reaction mixture due to the three-dimensional cross-linking of the sol particles, and the formation of an infinite cluster of them.

Therefore, as a rule, for obtained materials is carried out an aging procedure over time at an ambient or elevated temperature. During the aging, there is a further sealing of material, which is conditioned by the continuing hydrolysis and condensation reactions. As a result, the gel is shrunk. Removal of the solvent from the unmodified gel, for example, by its evaporation at an elevated temperature, generally results in destruction of the gel structure and ends with the formation of the powder. The reason for this is the high capillary forces that arise during the evaporation process of the liquid phase, which destroy the filigree gel network.

To eliminate or reduce this phenomenon, the liquid in the pores of the gel can be replaced by a solvent that provides a low capillary force. One such process is the exchange of solvent in the composition of the gel with a substance that is in its supercritical condition and thus may be introduced into the material directly in the gas phase. This technique is called drying in the supercritical region. Application of this method leads to the preservation of the gel structure. The thus obtained light materials are called aerogels. In addition to the above described morphology for the gel network, depending on the treatment conditions, can also be prepared different particles, fibers, and thin films.

If tetra-alkoxysilanes are the only precursors used in the formation of the structure of the silica, the obtained materials have a hydrophilic surface. Thus, these materials can actively interact with water and atmospheric moisture. This is particularly the case when the materials have high porosity, such as aerogels [13]. The hydrophilic properties of the surface can be changed if the silanol groups on the surface are replaced with hydrophobic organic

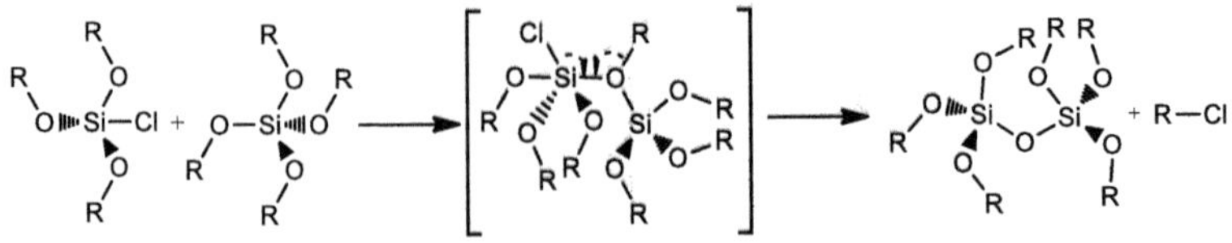

Figure 2.7 Example of one of the nonhydrolytic reaction mechanisms for the sol-gel process of obtaining inorganic oxides [27].

groups. This process can occur after the preparation of the material as well as the process of obtaining of the material. In the latter case it is possible if the material is prepared by co-condensation in the presence of a second functional organic substance.

Adding the gel structure of certain functional properties, using different ways, is also an important step in the formation of nanocomposites. This is because in these materials the interfacial interaction between the inorganic and organic components plays an important role in determining what kind of material is formed—the homogeneous or heterogeneous.

2.2.2 Nonhydrolytic Method of Sol-Gel Synthesis

Another method of obtaining organomineral hybrid materials is a nonhydrolytic method. This method is based on nonhydrolytic reactions of hydroxylation, or aprotic condensation reactions (Fig. 2.7 [27]). In a particular case, this method is based on the reaction of a metal halide ($MHal_n$) in an anhydrous medium with an oxygen donor, such as a metal alkoxide, ether, or alcohol [14]. As a byproduct, this reaction produces an alkyl halide compound.

In many cases, higher temperatures are required to carry out these reactions. Therefore, the applicability of this process to the organic groups is limited. A nonhydrolytic reaction or a nonaqueous sol-gel process has recently been given a lot of attention because it is a method for creating highly crystallized nanoparticles [27].

In the preparation of nanocomposites, the method is rarely used in connection with temperature limitations and consequently available only for certain types of polymers.

However, this method has several advantages:

- The absence of solvents
- Reduction in or elimination of the formation of silanol groups in the final product owing to another reaction mechanism in comparison with a hydrolytic sol-gel method synthesis
- Easy-to-achieve homogeneity of the mixture of starting substances, in particular for nonpolar molecules

At the same time it should be borne in mind that:

- It is necessary to take extreme caution when dealing with some highly reactive reagents used in a nonhydrolytic method.
- The interaction of oxygen-containing molecules can be complicated by their participation in other reactions as oxygen donor.

The nonhydrolytic method has been the subject of research in a number of studies that were carried out to identify its advantages in order to obtain inorganic oxides. However, it extremely small used for the synthesis of organomineral hybrids. In 1955 was described the synthesis of several alkyl- and aryl-modified silicates (and linear polyorganosiloxanes) by various combinations of di-methyl-di–chlorosilane, methylphenyldichlorosilane, phenyltrichlorosilane, phenyltriethoxy-silane, and feniletildietoksisilana in the presence of iron chloride (III) or aluminum chloride (III) at t = 95–100°C [15]. Checking the obtained results showed that these reactions occur by the mechanism of heterofunctional, stepwise polycondensation to form an insoluble branched organomodified silicate. Such passage of process is observed in the case when di- and trifunctional alkoxysilanes are used as the silicon-containing precursor. But additional studies of the samples were not conducted.

The nonhydrolytic sol-gel method of synthesis was studied in the formation of the organo-modified silicates (called ormosil) with various organic radicals [16]. For the formation of the silica lattice were used mono- and di-substituted alkoxy precursors with alkyl groups of various lengths, from $-CH_3$ to $-C_{10}H_{21}$. Although similar hybrids can be obtained with hydrolytic methods, the nonhydrolytic approach has some advantages, especially in the synthesis of hydrophobic hybrids. For example, there is a limitation to the introduction of the compound containing $-C_8H_{17}$ groups during the synthesis of the hybrid by hydrolysis. This is due to the fact that with increase in their concentration, phase separation of the mixture is observed in the system. Such problems do not arise in the process of nonhydrolytic synthesis of silicon-containing compounds having as substituents even $-C_{10}H_{21}$ groups. The only restriction of such interactions with bulky substituents is a steric effect that can affect the rate of the condensation reaction and the overall degree of condensation.

For example, hybrids of SiO_2 and polydimethylsiloxane can be obtained from a hydrolytic or nonhydrolytic sol-gel method. Using the method of hydrolytic sol-gel synthesis materials can be obtained that exhibit different degrees of hardness [27]. Properties of the resulting materials depend on the ratio of precursors, and may vary, ranging from solid to rubber products. With a nonhydrolytic process may be synthesized hybrid materials based on a silicon-containing compound and polydimethylsiloxane, using as catalyst iron chloride (III). Thus, reaction products do not have elastic properties, even when the content of siloxane is 50%.

2.2.3 Colloidal Method of Sol-Gel Synthesis

The first step in the preparation of nanocomposites of silica is a sol-gel process preparation of the gel. The colloidal sol-gel method of synthesis involves a preliminary synthesis of colloidal particles, the colloidal particles binding together and forming three-dimensional gel networks.

Benefits of a colloidal method in comparison with an alkoxide method are as follows:

- The use of ready-made, aggregately stable sols of polysilicic acid with different particle sizes, from 5 to 100 nm
- Low cost of the silicon-containing precursor

- The ability to use various modifying agents that promote changes in adhesion, strength, electrical, and other properties of the resulting material

The term "colloidal silica" refers to stable dispersions of discrete particles of amorphous silica (SiO_2). It is usually considered hydrophilic sol because particles are stabilized by "solvation" or "hydration." Such definition excludes from this group the solutions of polysilicic acids in which the polymer molecules or particles are so small that they are unstable. In aqueous solution, the silica at $t = 25^{\circ}C$ and pH 7 exists as $Si(OH)_4$ and its solubility is about 0.001 wt.%. At pH 2, it is increased by 1.5 times, and at pH 10 nearly 10-fold. When the monomer concentration in the solution exceeds the value corresponding to the equilibrium solubility, and there is no solid phase on which soluble silica might be precipitated, then the monomer is polymerized by polycondensation.

As a result, the polycondensation of low-molecular-weight silicic acid sol the germinal is formed, and takes place growth of its particles. Aggregation of the particles does not occur if the electrolyte concentration is less than 0.1–0.2 N, depending on the silica concentration. In silica sols, the free energy of interfacial interaction amorphous silica-water is 50 erg/cm^2 [17].

Silica gels are synthesized from molecular silicon-containing precursors. Two general methods are used to initiate the gelation of the water glass solution:

- Acidification or partial neutralization of a sodium silicate solution by adding Brönsted acids
- Replacement of the sodium ions Na^+ on hydroxonium ions H_3O^+ using an ion exchange resin in acid form, forming thereby a solution of silicic acid and initiating gelation by addition of a Lewis base (F^-) or a Brönsted base (OH^-)

The first method is a so-called single-stage process. Adjusting the pH to a value between 5 and 9 is equivalent to the partial neutralization of sodium silicate. Typically, for the description of this process use the term "acid catalysis." Strictly speaking, this is only partially true, because the addition of an acid serves as the primary reason for the partial neutralization of the alkaline solution

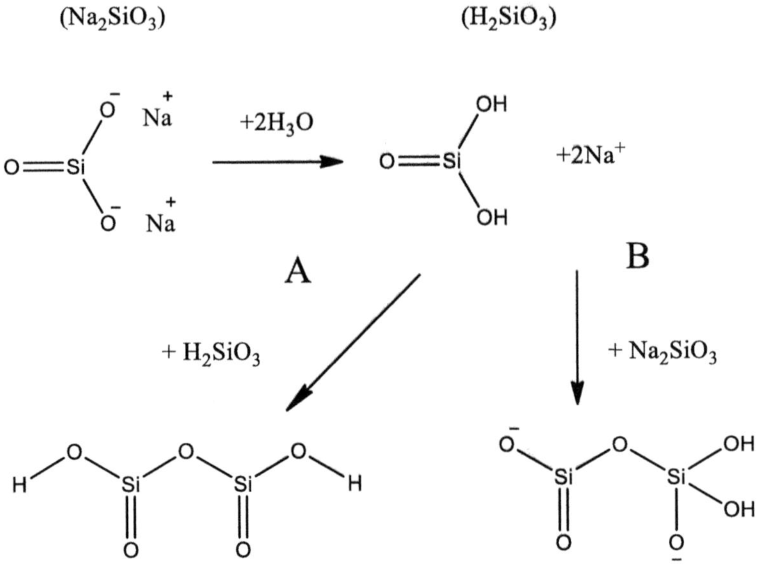

Figure 2.8 Acidification of the sodium silicate molecule to produce a silicic acid, and reaction with another molecule (A) of silicic acid and (B) of sodium silicate [27].

of sodium silicate and a decrease in pH. The second method is a classic two-stage process. The terms used in describing the various steps of preparing of the sol and the gel formation are often found in the literature for systems based on liquid glass. Subsequently, let us consider the formation of silica gel from a liquid glass. The two main steps in this process are neutralization and condensation. Figure 2.8 [27] shows the neutralization of the silicate with the formation of silicic acid H_2SiO_3. In the second stage, respectively, is shown the formation of dimeric particles by the reaction of the silicic acid (A) or sodium silicate (B) with one equivalent molecule.

The main step in the formation of a gel is the collision of two silica particles having a relatively low surface charge. When the particles come into mutual contact, between them are formed siloxane bonds, which irreversibly hold the particles together. For the formation of such a connection the catalytic action of hydroxyl ions or the dehydration of the particle surface at higher pH values is necessary. This is confirmed by the fact that the rate of the gel formation at pH 3.5 increases with pH and is proportional to the concentration of hydroxyl ions. At pH < 6, the lack of hydroxyl ions is not any more a factor that limits the rate of gelation. However, the aggregation rate is reduced due to reduction in the number of collisions between particles, due to the increased amount of charge on their surface.

The overall result of the simultaneous action of these two effects is the highest rate of the gelation at pH 5. As soon as siloxane bonds begin forming between the particles, further deposition of silica begins at the contact point due to a negative radius of curvature of the surface, the resulting particles. This process goes fast above pH 5 and slowly at pH 1.5. The rate of gelation appears proportional to the total surface area of the silica present in a given volume of the sol and increases with an increasing temperature. Substantial data relating to the activation energy of the particle aggregation can only be obtained when the particles have already completed their growth and stabilized at a higher temperature than provided for in the experiments. Below pH 3.5, the presence of salts weakly affects the rate of gelation, whereas water-miscible organic liquids like alcohols slow this process.

Once the sol turns into a gel, at first, the viscosity of the system increases, since the particles bond together to form branched chains that fill the whole volume, and then, the gel solidifies. In this case, you must always bear in mind that such a network, by the capillary structure, can hold a significant amount of fluid.

2.2.4 Soluble Silicates as Precursors Are in the Sol-Gel Technology of Nanocomposites

Soluble silicates of sodium and potassium (water-soluble glass) are substances that, in the amorphous glassy state, are characterized by particular oxides content—M_2O and SiO_2, where M is Na and K. The molar ratio SiO_2/M_2O is 2.6–3.5 when the content of SiO_2 is 69–76 wt.% of sodium water glass and 65–69 wt.% of potassium.

Liquid glass can be subdivided by type of alkali cations on the sodium, potassium, lithium, and organic bases. By mass or molar ratio in the glass: SiO_2 and M_2O, where M is K, Na, Li, or an organic base. In this case, the molar ratio SiO_2/M_2O is called "silicate module" of liquid glass –*n*. The secondary characteristic of liquid glass is the content of SiO_2 and M_2O in wt.%; content of impurity components: $A1_2O_3$, Fe_2O_3, CaO, MgO, SO_4^{2-}, etc., and its density (g/cm^3). The chemical composition of the liquid glasses is characterized by the content of silica and other oxides, regardless of the specific form of their existence in the solution. In some countries, in the characteristics of liquid glasses is also included the solution's viscosity.

Sodium liquid glasses typically produce within the silicate modulus values from 2.0 to 3.5, with the density of the solutions from 1.3 up to 1.6 g/cm^3. Liquid glasses based on potassium have the silicate modulus values in the range of 2.8–4.0, with a density of 1.25–1.40 g/cm^3 [18–21].

Acid-resistant building materials based on liquid glass are widely used in construction as silicate polymer concretes (SPCs), putties, fillers, etc. Soluble sodium silicates (liquid glass) are used as binders for the production of heat-resistant and chemically resistant materials. Liquid glass has high cohesive strength, is easy and safe to use, has a low cost, is not subject to corrosion, and is not inflammable (because the volatile components were evaporated) and does not adversely affect the environment on use.

A new trend in the technology of ceramics and inorganic composites is the use of sol-gel processes to form materials directly from solutions of sols. Naturally, in the first row of such materials are the products based on the silica sol; at continuing growth of a number of liquid glasses silicate module goes to infinity.

Agglomerated materials require a binder in order to achieve acceptable strength. In general, binders can be divided into three groups: matrix, film, and chemical. Sodium silicate is unique in that it can serve in all three of these capacities. For example, as a matrix binder, sodium silicate would be used in conjunction with Portland cement or pozzolan-blended cement binders [4].

Film forming binders are like glues and function by the evaporation of water or other solvents. Commercially available sodium silicates contain 45–65% water by weight. Loss of a small portion of this water, even under ambient conditions, will result in a strong, rigid, glassy film. The rate of drying will depend on ratio, concentration, viscosity, film thickness, as well as temperature and relative humidity. The silicate binder may be subject to dissolution depending on use conditions; however, some moisture resistance can be obtained by simply drying the silicate more completely through the addition of heat [27].

Good lubrication properties are important in briquetting because reducing the particulate friction and improving the flow onto the rolls provide for better and more efficient compaction. Almost all liquid binders will also act as lubricants; however, some chemicals are better at performing this dual task. High-quality unadulterated silicate has a very low coefficient of friction. Regrettably, this low coefficient of friction is lost with the addition of solids and/or water. Also investigated and found effective was the modification of sodium silicate by the addition of tetraalkylammonium compounds such as tetralkylammonium hydroxide (TMAH). It is thought that TMAH reacts with sodium silicate to form tetralkylammonium silicate with improved lubricity coming via the methyl groups [4].

Chemical binders function by reacting with the material being agglomerated or by formulating with multiple components that will react with each other. Sodium silicate has a long history of being used as a chemical binder. The best example is the use of sodium silicate with a soluble source of calcium. The reaction of calcium salts with silicate forms calcium silicate hydrate. Development efforts [28] have focused on nontraditional setting agents such as glycolic acid, sodium acid pyrophosphate, and calcium lignosulfonate. The goal was to achieve one or more of the following attributes: higher final strength, better control of set, longer set, and environmental benefits. A laboratory study was designed to investigate the interaction between calcium lignosulfonate and silicates over a range of concentrations. The interactive effects were observed with both sodium and potassium silicates. The results of this work showed that the setting characteristics of these systems were not typical of the rapid interaction seen with other calcium salts and allowed for much more flexibility in the binding process.

Processes that occur during the curing are complex and diverse. The modern view of the general idea of hardening liquid glass itself and in the various homogeneous and heterogeneous systems, the most commonly encountered in practice, is presented in several reviews [19, 21, 22]. The system is based on liquid glass, acting as adhesive or binding material, changing from a liquid to a solid state through many methods. They can be divided into three types:

- The loss of moisture by evaporation at ordinary temperatures
- The loss of moisture from the system, followed by heating above 100°C
- The transition to the solid state by introducing specific reagents, which are called curing agents

Naturally, these three types are used in combination.

In a solution the degree of polymerization of silicate anions is known to depend on two factors, the silica modulus and the solution concentration. Each solution has a distribution of degree of polymerization of anions. These two factors determine the distribution of the degree of polymerization of the anions and imposed on it charge distribution of the anions.

Sodium metasilicate Na_2SiO_3, sodium silicates, and high modulus, also known as water glass or liquid glass, are inorganic compounds that are readily soluble in water. Their saturated solution is a viscous liquid with a density of about 1.4 g/cm^3 and a pH of about 12.5. Liquid glass is synthesized by the reaction of commercial quartz sand with sodium hydroxide and/or sodium carbonate at an elevated temperature and pressure. Given the wide abundance and inexpensive nature of these reagents, liquid glass is probably the cheapest source of soluble silicon in the industry. The polar nature of the molecule (the presence of ionic pairs Si–O^- and Na^+), on the one hand, makes it readily soluble in water and, on the other hand, prevents the spontaneous formation of large silica polycondensate or gelling due to electrostatic effects. In addition, it is simple to use and does not constitute an ignition hazard, which can occur with silicon alkoxides such as TEOS or TMOS. It is chemically stable in the long term, under standard conditions of use and storage. Consequently, this type of precursor combines most of the key advantages needed to produce silicate-based nanocomposites on the industrial scale.

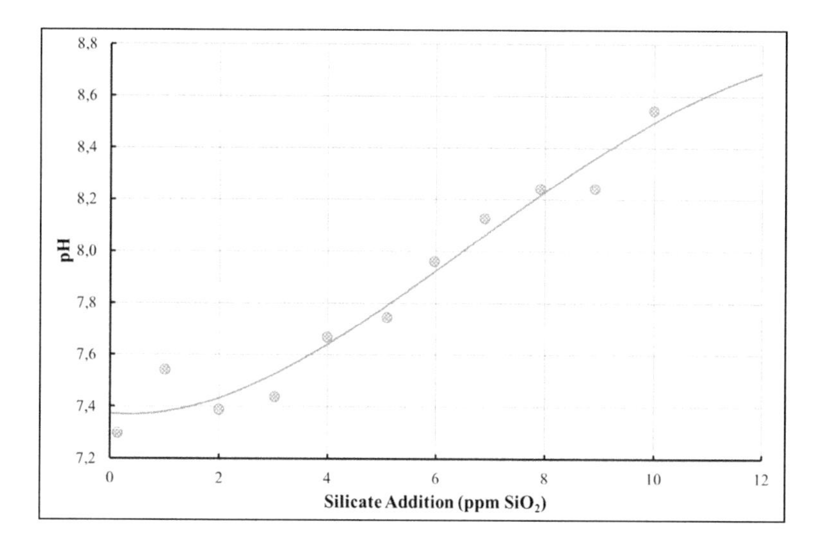

Figure 2.9 pH response with soluble silicate addition [33].

Sodium silicates are alkaline chemicals. Treating water at typical levels of 4 ÷ 24 mg SiO_2 may raise the water pH anywhere from 0.1 to 2.0 pH units or more. The actual pH increase will depend on overall water quality and silicate dosage (Fig. 2.9 [33]). Increases in pH will generally help minimize corrosion and will provide a synergistic effect along with the deposition of monomolecular silica film.

The use of sodium silicates for the control of corrosion in municipal water systems is approved by the American Water Works Association and the American National Standards Institute (refer to ANSI/AWWA Standard B404). Sodium silicate also has Food and Drug Administration (FDA) unpublished "generally recognized as safe" (GRAS) status as a corrosion preventative in water (at levels below 100 mg/L). The US Environmental Protection Agency (EPA) recognizes that silicate inhibitors may be effective in controlling corrosion of lead and copper in potable water systems [23].

Studies have shown that soluble silicates are reactive with cationic metals and metal surfaces [17]. This phenomenon is the basis by which silicates inhibit corrosion and is illustrated by the following scheme [33]:

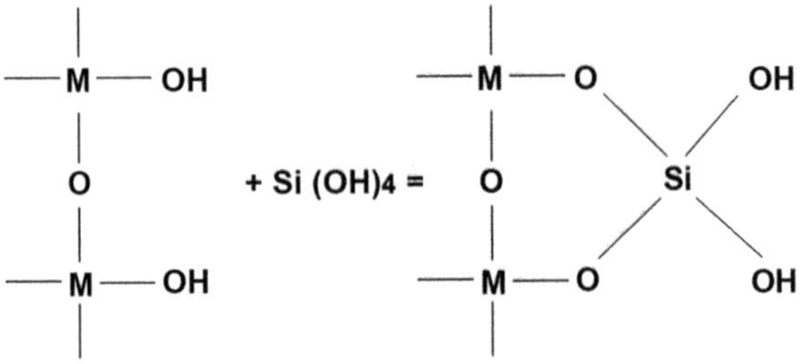

```
   |                          |
 —M— OH                     —M— O         OH
   |                          |     \     /
   O        + Si (OH)4 =      O       Si
   |                          |     /     \
 —M— OH                     —M— O         OH
   |                          |
```

Monomeric and polymeric silica, introduced into a water distribution system as sodium silicate solution, is carried by allowing water to flow to all parts of the distribution system. At the dilution levels used for water treatment, the majority of the silica depolymerizes to a reactive monomer form. The monomeric silica, which can be represented by $(SiO_3)^{2-}$, is adsorbed onto metal pipe surfaces at anodic areas, forming a thin monomolecular film on the interior of the pipe. This prevents any further corrosive reaction at the anode.

Corrosion will be inhibited when an anodic reaction of the type proposed between ferrous iron and silica (shown as Eq. (2) in Fig. 2.10 [33]) occurs in the place of the reaction that forms ferric hydroxide (shown as Eq. (4) in Fig. 2.10). Protective films can be formed by such a reaction.

Microscopic and x-ray examination of the film formed at the metal surface shows two layers, with most of the silica in the surface layer adjacent to the water. When the hydrous metal oxide film has been covered with a silica film, silica deposition stops. The film does not build on itself and, therefore, will not form excessive scale. The film is an electrical insulator and blocks the electrochemical reactions of corrosion, yet it is thin enough that it does not obstruct water flow.

Corrosion protection with sodium silicates can be achieved by modifying the SiO_2 content of the water. Therefore, the key water property is SiO_2 content, not pH or calcium level, as in other types of corrosion control practices.

Alkalinity, pH, and water hardness may influence the effectiveness of silicate treatment. In many cases, natural SiO_2 found in water probably has already "reacted" (i.e., adsorbed) with other metals in the water and may not be effective in reacting with metal

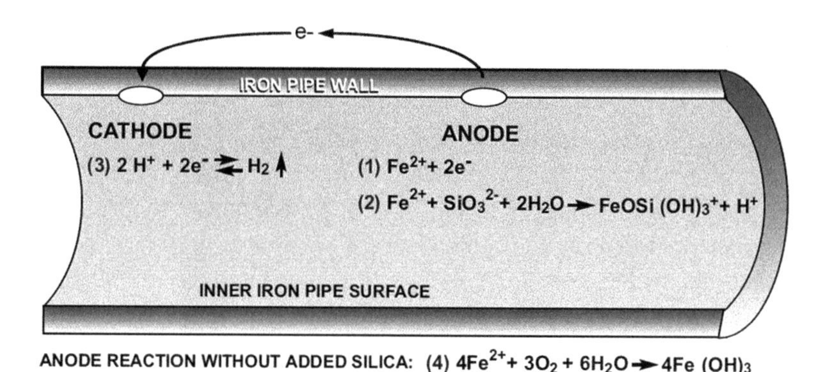

Figure 2.10 Iron electrochemical reactions scheme in water, with and without the addition of silica [33].

pipes. Therefore, a fresh source of "reactive" SiO_2 is needed (as from soluble silicate solutions).

Addition of silicate to water systems can protect cementation materials from long-term deterioration. The silicate reacts with available calcium to form insoluble calcium-silicate compounds. Studies have shown that silicate treatment may reduce the breakdown of asbestos-cement surfaces, thus prolonging the life of the material and minimizing the release of fibril asbestos [33].

Practical use of liquid glasses is realized in the following directions. The first direction is the manifestation in the liquid glass binding properties—the ability to self-harden to form an artificial silicate rock. The unique ability of the liquid glass is its high adhesive properties to substrates of different chemical natures. In these cases, the liquid glass is acting as a binder for a chemical gluing different materials used in coatings and production of the composite materials of wide application.

The second direction involves the use of liquid glasses as a soluble source of silica, that is, a raw source component for the synthesis of various siliceous materials—of silica gel, white carbon, zeolites, catalysts and carriers for them, silica sol, etc.

The third area relates to the use of alkali metal silicates, as chemical components in various substances. This direction provides for the use liquid glass in the manufacture of synthetic detergents, for bleaching and cloth dying, in papermaking, etc.

Liquid glass–alkaline solutions of sodium and potassium silicates are representatives of a wider class of water-soluble silicates and liquid glasses produced on an industrial scale. The group of water-soluble silicates includes crystalline anhydrous sodium and potassium silicates, crystalline and amorphous sodium and potassium hydrosilicates in the form of powders, etc. Amorphous powder hydrosilicates of alkali metals are characterized by compositions within the $SiO_2/M_2O = 2.0–3.5$ when the content of bound water is 15–20%. Such powders are usually obtained by spray-drying the concentrated liquid glasses and high hydration of glassy silicates. They are loose and quickly dissolve in hot and cold water. Manufactured crystalline hydrosilicates, usually represented by crystalline hydrate disubstituted sodium orthosilicate $Na_2H_2SiO_4$, contain from 4 to 9 molecules of crystalline hydrate water. This is also known as hydrated metasilicate formulas $Na_2O{\cdot}SiO_2{\cdot}5H_2O$ and $Na_2O{\cdot}SiO_2{\cdot}9H_2O$.

The above products—liquid glass, glassy silicates, hydro silicates in crystalline, and amorphous state—are so-called low-modulus silicates with a molar ratio $SiO_2/M_2O = 1–4$. The need to improve certain properties of composite materials based on them, such as water resistance and thermal properties, has led to the development of "high-modulus liquid glass"—polysilicates of alkali metals. The polysilicates group includes alkali metal silicates (silicate module 4 to 25), representing the transition region of compositions from liquid glass to silica sol stabilized by alkali [17]. Polysilicates have a wide range of polymerization degree of the anions, and they are colloidal silica dispersions in an aqueous solution of alkali metal silicate. Synthesis and practical application of polysilicates as the binder allow for the filling of the space existing among the alkali silicate binders, creating four groups represented by decreasing alkalinity: soluble (liquid), glass, polysilicates, and silica sols.

A relatively new field of water-soluble silicates, which found currently considerable practical output amounted silicates organic bases. The synthesis of this class of compounds is based on the ability to dissolve silica, at a pH above 11.5, in the organic bases of different nature and above all in the quaternary ammonium bases. Quaternary ammonium bases are sufficiently strong bases for the dissolution of silica in their solutions. Water-soluble silicates of this class—quaternary ammonium silicate (QAS)—are characterized by the general formula $[N(R^1, R^2, R^3, R^4)]_2O_{1-n}SiO_2$, where R^1, R^2, R^3, and R^4 represent H, alkyl-, aryl-, and alkanolgroups, respectively [24].

QAS solutions are usually highly siliceous lipophilic stable dispersion systems in which the silica is present in colloidal forms and forms specific to true solutions. They are often produced in those cases when the sodium or potassium analogues of such systems are not sufficiently stable. The dissolved silica in such systems is an oligomer with a polymerization degree of 10–25. The particle size of the colloidal silica increases from 2 to 100 nm depending on the value of the silicate modulus in the range $n = 2$–12. The greatest practical application was found among lower alkyl- and alkanolderivatives—tetrabutylammonium silicate (TBAS), tetraethyl silicate, and tetraethanolammonium silicate. Absence of alkali metal ions in this group of water-soluble silicates and the ability to control a wide composition of organic bases have opened up new areas of application, such as water-soluble silicates that differ significantly from traditional applications.

Thus, the group of liquid glass–alkali silicate solutions is very extensive. This group of silicate systems are classified by the following features.

By degree of polymerization (l) of silica the average number of silicon atoms forming the siloxane bonds continuous system ≡Si–O–Si≡ during polymerization. In the polymerization of silica occurs increase of its molecular weight (M), and at high degrees of polymerization of - increasing the size (d) of colloidal silica particles. At a certain degree of polymerization (l) in the alkali silicate systems appears colloidal silica as a sol or as highly dispersed hydrated silica [27, 29]:

Monomers →	Lower oligomers →	Higher oligomers →	Colloidal silica, sols
($l = 1$)	($l = 1$–25)	(polysilicic acids, $M < 10^5$)	($M < 10^5$ or $d > 2$ nm)

According to chemical composition with increasing alkalinity, the alkali silicate system is characterized by a certain molar ratio SiO_2/M_2O (silicate system module n) and forms a series corresponding to the four previously listed forms of silica [27, 29]:

Overbased systems	→	Liquid glasses	→	Polysilicates	→	Sols
($n < 2$)		($n = 2$–4)		($n = 4$–25)		($n > 25$)

The types of cation liquid glass are divided into potassium, sodium, lithium silicate, and silicates of organic bases. Synthesize mixed liquid glass inside these four groups [19].

Processes that occur during the curing are complex and diverse. The modern view of the general idea of hardening liquid glass itself and in the various homogeneous and heterogeneous systems, the most commonly encountered in practice, is presented in several reviews [17, 21, 22]. The system is based on liquid glass, acting as adhesive or binding material, changing from a liquid to a solid state through many methods. They can be divided into three types:

- The loss of moisture by evaporation at ordinary temperatures
- The loss of moisture from the system, followed by heating above 100°C
- The transition to the solid state by introducing specific reagents, which are called curing agents

Naturally, these three types are used in combination.

In a solution the degree of polymerization of silicate anions is known to depend on two factors, the silica modulus and the solution concentration. Each solution has a distribution of degree of polymerization of anions. These two factors determine the distribution of the degree of polymerization of the anions and imposed on it the charge distribution of the anions.

One method of curing of liquid glasses is a process of curing under ordinary temperatures due to the removal of moisture. Processes occurring in the silicate solution are regulated by two reversible reactions:

$$\equiv \mathrm{SiOH} + \mathrm{OH}^- \leftrightarrow \equiv \mathrm{SiO}^- + \mathrm{H_2O} \quad (2.3)$$

$$\equiv \mathrm{SiOH} + \equiv \mathrm{SiO}^- \leftrightarrow \equiv \mathrm{Si{-}O{-}Si} \equiv + \mathrm{OH}^- \quad (2.4)$$

Polymers formed by the second reaction are preferably spherical in structure and are formed during polymerization as colloidal particles charged negatively [3, 17]. Therefore, they do not come together to interact if the conditions for coagulation are not created. Dimensions of colloidal particles and thus their concentrations are regulated by an internal distillation process. It lies in the fact that the solubility of small silica particles in the solution depends on the particle size and if the particle size is increased, solubility is decreased. During the internal distillation process, the large particles grow due to the dissolution of the smaller particles. For larger particles, their solubility is not dependent on size. Therefore, internal distillation, at a certain point, slows down, and further, stops absolutely. This leads to some particle size distribution. This phenomenon is especially characteristic of cases where the formation of a silicate solution starts from monomeric particles. If a silicate solution is formed by dissolving large polymeric forms of silica, the internal distillation process may not be developed, or as a secondary process, to obtain a solution and the polymer distribution of the anions other than the first case. The internal distillation process, especially in the later stages, proceeds rather sluggishly, so old and freshly prepared solutions may be very different from each other, although the module and the concentration of the solutions are the same. Sharp dilutions of solutions or temperature change also lead to changes in the anionic composition.

If a dilute solution having a large silicate unit is evaporated, the liquid phase is represented only by ionic forms of silica. However, because of the hydrolysis caused by a lower concentration of hydroxyl ions in the first reaction, there will be a greater amount of ions of the type $\mathrm{HSiO_4^{3-}}$ and in much smaller quantities ions of $\mathrm{H_2SiO_4^{2-}}$. During evaporation, the solution will start to change in

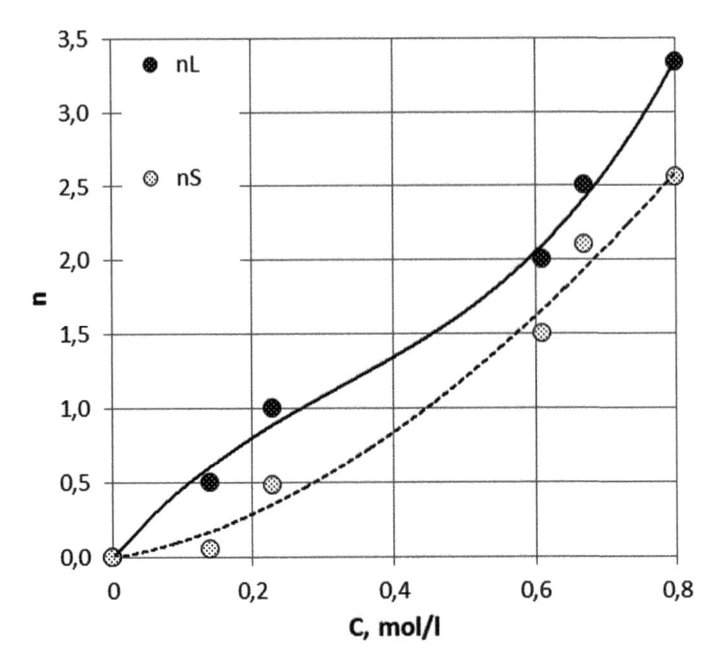

Figure 2.11 The initial solution concentration of tetrabutylammonium silicate and anionic composition of crystals obtained from it. n_L represents the composition of the liquid phase and n_S the composition of the solid phase [27, 29].

the direction of reducing the module, since the module in a solid phase is higher than the module source solution (Fig. 2.11 [27, 29]). Concentration of $HSiO_4^{3-}$ will be smaller and ions of SiO_4^{4-} more as it evaporates, will the emergence of new solid phases and, ultimately, would be to fall phase $Na_4SiO_4 \cdot mH_2O$ [27].

At some concentration of hydroxide ions in a solution of ionic forms of silica hydrolysis goes so far that there are completely hydrolyzed forms that have reached the uncharged molecular state $Si(OH)_4$. If the interaction between the two ions, the second type of reaction, it is unlikely due to electrostatic repulsion, then between the molecular and ionic forms, it is possible. So polymeric forms of silica may be obtained. They are already in the early stages and can take a three-dimensional structure where the silicon atoms are connected inside the Si–O–Si, and the outer atoms have at least one bond Si–OH. The latter may also exist in the ionic form SiO^-. With a length of chain equal to 4–5, there is a formation of ring structures, which subsequently acquire a three-dimensional structure.

Slow evaporation at elevated temperatures increases the degree of polymerization of silicates. Therefore, for obtaining readily soluble alkali silicate powders from the viewpoint of product quality, the process is advantageously carried out at a low temperature rapidly using not very concentrated solutions. Further conversion to the hardened silicate system associated with the slow loss of hydration water in the atmospheric conditions and the absorption of carbon dioxide

$$CO_2 + OH^- \rightarrow HCO_3^- \tag{2.5}$$

causes the migration of sodium ions to the surface to form the crystalline carbonate structure, forming a silica frame with a low water content. This leads to an increase of water resistance.

Another way of hardening of liquid glasses is a process of solidification by means of reagents. Alkali metal hexafluorosilicates have a special place among hardeners that increase the module of liquid glass. Their peculiarity lies in the fact that they interact with the alkali not only to reduce its content but also to form the silicic acid with its decomposition, which significantly plumps a hardening system, lowering its porosity. The reaction takes place between hexafluorosilicate ions and hydroxide ions according to the following conditional scheme:

$$SiF_6^{2-} + 4OH^- \leftrightarrow SiO_2 \cdot 2H_2O + 6F^- \tag{2.6}$$

This is a typical reaction of the ligand substitution in the complexes, but it is accompanied by a change in the coordination number of the silicon atom, and, as often happens in such cases, complexes with mixed ligands are very unstable. The reaction is reversible and takes place in acidic media in the opposite direction. Introduction of Na_2SiF_6 powder in sodium liquid glass, as in other cases, mixing with solid acidic hardeners immediately causes coagulation of silicate and then gelation occurs around the hardener grain surface. Therefore, usually sodium hexafluorosilicate powder is premixed with a filler and then a liquid glass.

Upon receipt of acid-resistant concretes and putties, hexafluorosilicate sodium is administered in an amount greater than needed to neutralize all the alkali of liquid glass [13, 22]. For example, to neutralize all the alkali contained in the sodium liquid glass ($n = 3$, $\rho = 1.45$ g/cm^3), sodium hexafluorosilicate requires slightly less than 16% by weight of the glass; when $n = 2$ and $\rho = 1.40$ g/cm^3, sodium hexafluorosilicate needs 18 wt.% by weight of the glass. Featured recipes offer 25–0 wt.% Na_2SiF_6 for acid putties [17–19, 28, 29]. After neutralization of all the alkali entering the liquid glass composition, decomposition of sodium hexafluorosilicate is completely stopped, and it is probable that the hardened system practicable simultaneous presence of Na_2SiF_6 and silica. It is also important to note that in an acidic environment, this reaction goes in the opposite direction when NaF, formed during the manufacture of putties, will be present in sufficient concentration in the system. Therefore, washing of NaF after solidification will increase acid resistance for three reasons, firstly because of the removal of NaF; secondly available moisture because of Na_2SiF_6, which remained in the system and enters into the reaction; and thirdly due to plugging pores in the material, with the help of the resulting silica gel. To hardeners of liquid glass relate esters of light organic acids, and esters of carbonic and silicic acids are saponification by alkali of liquid glass:

$$\mathrm{RCOOR'} + \mathrm{OH^-} \rightarrow \mathrm{RCOO^-} + \mathrm{R'OH} \qquad (2.7)$$

Different esters have their rate constants for the reaction. However, most of the ester hardeners used have very limited solubility in water and form a separate phase in the form of emulsion droplets. Around these droplets is formed silicate semipermeable membranes that are broken off under the action of osmotic pressure, and the mechanism of action of such hardeners is quite complicated. The hardener composition for each technological object must be selected by mixing various esters that slow down or speed up the process, as well as there is the need to experimentally select hardener dosage.

The preferred hardener is aluminum triphosphate, which is a kind of solid acid [25] having the following formula:

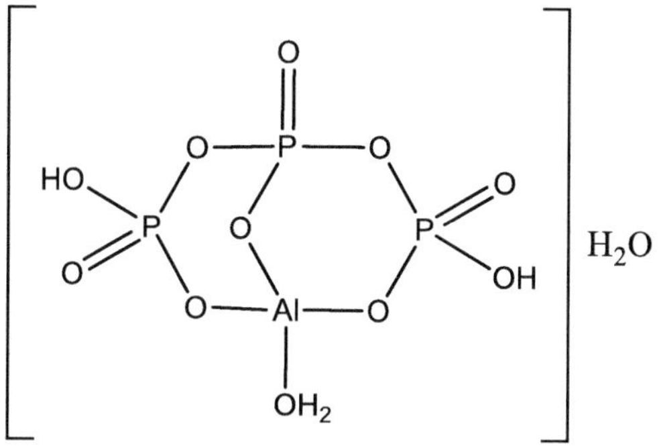

This material has been found to have no oral toxicity and no skin irritation. Aluminum triphosphate reacts with soluble sodium silicate as follows:

$$Na_2O \cdot xSiO_2 + H_2AlP_3O_{10} \rightarrow Na_2AlP_3O_{10} + H_2O \cdot xSiO_2 \quad (2.8)$$

The time necessary to initially form a gel after addition of the hardener to the soluble alkali silicate decreases as the amount of hardener used is increased. When aluminum triphosphate is used as the hardener, amounts from about 3 parts by weight to about 8 parts by weight of triphosphate per 100 parts by weight of soluble alkali silicate will give initial gelling times from 1 to 12 h. The higher the content of hardener, the shorter the useful working life of the composition will become. The amount of hardener included should, therefore, be chosen to provide a convenient initial gelling time consistent with the circumstances under which the composition is to be used [29].

Hardening of liquid glass may also be effected by its interaction with neutral electrolytes and water soluble organic compounds [18–21]. This process is widely described in the technology of silica gels, but not directly used in binding systems. There are many technologies that produced structures with very different porosity and strength in the hardened state. These processes are controlled by the temperature change of the process, the type and concentration of added salt and silica modulus liquid glass solution, and the exposure time of the system at a pH in the range of weakly alkaline solutions. These studies are described in detail in some review articles [17, 28].

Hardeners of liquid glass are the compounds of calcium and other divalent metals [27]. Interaction of silicate solutions with calcium compounds is important in applied chemistry. Calcium silicates, which are precipitated using calcium salts from liquid glass solutions, at ordinary temperature, are amorphous substances. The crystalline products may be formed at elevated pressure and temperature in autoclaves or very dilute solutions of low alkalinity, as well as in aging. Deposition of silicates of alkaline earth, polyvalent, and heavy metals is possible, as a rule, at a pH slightly lower than the pH of precipitation of the corresponding hydroxides. Therefore, when mixing the two solutions, besides metal silicate, there is always the formation of metal hydroxides and the silica gel. Their formation always occurs to a greater or lesser amount depending on the mixing intensity. The procedure for their formation depends on the nature of the reactants. The result of the interaction of solutions of divalent and trivalent metal salts with a solution of liquid glass is silicate solution coagulation [33]. The composition of precipitated amorphous oxide flocks depends substantially on the order of draining reagents from the mixing intensity, the concentration of the solution used, and the pH of the resulting reaction mixture. It may include hydroxides of silicon and corresponding metal and its silicates, with the captured anions. Such nature of the interaction is observed with the majority of salts of divalent and trivalent metals. This process is called coprecipitation or cocrystallization of the hydrated metal oxide and silica, or metal hydroxide adsorption on colloidal silica, or conversely, the deposition of silica on metal oxides and hydroxides. Such interactions are widely used in hydrometallurgy and radiochemistry for the isolation and separation of radioactive elements.

2.2.5 Preparation of Nanocomposites through Aerogels

Composite materials are obtained by combining two different materials. In general, the composites are developed precisely in order to use to the maximum advantages of each of the types of materials used and to minimize their disadvantages. For example, silica aerogels are brittle substances. Thus, the other component in the obtained material can increase the strength of the material and, in turn, has, for example, the desired optical properties, high surface area, and low density, as a silica airgel.

In addition to these methods of synthesis and processing, it should be emphasized that the flexibility of the sol-gel processes can increase the variety of aerogels, except for silicon dioxide, aerogel-based materials such as that are at the moment still available. Architecture of bulk materials can be adapted by using the template method. The chemistry of the gel may be modified by grafting, either before or during or after gelation.

Composites and nanocomposites can be created by impregnating the foams or fibrous meshes, dispersing particles, powders, or polymers or the synthesis of mixed oxides based on silica or other metal oxides. Organic silica hybrids can also be produced using plurality techniques such as co-gelling and cross-linking or by reaction with functionalized particles [20, 27].

In the recent years there has been a large body of research in the field of preparation of energetic materials. Work was carried out for the application of aerogels and sol-gel derivatives for the preparation of nanostructured composites of energetic (e.g., explosives, propellants, and pyrotechnics) and their characteristics studied. Aerogels have a unique density, composition, porosity, and particle size and are created through chemical synthesis techniques using low temperatures and mild conditions, all of which makes them attractive candidates for creating energy nanomaterials.

Using these materials and methods in this field of technology has led to three principal types of energy sol-gel materials [13]:

- Pyrotechnics-inorganic sol-gel oxidants/metallic fuel (thermite composites)
- Sol-gel derivatives of porous pyrophoric metal powders and films
- Organic sol-gel fuel/inorganic nanocomposite oxidants (composite solid propellants and explosives)

The behavior of all sol-gel nanoenergy materials to a large extent depends on several factors, including the surface area, the degree of mixing between phases, the type of mixing (a sol-gel or physical mixing of solids), ways of loading solids, and presence of impurities. Sol-gel methods are attractive for the preparation of nanostructured energetic materials. These methods offer many options for the form of the obtained materials, such as monoliths, powders, and films, and have a wide compositional flexibility. These attributes, combined with the severity of the synthetic control of microstructural properties of sol-gel matrix, ensure the preparation of energetic nanocomposites with reconfigured characteristics.

Energetic materials are divided into three classes [30]:

- Explosives
- Solid rocket propellants
- Pyrotechnic materials

Thus, materials may be classified based on the speed of interfacial interaction of reactants and the type of the energy output. Explosives are materials that react to a supersonic velocity (detonation) and whose reaction products primarily are gaseous substances. Rocket propellants also react quickly and give mainly gaseous reaction products but react, unlike explosives, at subsonic speeds. Pyrotechnic materials tend to react most slowly among the three types of energetic materials and generate high-temperature, solid reaction products and a few gas products, thereby generating an intense visible light output.

At least in the past two decades, the field of nanoresearch was one of the most active areas of research in various scientific disciplines, and energetic materials were not an exception to this [26, 30]. Nanoenergy composites were synthesized through the use of nanomaterials and advanced manufacturing techniques, which are promising opportunities. Nanoenergy composites are defined as a mixture of oxidizer and fuel particles that have dimensions or at least one critical dimension of less than 100 nm [30]. Reducing the size increases the surface area of contact between the phases of reactants. This has been achieved using a variety of methods, including vapor condensation, micellar synthesis, chemical reduction, ultrasonic mixing, as well as mechanical mixing methods. Very good results have been received. For example, for the pyrotechnic nanocomposites Al/MoO_3 burning rates almost 3 orders of magnitude higher than those of conventional mixtures were fixed. Energetic materials with such properties may be sensitive to impact or shock depending on particle size. Energetic materials with smaller particle sizes may be less sensitive to ignition and thus have better properties in terms of safety. These examples provide a good stimulus for the use of nanomaterials and technologies in the energy fields. With this in mind, aerogels and other gelatinous materials obtained from sols were investigated in the last decade as nanostructures of energetic materials.

Along with good miscibility energy nanocomposites have an extremely high surface area interface. The sol-gel method of obtaining these materials enables more large interfacial contact areas. All these favorable attributes have led to active research on the use of sol-gel chemistry in the research and development of energetic materials.

Organomineral nanocomposites based on silica aerogels possess a complex of unique optical properties. The refractive index of the airgel modified with tri-methylsilyl groups may be in the range of 1.008–1.06, depending on their densities. Figure 2.12 [27] shows the relationship between the density and the refractive index of TMSA aerosilica gel. The relative value of the index of refraction n is almost proportional to the density of the airgel material in a range of high porosity. This result corresponds to the theoretical ratio of Maxwell-Granat as applied to nanocomposites formed from organically modified silica and air [27, 31].

Since modified aerogels have excellent optical properties, transparency, an extremely low index refractive index, and moisture resistance, they are often used as media in Cerenkov counting. When a charged particle passes through a transparent medium at a speed faster than the speed of light in the material, there is a glow of Cherenkov radiation. Although monolithic blocks of silica airgel produced by supercritical drying methods are quite expensive for industrial applications, they have greatly contributed to progress in such scientific fields as high-energy physics. The progress of science has always contributed to the improvement of research and development in the industrial world, so we can expect that airgel can

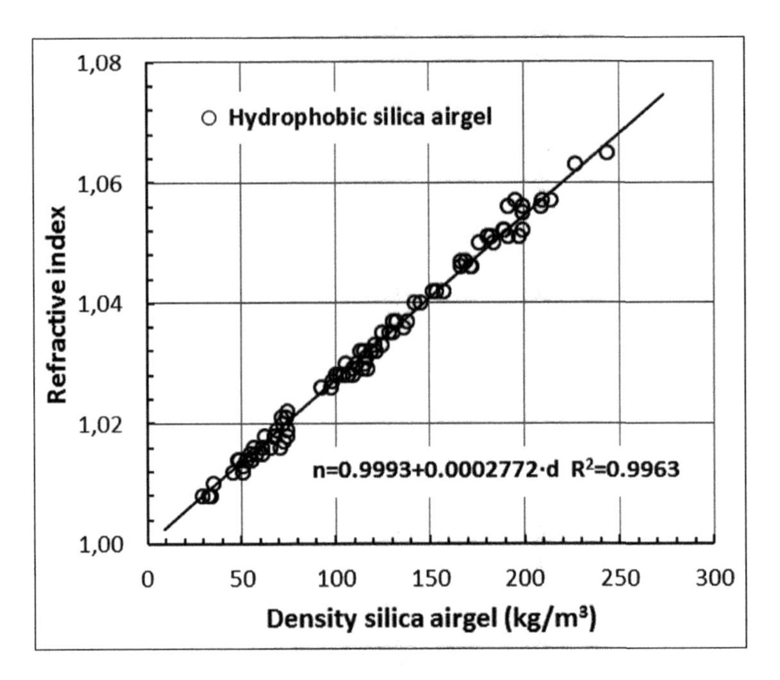

Figure 2.12 The correlation between the density and the refractive index of the hydrophobic silica aerogels [27].

be a pioneer of new technologies, such as nanocomposites, optics, space exploration, and energy devices [13, 27].

Active work is being done in the field of nanoscale engineering of composites based on silica to create a variety of sensors [23]. Authors have described "composite silica—modified silica" prepared by the modification of the silica gel after gelling, base-catalyzed, with another silica sol, this time prepared using acid catalysis. This base-catalyzed, acid-modified gel is then treated with the carbon dioxide supercritical extraction method to obtain the airgel. Airgel monoliths obtained as a result of this process have the bulk properties of silica aerogels prepared base catalysis, including a high level of transparency. However, at the same time the surface properties are more typical of the airgel obtained by acid catalysis. Consequently, it is possible to catch various kinds of strongly polar molecules, including acid-base indicators, and use them as an interface to the respective sensors.

In Ref. [32] is reported composite aerogels silica containing colloidal metal particles (gold or platinum) that have optical transparency of silica aerogels, combined with the surface and optical properties of the metallic colloid. Metal colloidal particles are uniformly distributed throughout the volume of the mixture and hence are isolated from each other. At the same time, the porosity of the silica matrix makes these metal colloid particles available for the particles that pass through the matrix. The surface of the metal colloid may be modified either before or after gelling, in order to adapt it to the optical properties of the material.

Subsequently, this method was applied to the preparation of airgel monoliths doped protein cytochrome c [27]. In the buffer, the protein forms a superstructure containing thousands of individual protein molecules around a colloidal gold particle. The modified particle of gold is reacted with a TMOS-catalyzed base sol to obtain a composite material prepared as described in Ref. [32]. Despite the fact that fragments of cytochrome c in the outer part of the superstructure are damaged during the process of exchange and solvent extraction, most of the internal proteins that survive the extraction process without change remain in the environment as a buffer around the gold particles. These monoliths airgel retain some reactivity of cytochrome c, as shown by their response to the presence of NO in the gas phase, the presence of which is monitored by a change in optical density over time.

It should be noted that the relatively low temperature process using carbon dioxide supercritical extraction is of great importance for the conservation of protein function in this application. When using a fast process, supercritical extraction, one should not expect comparable results, because this protein cannot withstand the higher temperatures required for the implementation of such a drying process.

The inclusion of nanofibers of polyaniline in silica aerogels obtained on the basis of TMOS and carbon dioxide supercritical extraction leads to an increase in the strength of materials. Thus, there is the possibility of their potential application for the detection of gaseous acids and bases. It has been found that including in the introduction, only about 6% of polyaniline by weight of the material increased strength airgel 3 times, in obtaining a material such as low

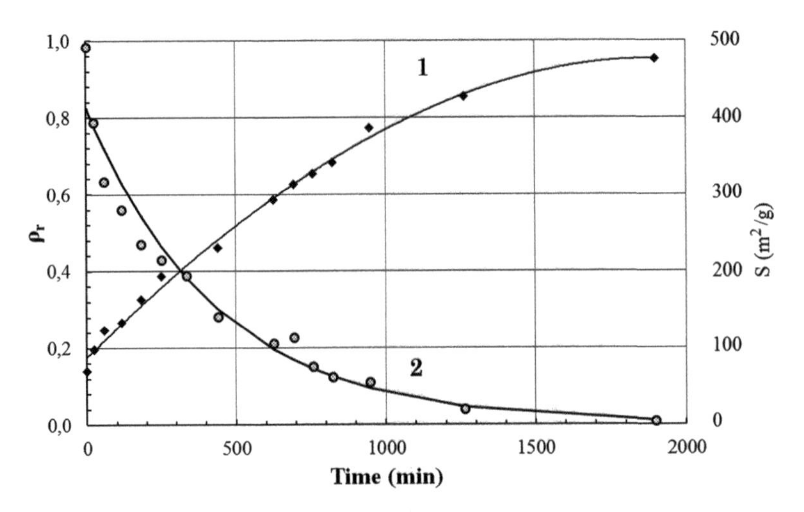

Figure 2.13 Change in the relative density ρ_r (1) and the specific surface area S (2) for the airgel, depending on the time of sintering at 1000°C [27].

density (0.088 g/cm^3). When using a gold electrode on the surface of the airgel composite, there is a strong decrease in resistance when the airgel is exposed to a vapor HCl.

After carrying out all stages of the synthesis process, airgel is solid and amorphous but it is an extremely porous (75–99% porosity) material. The last step in the transformation is its densification by thermal treatment. It is often necessary to convert the material by the sintering of the airgel, a solid glass devoid of porosity, that is, having a relative density equal to 1. Relative density is the ratio between the bulk density of the airgel and the density of quartz glass (2.2 g/cm^3). Figure 2.13 [27] shows a typical evolution of the relative density and the specific surface area during sintering by heat treatment. These curves are strongly dependent on the temperature of heat treatment and the content of hydroxyl groups in the airgel structure, which affect the viscosity of the airgel [27, 31].

Gels that are initially noncrystalline may crystallize during subsequent heat treatment. The successful formation of a glass is the result of competition between the processes that lead to the densification of the material and those that promote the crystallization.

What follows from these data is the importance of the use of nanocomposites based on silica aerogels, which is the sealing of radioactive waste from nuclear power plants. The actinides and other radioactive nuclides generated in the nuclear fuel cycle are present as salts in aqueous solutions. Using a fully open pore structure of the airgel can be filled with solutions of these salts, the entire volume of the airgel. Then, the liquid phase was removed by evaporation and the porous composite material (airgel + salt) completely sintered, leading to the synthesis of a multicomponent material. The porous structure of the airgel is used as the receiving vessel. In accordance with the small pore size of the airgel, the preparation of such nanocomposites is a very simple process. The size of the domains being formed will depend on the size of the pores in the airgel and the content of actinides and other radionuclides in the liquid phase.

However, if on trying to fill an airgel with a liquid such as water, capillary forces may cause destruction of the airgel. Due to the complexity of the texture of the airgel, a detailed calculation of the local stresses on filling it with liquid is difficult; it depends on the surface energy of the liquid vapor and pore size.

Thus, to avoid cracking of the material during filling, different strategies can be offered:

- Synthesis of airgel with large pores that reduce the magnitude of capillary forces
- Increase of the mechanical strength of the airgel through aging and its partial sintering
- Surface functionalization due to the imparting surface of the airgel of chelating groups

2.2.6 Modification Products Sol-Gel Synthesis by Polyurethanes

New high-strength organomineral products were prepared by joint gelation of polyisocyanates, aqueous alkali silicate solutions, and cement. These products, as compared with conventional polyurethane formulations and alkali silicate masses differ by more rapid curing, in their preparation, improved hardness, and incombustibility [29].

They are suitable for use as putties, bonding agents, thermal insulating and sound-proof materials, and also waterproofing. They differ from previously known products by high mechanical strength.

For these organomineral products interaction of the polyisocyanate and the alkali silicate solution is carried out in the presence of a polyisocyanate cross-linking catalyst. This leads to the formation of, closely related with each other, inorganic and organic three-dimensional bodies having improved mechanical strength.

Reference [27] describes a method for hardening and sealing coal mining, ground, or brick structures in mining, tunnel construction, and construction industries. This method is the reaction of the polyisocyanate and the alkali silicate solution, leading to the production organomineral products. When using this method, the mixture of the starting compounds is introduced into the reinforcing formation through the boreholes or injection tubes. In some cases, to achieve the purpose, high pressure is used. The starting components can be located in separate multichamber cartridges and putting them into the formation, to mix with each other, results in the destruction of the cartridge.

The method described in Ref. [27] leads to significant strengthening of the treated rock compared to uncontrolled conversion of components. This is controlled interaction of the polyisocyanate and the alkali silicate solution, which yields a linked-together three-dimensional network of inorganic and organic polymer frame. However, organomineral products formed by this process, because of their organic components, are not completely noncombustible, although their flammability is significantly reduced compared to polyurethane systems used for the same purpose.

Reference [28] discloses an organomineral synthetic material with improved strength, elasticity, resistance to deformation at higher temperatures, and the lack of flammability. This material is suitable for use as putty for repairing cracks and hollow spaces and also for the manufacture of building materials. A material is obtained by mixing an organic compound that contains two or more reactive hydrogen atoms and one or more nonionic hydrophilic group with an aqueous silicate solution and organic polyisocyanate. From the resulting mixture, a colloidal dispersion is produced. For these synthetic materials may be used catalysts that promote the reaction of isocyanates with a reactive hydrogen atoms. For this purpose are tertiary amines, silaminy, nitrogen bases, and organometallic compounds. These synthetic materials can be administered to various auxiliary and additional substances. Such additives may be the following substances: surfactants, additives, foam stabilizers, reaction inhibitors, enhancers, organic or inorganic fillers, or diluents of various types.

In some examples is also described a cement additive in the preparation of synthetic materials. In these examples has been described the joint interaction of three components—a polyisocyanate, a silicate, and an organic compound—with reactive hydrogen atoms. The first component is introduced into the prepared mixture of the second and third components. Triethylamine was used as a catalyst for this reaction.

These studies indicate that the resulting composite material may be used with fillers, in fairly large quantities, without losing its valuable properties. However, in the case of cement additives, even at a low-volume fraction of filling increased compressive strength is achieved. Furthermore, the duration of the setting, that is, the time between the start of mixing and curing, is significantly increased by the addition of cement.

When using certain catalysts, the presence of cement in the compositions provides an interesting synergic effect. This effect is that the reaction of polyisocyanates and aqueous alkali silicate solutions and the cement setting reaction with the available water lead to a hardening of the material, which occurs in a very short time. As a result of this reaction are obtained products with high mechanical strength. Furthermore, due to the high proportion of inorganic substances, they are noncombustible products.

Reference [29] indicates that the polyisocyanates in an aqueous alkaline solution containing SiO_2 can be significantly inclined to cross-linking. Moreover, the reaction between NCO and water is significantly inhibited and forms a controlled amount of gaseous CO_2, which is used to interact with the liquid glass. The reaction simultaneously produces two polymeric frames mutually intertwined with each other.

In the first stage of the reaction a part of the polyisocyanate reacts with water to form a polyuria and cleavage of gaseous

$CO_2 \cdot CO_2$, formed in situ, reacts with an alkali component water glass solution to form $Me_2CO_3 \cdot H_2O$ (where Me is an alkali metal, especially sodium or potassium). It increases the silicate module water glass solution and formed polysilicon acid. During this reaction, a significant amount of heat is released.

As a result of this process were developed high-strength organomineral products. The optimal mode of their preparation by reacting a polyisocyanate, an aqueous solution of alkali metal silicate, and cement is molar ratios $NCO/SiO_2 = 0.8–1.4$, $SiO_2/Me_2O = 2.09–3.44$ (where Me is an alkali metal), and the ratio of NCO/cement = 10–0.5. The process is carried out in the presence of a dispersed catalyst, is stable in polyisocyanate-heterocyclosubstituted ether having the formula B–A–O–A–B, where A–C_1–C_4- alkylene, B- 5–8-membered N- and/or O, and/or S-containing mono- or bicyclic heterocyclic radical. Preferred is the use, as a catalyst, of a simple dimorpholinoethyl ester that is stably dispersed in the polyisocyanate. Moreover, the dispersion of cement and the catalyst must be added to the mixture of polyisocyanate and aqueous alkali metal silicate solution.

According to the work in Ref. [18], by reacting a polyisocyanate and water glass in the presence of cement, and also in the presence of a catalyst that is stable in polyisocyanate and dispersed therein, there is an immediate gelling and hardening of the mass begins almost instantaneously.

For more alkaline, liquid glasses, in which the silicate modulus is significantly lower than the standard range of conventional liquid glass in the formulation, it is necessary to provide a higher proportion of components giving CO_2, by reaction.

For optimal product hardness, the composition and amount of used liquid glass must be considered in determining the amount of other components of the reaction and, under certain conditions, the amount of catalyst. Organomineral products with excellent bending strength are obtained according to the invention in Ref. [18], in the event that the polyisocyanate and water glass solution are used in a molar ratio NCO/SiO_2 from 0.8 to 1.4 (preferably from 0.85 to 1.15). In this the preferred molar ratio of NCO/SiO_2 is equal 1.0.

The use of concentrated solutions of liquid glass is preferred as it reduces the total water content in the products, which has a negative influence on their strength properties. Concerning the cement used, the proportion of water should be 20–80%, preferably from 40–60%. This value expresses the water-cement ratio, that is, the ratio of the mass of water and the mass of cement. The required water fraction comes mainly from water glass. The lower boundary share of water glass is determined by the factor that its amount should be enough for the construction of the inorganic framework. For this are required at least 0.2, preferably up to 0.5, parts by mass of water glass, on mass part of polyisocyanate. The upper limit of an admissible proportion of liquid glass in this composition is achieved if the allocated amount of CO_2 is not enough to bind the entire amount of Me_2O contained in the liquid glass. Similarly, at too-high water content, it is impossible to achieve complete cure. When using liquid glass with a molar ratio of SiO_2/Me_2O at 2.85, the upper limit of the content of liquid glass is, for example, from 1.6 to 1.7 parts by mass of water glass, on mass part of polyisocyanate. These limit values, using water glass, with a different composition may shift slightly.

For organomineral products according to the invention polyisocyanates conventionally used in this field, such as those described in Ref. [39], can be used. In addition are suitable other products of the prior attachment of NCO, are known in the preparation of polyurethanes.

For obtaining organomineral products according to the invention [10, 18], preferred are polyisocyanates, which can easily react and lead to the formation of cross-linking when creating an organic, three-dimensional framework. These are compounds that have no steric hindrance for groups of NCO, which are involved in the interaction. A specific example of such sterically unhindered polyisocyanate is 4,4′-diphenyl-methane-di-iso-cyanate.

At present the most preferred catalyst is a simple 2,2-di-morpholino-di-ethyl ester of the following formula:

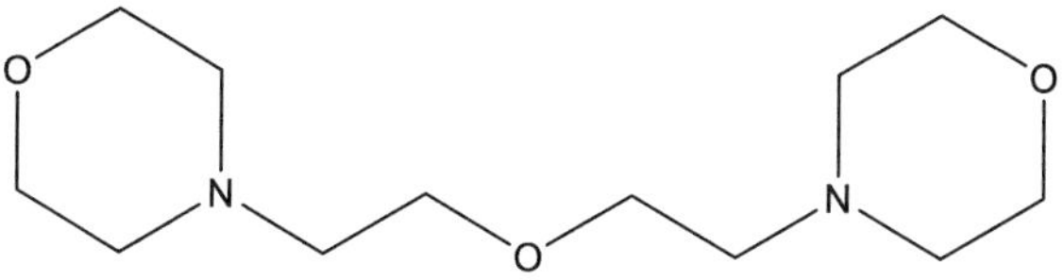

4,4'-(oxybis(ethane-2,1-diyl))dimorpholine

The catalyst of this type has a number of technological advantages, for example a stable, resistant dispersion. In addition to the catalyst, to accelerate the interaction further, can be used the catalysts known from polyurethane chemistry.

Examples of such catalysts are tertiary amines (triethylamine; tributylamine; N-methylmorpholine; N-ethylmorpholine; N-kokomorfolin; N,N,N′,N′-tetramethylethylenediamine; 1,4-diazabicyclo(2,2,2)octane; N-methyl-N′-dimethyl-amino-ethyl piperazine; N,N-dimethyl-benzyl-amine; bis-(N,N-diethyl-amino-ethyl)-adipinate; N,N-diethyl-benzyl-amine; pentamethyl-diethylene-triamine; N,N-dimethyl-cyclo-hexyl-amine; N,N,N′,N′-tetramethyl-1,3-butanediamine; N,N-dimethyl-β-phenyl-ethyl-amine; 1,2-dimethyl-imidazole; 2-methylimidazole; and derivatives of hexa-hydro-triazine); silamins with carbon-silicon bonds, as described in Ref. [33] (2,2,4-trimethyl-2-silamorfolin and 1,3-diethyl-amino-methyl-tetramethyl-disiloxane); nitrogen bases (tetra-alkylammonium hydroxides, alkali metal hydroxides, alkali metal phenolates, or alkali metal alkoxides); organic metal compounds, in particular, organic tin compounds, such as tin (II) salts of carboxylic acids (tin (II) acetate, tin (II) octoate, tin (II) 2-ethyl-hexanoate, and tin (II) laurate); dialkyltin salts of carboxylic acids (dibutyltin diacetate, dibutyltin laurate, dibutyltin maleate, or dibutyltin diacetate); ε-caprolactam, aza-norbornane of formula

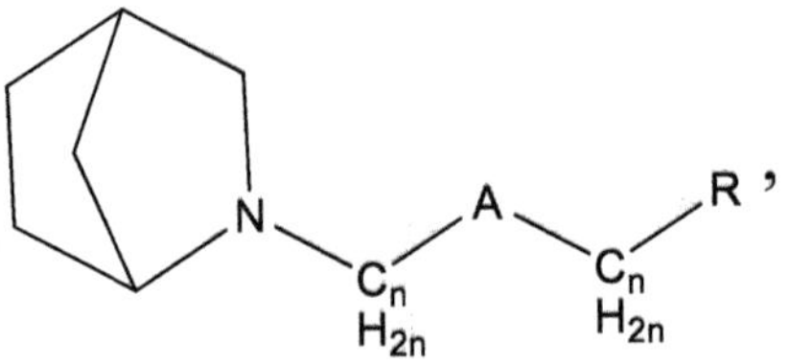

where A = O or CH_2, R = H or OH, and m and n are whole numbers; and catalysts, which catalyze the addition reaction and polymerization and cross-linking of isocyanates, such as 2,4,6-tris-(dimethyl-aminomethyl)-phenol.

To produce high-quality organomineral products by reacting polyisocyanate and a water glass solution, preferably a uniform distribution of the catalyst in the reaction mixture is required. When adding the catalyst to a solution of water glass, it is impossible to achieve a uniform dispersion as the resulting mixture is inclined to spontaneous delamination. Therefore, the catalyst according to the invention in Ref. [18] is added to the isocyanate, which is a particularly suitable (mentioned above), heterocyclic, substituted ether.

For foaming of the product suitable volatile substances are used that at room temperature are liquids and during interaction of liquid glass with a polyisocyanate evaporate as a result of heat produced from the reaction. Examples of suitable volatile substances are alkanes and halogen alkanes.

2.3 Mixing Technologies of Nanocomposites

A method for producing polymer nanocomposites in the melt, the so-called extrusion method, consists of mixing the molten polymer with nanosized material particles, surface modified with organic compounds. During the intercalation, the polymer chains substantially change their shapes and lose conformational entropy. As the driving force of this process is likely when mixed, the most significant contribution is made, the enthalpy of interaction in the system, the polymer-nanoscale filler. It should be added, for example, that the polymer nanocomposites based on clay materials were successfully produced by extrusion. The advantages of an extrusion method are the absence of any solvents, which eliminates the presence of hazardous effluents; the considerably higher process rate; and the considerably easier technological design of manufacture. That is, the most preferred method for the preparation of polymer nanocomposites is the industrial-scale extrusion method is, which involves less costly raw material and maintenance of the technological scheme [27].

The polymer-silicate nanocomposite silicate modified with organic substances, in obtaining, swells in a solvent such as toluene or N-dimethylformamide. Then, it is added to the polymer solution that penetrates into the interlayer space of the silicate. Then, solvent removal is carried out by vacuum evaporation. The main advantage of this method is that the "polymer-layered silicate" can be obtained based on a polymer of low polarity or nonpolar material.

Nevertheless, this method is not widely used in the industry due to the high solvent consumption [27].

By using mixing technologies, an important factor is the viscosity and fluidity of the reaction mixture. This is especially important for highly filled compositions, where there is low binder content that is in the liquid phase. In industrial practice, a relatively long-known way to increase mobility and moldability of the composition is based on the use of hydrophilizing, or hydrophobizing, surface-active agents (surfactants). They can significantly reduce the amount of solvent or liquid binder phase in the compositions while maintaining or even improving their rheological properties. Thus, organic solvents such as ionic and nonionic surfactants (alkyl sulfonates, alkylbenzenesulfonates, and the fatty acid content greater than C_9) allow obtaining compositions based on liquid glass with a high fluidity. These substances have lubricity and reduce frictional forces between the particles of the composite material. Thus the use of surfactants in the compositions based on liquid glass allow to further reduce the binder content in the composition and bring it up to $5 \pm 0.5\%$. Such a mixture has good mobility, adequate for the manufacture of complex-shaped products by free casting.

A method for producing polymer nanocomposites in the melt consists of mixing the molten polymer with the particles modified with organic substances, nanosized dispersed material. Preparation of polymer nanocomposites such as clay fillers in the synthesis of the polymer is the intercalation of the monomer in the clay layers. The monomer migrates through the galleries organo-clay material, and polymerization takes place within the layers. The polymerization reaction may be initiated by heating, irradiation, or appropriate initiator [27].

The nature of the rheological properties of the compositions is influenced by various factors—the size of the starting particle agglomerates, microstructures formed by them, forces acting between the particles and contributing to the formation of agglomerates, and processes that take place in a structured composition on flowing. After shear forces are applied, macroscopic agglomerates of the filler are destroyed; this is reflected in the results of rheological measurements. The transitional period, corresponding to the period of time from the beginning of the destruction of macroscopic agglomerates until the equilibrium, is characterized by a gradual change in the intensity of the scattered light. In the systems studied, the destruction of the agglomerates, to form separate particles and their stable compounds and to achieve their uniform distribution in volume of binder, was confirmed by microscopic observation of the sample's dispersion compositions.

Particles of fine fillers are also capable of forming a continuous structure, penetrating the entire volume of the system, and lead to changes in the macroscopic properties of the oligomeric composition. Primarily there is an increase in energy dissipation during flow, and hence the change of rheological properties, namely to increase the viscosity of the composition. In case of use, electrically conductive fillers such as carbon black, when creating continuous spatial structures, will significantly change the properties of the composition, making the material electrically conductive. Obviously, the length of the continuous chains must be commensurate with the distance between the electrodes. Consequently, the conductivity of filled compositions is a more sensitive parameter in the creation process of continuous spatial structures of particles than the viscosity of the composition. The sharp increase in viscosity appears to occur during the formation of the local structures of the filler particles. The dimensions of such structural formations are not commensurate with the size of the space between the electrodes. Hence, the composite retains dielectric properties.

Thus, to obtain conductive compositions on the basis of highly dispersed components necessary to construct chains of particle sizes which are many orders exceed the sizes of the particles themselves, and the number of particles in the chain of thousands or tens of thousands. It is obvious that such complex associates are easily susceptible to damage. It is possible to assume that these large particles, in turn, could become centers of structure formation carbon black particles. Dimensions of continuous chains of carbon black particles required to produce an electrically conductive system in this case would be dramatically reduced.

Figure 2.14a [27] shows the relative change of the composition flowing through the electric current, showing how with all equal conditions, the electrical conductivity of the composition with

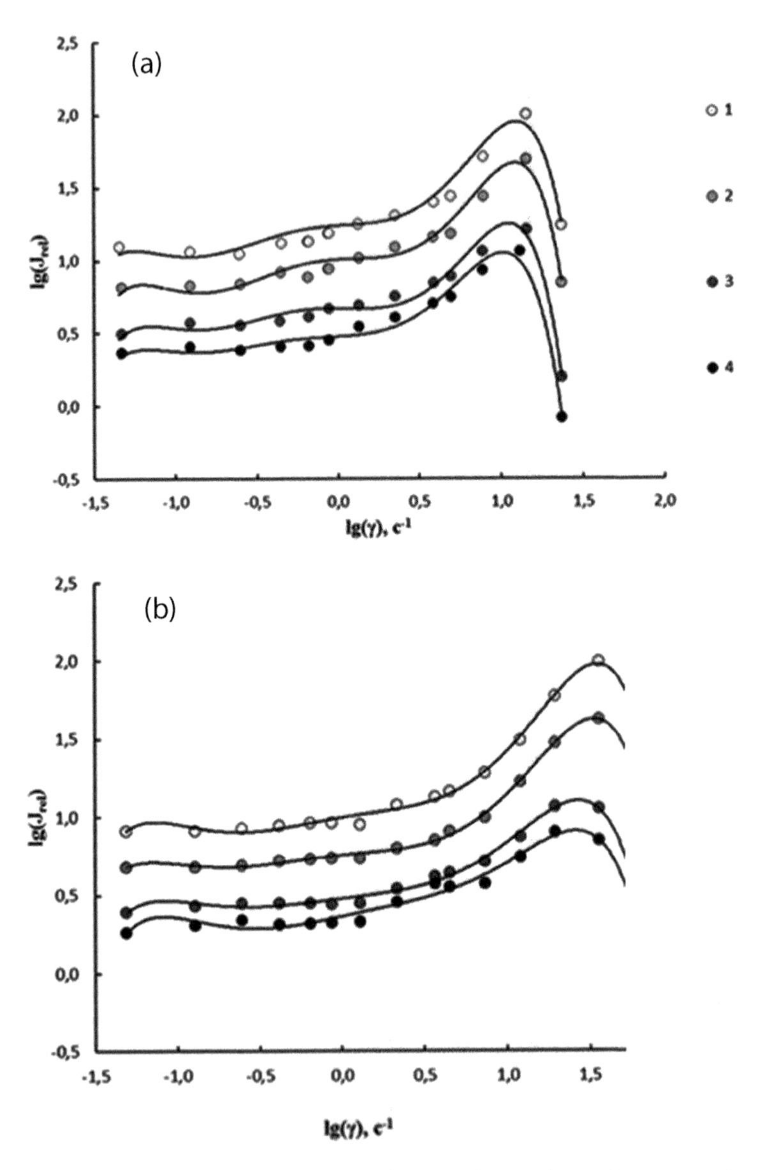

Figure 2.14 Dependence of the relative electrical conductivity of a composition (comprising 3 wt.% carbon black) on the shear rate at 60°C (a) and 80°C (b). Graphite content (wt.%): 1 = 20, 2 = 10, 3 = 5, and 4 = 2.5 [27].

graphite and carbon black is more than the electrical conductivity of the composition with only carbon black.

These studies showed that introduction in an oligomer of 22 wt.% graphite does not produce electrical conductivity of the test composition, but an oligomer containing 3 wt.% carbon black has conductive properties (Fig. 2.14a). However, the introduction of graphite into the composition containing 3 wt.% carbon black leads to a sharp increase in the electrical conductivity of the whole system. The increase of the shear rate leads to the destruction of the internal structure of the composition and reduction of its electrical conductivity. Figure 2.14b [27] shows similar curves for elevated temperatures. As follows from the data presented the effect of graphite increases with an increasing shear rate. The maximum effect of graphite at a temperature of 60°C corresponds to a shear rate of 10 s^{-1}. Increasing the temperature to 80°C leads to an increase in this value to a shear rate of approximately 20 s^{-1} [34].

2.4 Different Types of Nanophases

2.4.1 Nanosized Filler

For a long time to obtain nanoparticles was a core part of nanotechnology. To the same subject belonged basic postulates of nanotechnology, which were laid down at the beginning of the last century, in the ideas of R. Zsigmondy and T. Svedberg. However, the essence of nanotechnology and nanomaterials was formulated by R. Feynman, who offered two diametrically opposed approaches to the creation of nanomaterials" "bottom-up" and "top-down." The first of these methods is linked to the implementation of processes in atomic and molecular self-assembly or assembly technology of the final products. The second is related to the processes of mechanical disintegration, severe plastic deformation, decomposition of solid solutions, etc. [35, 36]. Actually other technological approaches do not exist.

The top-down approach is based on reducing the size of the physical bodies, by mechanical or other treatment, up to objects with ultramicroscopic, nanometric parameters. This method of obtaining nanomaterials has fundamental physical limitations. For example, when using photolithography techniques, there are the dimensional limitations associated with the wavelength of incident radiation. A mechanical action on the material requires the application of sufficiently high energy to create a new surface when the particles break down to the submicron level. In addition, any mechanical impact is bilateral in nature by virtue of Newton's third law. If you destroy some material, you (the instrument of destroying) are destroyed simultaneously. Thus, you bring to the product obtained as a result of this impact the particles of the instrument itself. On the nanometric scale, this contamination even in an amount of about several parts per million can prove fatal.

The idea of the bottom-up technology is that the assembly of produced "construction" is carried out directly from the lower-level elements (atoms, molecules, etc.), which are stacked to the desired order.

An important aspect of this approach is that it has been included in the essence of the nature of the chemicals. For centuries, chemists were engaged in obtaining new, more complex chemicals, by assembling them out of individual atoms or groups of atoms, creating complicated molecules from more simple molecules. Giant crystals were obtained by laying in the correct order individual atoms or molecules. The main problem is, how do you control this process to stop it at the desired level, which corresponds to the desired level of nanometer dimensions?

Of course, the first path is not closed today. There are a lot of technological methods, and new methods are constantly created of crushing materials without contamination of them by foreign impurities. There are methods involving the use of explosive technology, high-energy flows, plasma, radiation, laser, acoustics, etc.

But nevertheless, bottom-up assembly is the most natural process, in which there are no fundamental limitations and physical prohibitions on the possibility of obtaining structures in the range from the atomic to the macroscopic level.

In connection with this, we consider the chemical methods for the preparation of nanoparticles that may be the fillers in nanocomposites.

There are many chemical methods that can be used to obtain nanoparticles of various materials. Naturally, different classes of chemical substances and compounds obtained by various methods appropriate to the chemical properties of substances from which are obtained nanoparticles.

For example, to obtain the metal nanoparticles several types of reducing agents can be applied, for example, complex hydrides and alkyl hydrides. Thus, molybdenum nanoparticles can be prepared by the reduction of a molybdenum salt using sodium tri-ethylhydroborate, $NaB(C_2H_5)_3H$, dissolved in toluene. This reaction gives a good yield of molybdenum nanoparticles with dimensions of 1–5 nm. The reaction equation is as follows:

$$MoCl_3 + 3NaB(C_2H_5)_3H \rightarrow Mo + 3B(C_2H_5)_3 + (3/2)H_2 \quad (2.9)$$

Nanoparticles of alumina can be obtained by the decomposition of adduct, tri-ethylamine and aluminum hydride $(C_2H_5)_3N{:}AlH_3$, dissolved in toluene, followed by heating to 105°C for 2 h. As a catalyst for this reaction is used titanium iso-propoxide. The choice of catalyst determines the size of the nanoparticles formed by this reaction. When using titanium iso-propoxide it is possible to obtain particles with a diameter of 80 nm. To prevent agglomeration of nanoparticles to the solution surfactants such as oleic acid may also be added.

Obtaining metal nanoparticles in the conditions of high energy impacts on the chemical system is associated with the appearance of strong reducing agents with high activity, such as hydrated electrons, radicals, and excited particles.

Photochemical (photolysis) and radiation-chemical (radiolysis) recovery differ by energy. For photolysis, typically, energy less than 60 eV is required, and for radiolysis, more than 100 eV. The main features of chemical processes under the influence of high-energy radiation are nonequilibrium in the distribution of particles by energy and the overlap of the characteristic times of physical and chemical processes. Decisive role for the chemical transformations of the active particles has a multichannel, and nonstationary processes occurring in reacting systems.

Photo- and radiation-chemical reduction processes, when compared with the chemical method, have certain advantages. These processes are different in that higher-purity nanoparticles are produced by such recovery, as there are no impurities, which is a disadvantage using conventional chemical reducing agents. Moreover, using photo- and radiation-chemical reduction techniques synthesis of nanoparticles in solids at low temperatures is possible.

Photochemical reduction in a solution is most often used for the synthesis of noble metal particles. Upon receipt of such particles from the corresponding salts, use their solutions in water, alcohol, and organic solvents as the medium. In these media active species are formed by exposure to light. However, not only is photoreduction a process of preparation of nanoparticles of a certain size but there is a formation of larger aggregates as well.

As an example of photochemical production of films of metals, nanometer thickness, discloses a method for applying a metallic coating to a substrate. The substrate is placed in a reactor with a pressure of about 10 Pa. Next, the substrate is heated to a temperature of 50–70°C, a mixture of hydrogen and precursor vapor is fed into a reaction zone, and the substrate is subjected to UV radiation in the wavelength range of 126–172 nm. Here the flux of hydrogen is fed at a rate 0.3–4.0 l/h, and vapors of the precursor at a temperature of 25–50°C is fed at a rate of 2–4 l/h, with the help of the carrier gas argon. As a result, a chemical reaction of reducing the corresponding metal from the precursor to the metallic state takes place. In this case, on the surface of the substrate, a film is formed, consisting of grains of metal the size of 100–150 nm.

Decomposition of the precursor is achieved through the photo dissociation of hydrogen and oxygen, with the formation of highly reactive particles, such as the H, O, O_3, which sharply reduce its decomposition temperature. The metallic coatings can be prepared from Pd, Pt, Ni, and Cu. As the substrate, use copper, polymers, ceramics, or silicon. As precursors, use volatile organometallic compounds such as fluorinated β-diketonates (hexa-fluoro-β-diketonate [HFA]), palladium ($Pd(HFA)_2$), platinum ($Pt(HFA)_2$), nickel, and copper, as well as volatile trimetilnyh derivatives Pt (IV) ($(CH_3)_3Pt$ (HFA)Py) [27].

Radiation-chemical reduction for the synthesis of metal nanoparticles, due to their availability and reproducibility, is becoming more widespread. In the synthesis of metal nanoparticles in a liquid phase, issues related to the spatial distribution of the primary and intermediate products are of great importance. When radiolysis, unlike photolysis, for the intermediate active particles or radicals observes uniform distribution by volume, it promotes the synthesis of nanosized particles with a narrow size distribution.

Currently, for the formation of metal nanoparticles, porous inorganic materials such as zeolites are widely used. Solid zeolites having pores and channels strictly of a certain size are convenient matrices for stabilizing the nanoparticles with desired properties. Typically, two basic methods are used for the preparation of nanoparticles in the pores of the zeolite. One of them is related to the direct adsorption of metal vapor in the pores of thoroughly dehydrated zeolites.

Another more widely used method is based on chemical transformations of precursors introduced into the pores in the form of metal salts, metal carboxylates, metal complexes, and organometallic compounds. In a similar way, for example, in the channels of the molecular sieves were obtained nanowires with a diameter of 3 nm and a length hundreds of times larger. Zeolites of high thermal and chemical stability introduced inside the nanoparticles can be considered as the most promising catalysts.

To chemical methods for the synthesis of oxides and sulfides nanoparticles relate different versions of the sol-gel synthesis. When implementing the process of the sol-gel synthesis various precursors, such as salts, alkoxides, and chelate compounds, are also used. The process is catalyzed by a change in the pH of the initial solution. In acidic media are formed linear chains, while in alkaline media branched chains. Nanoparticles of metal sulfides can be obtained by replacing metal alkoxides of the corresponding compound of thioalcohols $M(SR)_n$ and reacting it with hydrogen sulfide.

A separate group of methods for the synthesis of nanoparticles involves reactions in micelles, emulsions, suspensions, and dendrimers (nanoreactors). Micelles are particles in colloidal systems consisting of insoluble in this medium of the nucleus, which is very small in size, surrounded by a stabilizing shell of the adsorbed ions and solvent molecules. They are associates of the characteristic structure, the shell of which is constructed of a plurality of amphiphilic molecules consisting of long-chain radical lyophobic and lyophilic polar groups.

For the preparation of nanoparticles of metals and their compounds micelles, emulsions, and dendrimers are used, which can be regarded as kinds of nanoreactors it possible to synthesize particles of a certain size.

Issues related to obtaining solutions of colloidal particles SiO_2 – silica sols, are discussed in detail in Refs. [20, 27, 29, 37, 43]. However, for various fillers for nanocomposites, it is important to obtain a wide range of particle sizes, from a few nanometers to hundreds of nanometers. Therefore, to obtain concentrated and stable silica sols enlargement of particles and ensuring a monodispersed system are necessary.

Uncontrolled growth of particles ultimately is carried out due to the change of amount or the disappearance of the other portion of the particles. To increase the rate of particle growth, it is necessary to increase the solubility of silica; it can be achieved by increasing the temperature and increase in pH to 9–10.

The most versatile is the so-called method with a feeder. This method is intended to produce monodispersed silica sols with particle sizes up to 150 nm. It was proposed by Bechtold and Snyder and developed by N. A. Shabanova [37]. Some of the important conditions for the success of the implementation of this method are the quality of the feeder, the feeding rate, and reduction in the polydispersity of the sol.

Effective implementation of the growth process is achieved by strict regulation of the feed rate of the feeder—active polysilicic acid. It has been established that taking the feeder feed rate proportional to the area of grown particles makes it possible not only to lower the polydispersity of sols but also to significantly accelerate the process. At high speed of filling a feeder, there will be a significant super saturation in silicon oxide. This leads to the formation of new fine particles of silica. On these particles subsequently would be delayed active silica that enters the system with a feeder. Thus, increase of the particle size will not be substantial. Supplying a flow of the feeder into a reaction vessel is necessary so that the supersaturation created in the system is not significant and the whole polysilicic acid is spent on growing of primary particles.

The following impede the normal course of the process:

- Irregularity of filling of the feeder (increased polydispersity of the sol).
- The duration of the synthesis (polydispersity increases, the average size does not increase significantly).
- Lack of vigorous stirring, perhaps the local formation of a new phase, the particle growth is slowing due to weak convection, as during the process, the number of particles remains constant, while the volume of the system increases to 20 to 25 times.

The average particle size of these papers is determined by the method developed by Sears [19] as well as by ultracentrifugation, by which is determined the distribution function of particle size.

All experiments were performed on samples of sols with a baseline pH of 7.5–10.0. The choice of this pH range is due to the fact that in this range the polymerization of silicic acid leads to the formation of spherical colloidal particles of SiO_2. Increasing the particle size of SiO_2 to more than 10 nm is accomplished by carrying out the process of heterogeneous polycondensation of active polysilicic acids on preformed particles.

The following processes are possible when adding polysilicic acid in the sol:

- Condensation on the particles
- The formation of new particles
- Losses on the walls of a chemical reactor

Given the heterogeneous nature of the polycondensation, and a material balance obtained the relation [37], which connects critical feed rate of the silicon acid in the solution (W_{kr}) with a concentration of SiO_2 in the system. It was obtained for conditions precluding the formation of new particles:

$$W_{\kappa p} = K\, S_{oud}\, (C_o V_o)^{1/3}\, (C_t V_t)^{2/3}\, C_{\Pi}^{-1}, \qquad (2.10)$$

where S_o, V_o, and S_{oud} are concentration, volume, and surface area of the germinal sol, respectively; C_t and V_t are total concentration and volume of the system, respectively; and C_f is the concentration in the feed solution SiO_2 ("feeder"). The value of the constant K is approximately equal to $5 \cdot 10^{-3}$ g / (m$^2 \cdot$ h).

As can be seen from the equation, with the increase in the total concentration (C) of the system or volume (V_t), the feed rate of the silica can be increased.

Before carrying out the process, the initial solution is prepared. It is prepared as follows: a certain volume of just-prepared nucleated sol having a concentration of 3.0–4.0%, stabilized by the addition of an alkali of a pH of 7.5–8.0 and boiled for about 1 h. In this case there is an acceleration in the formation of primary particles, which then act as the nuclei of condensation. Their diameter increases during this period up to 4–6 nm. The thus obtained sol at boiling treatment is fed off by fresh sol. Friable particles of the fresh sol getting in the heat-treated sol dissolve faster and stand out silicic acid, is deposited on the previously formed sol particles.

It is expedient to carry out the process of growth of particles at a certain increase of volume, that is, with partial evaporation. This technology is particularly preferred for the production of concentrated silica sol having a particle size of 15 nm and above. The following factors influence the efficiency of particle growth: feeding speed of the feeder and the ratio of the volume of the feeder to the original sol volume.

The first experiments investigated the effect of feeding speed of the feeder on the growth of the colloidal particles. Parameters such as the concentration of the starting sol, the concentration of the feeder, and the feeder volume and pH were constant. The feed rate of the feeder changed in the range from 2 to 25 mL/min. The value of the specific surface of sols increased from 256 to 412 m^2/g (Fig. 2.15 [27]). The minimum specific surface correspond feed rate of feeder 3–5 mL/min, was selected as the optimal rate of 5 mL/min.

In further experiments a constant speed of 5 mL/min was set, which changed the ratio of the feeder to the volume of the original sol. This ratio was varied in the range of 5–30, with all other things being equal. Data from these experiments are presented in Table 2.1 [27] and Fig. 2.16 [27]. In this range, V_f/V_o particles increased by 3 times. Further increase in the particle size requires a considerable lengthening of the time of the particles' growth, which can be measured by several days and more, depending on the desired

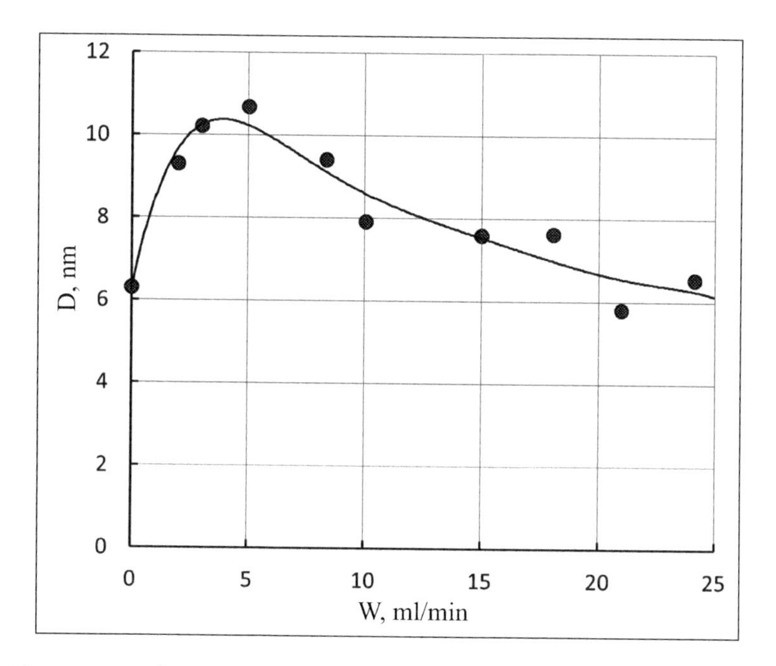

Figure 2.15 The dependence of the silica sol nanoparticle size on the feed rate of the feeder. The ratio $V_f/V_o = 10$ [27].

Table 2.1 Growing of particles with partial evaporation

No.	V_o, mL	C_0, g-eq/l	S_{oud}, m^2/g	W, mL/min	C_t, g-eq/l	V_t, mL	S_{tud}, m^2/g	D_t, nm	V_f, mL
1	200	0.5	500	5	1.6	350	256.0	10.7	1000
2	200	1.6	256	5	3.1	350	199.0	13.6	2000
3	250	3.1	199.0	5	3.1	500	179.0	15.2	3000
4	200	3.1	179.2	5	2.1	860	170.2	16.0	4500
5	200	2.1	170.2	5	–	–	173.4	15.7	5700

Source: [27]

nanoparticle size. Thus, for a given system, there are experimentally determined conditions for growing colloidal nanoparticles with sizes >10 nm.

Consequently, the nanoparticles of silica or other nanosized materials can interact both among themselves as well as with monomer and polymer components in the medium around them. An

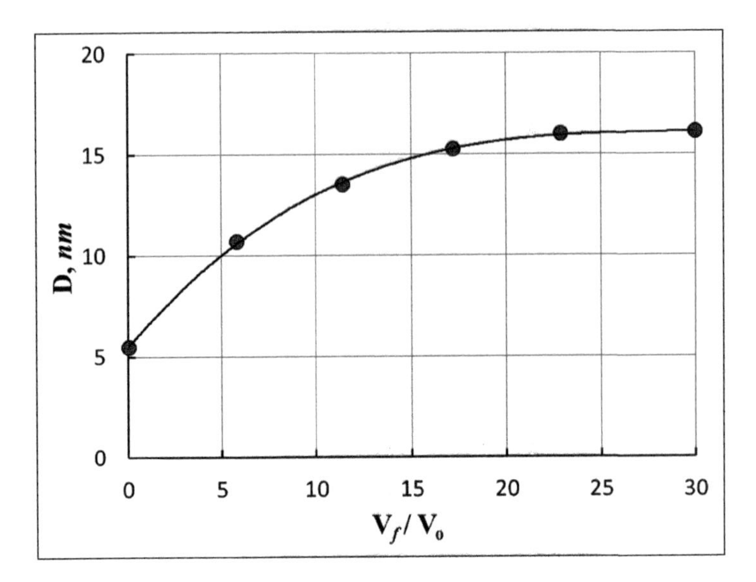

Figure 2.16 The dependence of the nanoparticle size silica sol on the ratio V_f/V_o [27].

important aspect of this issue is their interaction with the polymer matrix at their introduction in the composition with a polymeric binder. Interactions of the polymer chains with nanoparticles are very diverse. They differ in nature and intensity and often occur simultaneously. Polymer chains with chemisorption can form covalent, ionic, or coordination chemical bonds with surface atoms of the metal layer or of the corresponding compound. In recent years, covalent bonds caused the greatest interest in the design of nanomaterials from molecular building blocks [40, 41]. For this purpose, the particles were further functionalized for subsequent covalent binding of them with other components.

2.4.2 Nanosized Binder

Quaternary ammonium silicates (QASs) are organomineral surfactants with the general formula $([R^1R^2R^3R^4]N)_2(SiO_2)_{n-1}SiO_3$, where R^1, R^2, R^3, and R^4 are various organic radicals. In comparison with sodium/potassium silicates, stable aqueous solutions of QAS may easily have n (modulus) 5–10 and more. Variations of modulus and radicals allow changes in properties of QAS in a very large spectrum. QAS are compatible with many of mineral and organic compounds with pH values high enough to avoid precipitation of silica [38]. QASs have good adhesion to hydrophilic and hydrophobic surfaces and have the following applications:

- As quaternary ammonium compounds:
  - As biocides; as catalysts
  - In textiles (if two long chains) as textile softeners for home use
  - As the final rinse in the washing machine; as a rinse after shampooing; as emulsifiers
  - In metal working as additives to acid used in the cleaning and pickling of steel to prevent hydrogen corrosion
  - In road building, bentonite treatment, and oilfields
  - As antistatic in polymers, for example, in PVC belting
  - For the preparation of excellent-quality toner
  - As components in special systems of water purification
  - As components in self-setting aqueous mixtures for the manufacture of chemically resisting materials as additives in concrete and coatings
  - In structure-directing agents, for example, for the synthesis of molecular sieves with high-modulus silica
  - As raw material for the preparation of organosiloxanes
  - With aggregated titanium dioxide pigment products containing QAS, for pigment preparation
- As silicates:
  - For blends of hydrophilic medical use
  - As binders for concrete
  - For reinforcement of concrete; other building destinations
  - For coatings, linings, and claddings

QAS-based compositions provide coatings characterized by excellent adhesion, thermal stability, and fire- and corrosion resistance. Depending on additives, they may be insulators or electric conductors.

However, elaboration of QAS-based materials is very complicated because of difficulties in modeling of the QAS system. Silica and solid QAS have branched cross-linked structures not available for traditional methods of simulation.

Existing methods of study of such processes comprise two principal approaches: modeling by Monte-Carlo random simulation and the thermodynamic description.

The Monte-Carlo approach considers the solid-phase process of microporous cluster formation as a random process in which the initially empty space is divided into cells, every one of which can be eventually filled with a solid particle (silica). The probability of such an event is assumed independent of the prehistory of the process. The main drawback of this approach is the ignorance of the determination of this process. Therefore, such approximation is applicable only to solid structures with very low density, in which the probability of the neighborhood of two or more empty cells is negligible. The applicability of the Monte-Carlo approach is limited by three principal factors: finite size of the studied system (for various capacities of computers, from some thousand to some million cells), negligibility of surface tension, and absolute randomness of micropore formation processes. Therefore, the validity of the Monte-Carlo approach for the major part of real systems is very doubtable.

An available alternative approach may relate to the thermodynamic definition of the considered process, which does not take into account the possibility of deposition of silica [38]. The thermodynamic approach uses a macroscopic description based mostly on such or such distribution of structural elements in energy. This approach is well applicable to a system consisting of a very large number of cells and allows obtainment of very important information about a microporous system and the dependence of its properties on preparation conditions. However, since energy distribution is not directly related to internal surface area, this one cannot be found from equations of the thermodynamic model [24].

The problem of silica aggregation from a quaternary ammonium solution in a limited volume has been considered. The system has been described by a model combining the Monte-Carlo approach with thermodynamic limitations. Thermodynamic characteristics have been estimated by the statistical polymer method (equilibrium version). The model has been employed for the estimation of such properties as monomer concentration and tortuosity. The results of a computer simulation have been used for forecasting of properties of coatings and practical preparation of samples coated with a QAS-based composition [26].

In recent years begun to develop application silicate polymer composite materials, which are water soluble silicates with additives of active substances furan series. They operate under acidic and neutral media and under the influence of elevated temperatures. The materials are cheap, easy to manufacture, nontoxic, and non-flammable. The cost of polymer silicate materials is commensurate with the cost of cement concrete and several times lower than the cost of polymer concrete. Silicate polymer materials such as concrete, mortar, and putties used for making constructions for various purposes, monolithic lining, and tile lining. There is a certain perspective in composite materials based on liquid glass binder modified with furfuryl alcohol (FA).

A significant increase in strength and thermal and fire resistance of the silicate matrix is achieved by introducing into the composition esters of orthosilicic acid and FA (tetrafurfuryloxysilanes [TFSs]). The effect is achieved by strengthening of contacts between the globules of silica gel and modification of the alkaline component, due to the "inoculation" of the furan radical. Introduction to the binder of additive TFS leads to the formation of nanoparticles of SiO_2 and FA, which fills the matrix and forms the cross-linked polymer. These particles act as centers of nucleation and crystallization. Adding TPS increases the mechanical and chemical resistance of the binder, and this approach began to be widely used for the preparation of acid-resistant concrete and coatings [38].

This effect can be explained by the following considerations. The thermal stability of oxo-compounds can be judged by the relative strength of the interatomic bonds M–O and C–O in their crystal-structure. The lengths of the M–O and C–O, within the coordination polyhedron, can vary significantly, indicating their energy nonequivalence. During the dehydration and with thermal influence, the denticity of a certain part of the ligands may change. In the forming structure, they can begin simultaneously to perform the function as a ligand and a solvate that is absent in the system. Increasing denticity of ligands leads to distortion of the oxygen environment of the matrix element, or filler, with a corresponding change in the distance of the M–O and C–O in the structure, and hence to changes in their strength.

For increasing the strength, acid resistance, heat resistance, and flame resistance of construction materials and structures of them, TFSs are input to the binder composition. They are synthesized by transesterification of tetra-ethoxysilane and FA.

The composition of the resulting binder is as follows: a liquid glass 80–95 wt.%, TFS 2–7 wt.%, a hardener, and sodium hexafluorosilicate 13 wt.%. Thus as part of the liquid glass, organic alkali liquid glass is used, wherein the organic cation taken is 1,4-diazabicyclo[2.2.2]octane-1,4-diium or 1,5-diazabicyclo [3.3.3]undecane-1,5-diium silicate of 2–4 wt.%. (Fig. 2.17 [27, 29]).

Water-soluble silicate containing an organic alkali cation was prepared by the reaction the salt of an organic quaternary ammonium derivative with amorphous silica. A soluble organic alkali silicate, such as tetrabutylammonium silicate (TBAS), was used as a binder component for self-extinguishing.

A nanostructuring binder was prepared through laminar mixing of liquid glass containing cations of alkali metals such as sodium; TFS; and a water-soluble silicate containing an alkali organic cation such as diazabicyclo[2.2.2]octane-1,4-diium or 1,5-diazabicyclo[3.3.3]undecane-1,5-diium [18].

After mixing all components of the binder, it is necessary to use for 2–3 h. Adding of a hardener is carried out together with a fine-ground mineral filler.

Introduction of tetrafurfurylsilane, which is a nanostructuring component, in the binder leads to the formation of nanoparticles of SiO_2 and FA. SiO_2 nanoparticles act as centers of nucleation and crystallization. FA fills the silica matrix, which is destroyed and thus is polymerized. Adding TFS increases the mechanical and chemical resistance of the binder and is widely used for the preparation of acid-resistant concrete and fillers. Liquid glass with an additive of 1,4-diazabicyclo[2.2.2]octane-1,4-diium silicate is compatible with an aqueous dispersion of chloroprene rubber and polyurethanes, as well as most synthetic latex-based rubbers.

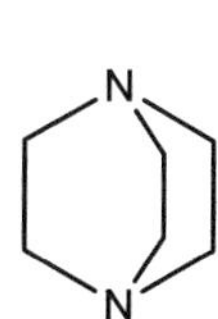

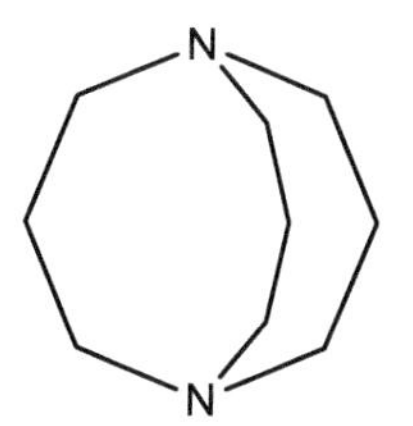

1,4-diazabicyclo[2.2.2]octane 1,5-diazabicyclo[3.3.3]undecane

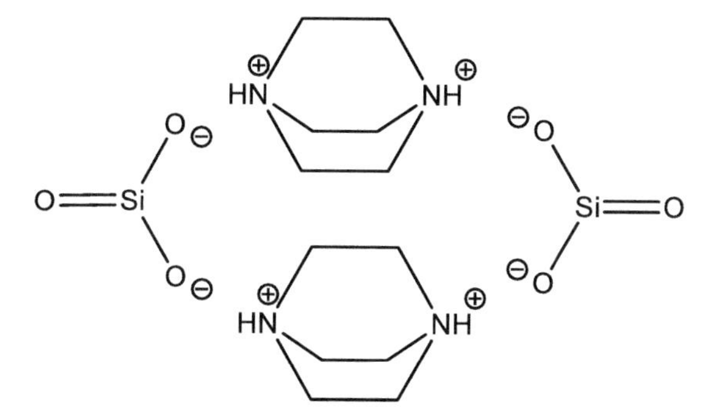

1,4-diazabicyclo[2.2.2]octane-1,4-diium silicate

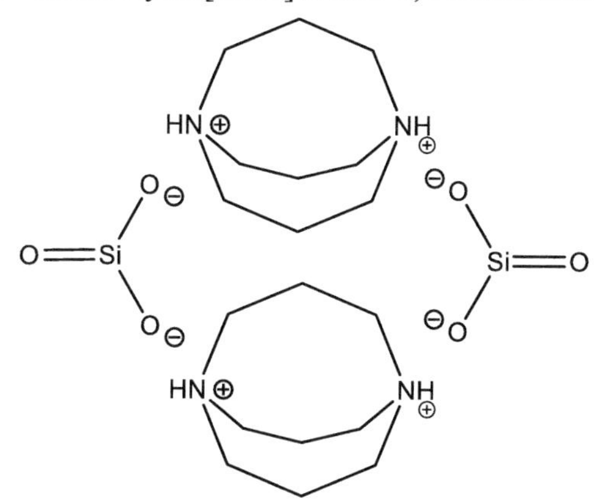

1,5-diazabicyclo[3.3.3]undecane-1,5-diium silicate

Figure 2.17 The structural formulas of 1,4-diazabicyclo[2.2.2]octane-1,4-diium and 1,5-diazabicyclo[3.3.3]undecane-1,5-diium and their silicates [27, 29].

Composite materials of interpenetrating inorganic and organic polymer networks are disclosed in Ref. [27].

2.4.3 Synthesis of Nanophase in the Matrix of the Composite Structure

Of interest is the introduction of the carbon material in a sol-gel process for obtaining nanocomposites as additives for modifying the functional properties of the nanocomposites as well as kind of pore-forming components (see Fig. 2.4). Pore-forming components are substances that disappear at various stages of forming and processing of materials to form pores. Carbon materials in the process of annealing in an oxygen atmosphere are converted into gaseous products. The ability to control the nanosizes of the objects consisting of carbon materials during their introduction to the structure of the nanocomposite determines the feasibility of their use to obtain the structures with specified pore sizes.

For these technological methods, the term "template synthesis" is used and input particles are called a template. New materials with a narrow pore size distribution are necessary to create semiconductor gas sensor devices, new-generation composite membranes, sorbents for chromatography, and others.

The role of porous materials obtained by the sol-gel method is highlighted in the development of technological operations applying dielectric coatings with low values of relative dielectric permittivity, in micro and nanoelectronics [26].

At the present stage of development of silicon nanoelectronics, intensive works are carried out to decrease the values of resistance of materials by replacing aluminum in conventional conductive materials on the copper conductors (or copper in a shell from TiO_2). To reduce the capacitance values there is intensively developing research on materials with low values of the effective relative dielectric permittivity. One of the main technological methods to reduce the average value of the dielectric permittivity is the formation of an insulating layer with nanopores of the sol-gel method.

Another promising area of technology is the formation of the catalytically active nanoparticles on the pore structure. The use of such substrates gives the material in the form of "nano grass" from the silicon, or heterostructure nanowires. Nanowire-manufactured devices obtained in this manner are used for nanoelectronics and nano-opto-electronics, for sensor techniques, and for photonics. To date has been developed the technology of heterostructure nanowires (nanofilaments) with a change in the composition both along the growth axis (axial nanowire heterostructures) and in the radial direction (radial nanowire heterostructures). The review summarizes the data on such structures on semiconductor compounds A_3B_5 and A_2B_6 and solid solutions based on them. As the catalyst, Au nanoparticles are usually used and nanofilament growth occurs by a "vapor-liquid-crystal" mechanism.

Interest in metal-polymer nanocomposites obtained by the method of synthesizing nanophase in the structure of the composite matrix is caused by a combination of unique properties of nanoparticles of metals, their oxides, and chalcogenides. Furthermore, these materials possess unique mechanical, film forming, and other properties and possibilities of their use as magnetic materials for recording and storage of information, as sorbents, as catalysts, and as sensors. Monomeric and polymeric metal carboxylates are classic examples of such starting materials [39]. Carboxylate compounds such as monomeric or polymeric structures may be used as molecular precursors of nanocomposite materials. In addition, the carboxylate groups belonging to macroligands are effective stabilizers of nanoscale particles. Sometimes, these properties are manifested in the form of a polyfunctionality of one system.

The amphiphilic nature of carboxylated polymers and copolymers allows encapsulating the metal nanoparticles and combining them with the polymer and inorganic matrices, as well as with biological objects. It also allows you to attach to the nanoparticles properties such as solubility in different environments and the ability to self-organization.

Processes of steric stabilization, flocculation, phase separation, electrostatic, and van der Waals interactions determine the aggregative stability of the particles in the polymer matrix. The van der Waals attractive force acts between the two surfaces of the nanoparticles that are not coated with a polymer, up to 200 nm, causing their aggregation. At the same time, the presence of an adsorbed layer of polyacrylate or polymethacrylate of ammonium on the surface of nanoparticles leads to the appearance repulsive forces between the two surfaces at a distance of 35 nm. Moreover, ammonium polymethacrylate provides a stronger repulsion than ammonium polyacrylate. This effect is apparently connected with an additional steric barrier due to the CH_3 group. Conformational effects of polymer chain are particularly sensitive to reaction conditions such as pH of the medium.

Accordingly, the thermal conversion processes of unsaturated carboxylate metal complexes allow combining the synthesis of nanoparticles with their simultaneous stabilization in a polymer matrix formed by decarboxylation. It is possible to assume that the almost complete homogeneity of thermal conversion processes of monomeric carboxylates is the cause of a fairly narrow character size distribution of metal nanoparticles and also their morphological features associated with the spherical shape. In this case there is only a partial heterogeneity in the macroscopic defects area.

Other monomeric carboxylates most frequently used as starting products in the preparation of nanostructured metals are metal octanoates and oleates. Thermolysis of their complexes, in combination with surfactants and other reagents, is usually carried out in a solution of high-boiling solvents (octadecane, octadecene, docosane, octyl esters, etc.). The undoubted advantage of the process of thermal decomposition of the carboxylate compounds in an inert solvent is the possibility of controlled synthesis of monodisperse nanocrystals, with a high yield, a narrow size distribution, and a high degree of crystallinity.

A promising direction is to conduct these processes in the pores of silica gel matrix, which will allow obtaining materials with unique optical and sensory properties. There is the possibility of combining these processes with processes of the sol-gel synthesis.

Thus, methacrylate-substituted metalloclusters Hf_4O_2 $(OOCC(CH_3)=CH_2)_{12}$ and methacrylo-yl-propyl-tri-methoxysilane $CH_2=C(CH_3)$ $(COO)\text{-}(CH_2)_3Si(OCH_3)_3$ are used for hybrid thin films based on silica gel with embedded oxo-clusters of hafnium [27]. Chemical bonding of the components is carried out with the use of photochemical polymerization of methacrylate groups. Alkoxy silane undergoes hydrolysis and condensation, with the formation of oxide grid SiO_2. Heat treatment at $\geq$800°C of the hybrid nanocomposite is accompanied by pyrolysis of the organic part and sealing of the oxide network. This leads to the formation of a nanostructured oxide material. Thus it is possible to obtain monolithic gels and thin films. Carboxylate ligands in organomineral composites offer a high degree of cross-linking, through coordination bonds between the polymer and the mineral component.

In Ref. [26] is proposed a new synthesis method of organomineral nanocomposites on the silicon oxide matrix, using borane compounds as homogeneous catalysts that promote Si-H bond activation. The reference describes an innovative method for modifying the surface of silica, using hydrosilanes as modifier precursors and tris-(penta-fluorophenyl)-borane ($B(C_6F_5)_3$) as the catalyst. The proposed mechanism of this process is shown in Fig. 2.18 [27].

Since the surface modification reaction between surface silanol groups and hydrosilanes is dehydrogenative, passing and stopping of the reaction can be easily confirmed with the naked eye. This new process, without participation of the metal compounds, can be carried out at an ambient temperature and requires less than 5 min to complete. Hydrosilanes carrying different functional groups, including alcohols and carboxylic acids, were immobilized by this method. The ideal preservation of sensitive functional groups that are easily destroyed when using other methods of synthesis makes this technique attractive for diverse applications.

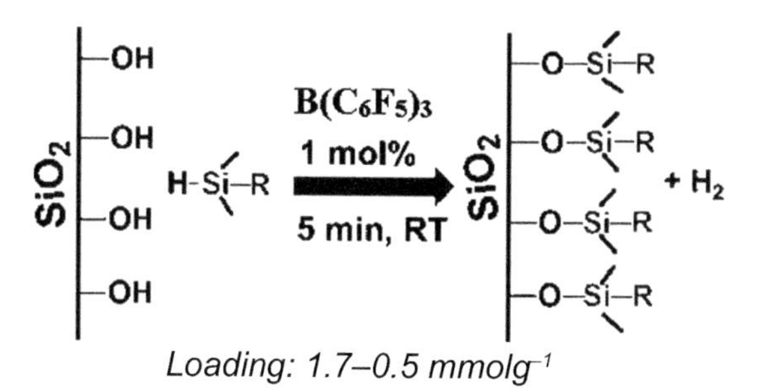

Figure 2.18 Schematic diagram of the silica gel surface modification, using the dehydrogenative reaction between the silanol groups and hydrosilanes [27].

2.5 Influence of Various Factors on Structure and Properties of Hybrid Materials

2.5.1 Packing of Spherical Nanoparticles of the Filler

The system of pores in a solid is entirely determined by the nature of the packing of the primary particles. For example, when the hexagonal closest packing of spherical particles between disposed in a plane, the three nearest tetrahedral cavities is octahedral cavity. In subsequent accretions over the octahedral cavity occurs again an octahedral cavity. Superimposed on top of each octahedral cavity is a formed channel that permeates the whole body of the matter in the orthogonal directions [31]. In cubic structures, the closest packing can also point out to the formation of interconnected channels. In a package of this type, the void volume is 25.95%. From here, it is possible to calculate the bulk density of the material (ρ):

$$\rho = \rho_{\mathrm{K}} \, (1 - L), \tag{2.11}$$

where ρ_{K} is the crystallographic density of the material used as a basis and L is the void fraction in the structure of the package.

If the lightest of refractory materials have a crystallographic density close to 3 g/cm^3, the bulk density of the product with the densest packing is 2.2 g/cm^3. From here, to obtain a material with a bulk density of less than 0.5 g/cm^3, it is necessary to achieve a greater porosity of 83%.

Let's try to analyze whether the particle packing can achieve the desired porosity.

Typically, with a decrease in the coordination of number of particles packed in the packaging structures, the pore volume of particles increases. In the case of the densest packings, the coordination number of the package is 12. Consider the following options for the coordination numbers of the package. Classically, such a packing density was 74%; the porosity of such a package is 26%. For monodisperse materials denser packaging does not happen.

If the coordination number is 8, there is cubic volume-centered packing. In contrast to the dense packing it has a loose structure. Here, the balls are in a flat layer, not touching each other, and are at a distance 0,155 R. The motive construction of this structure is characterized by the fact that the balls of the second layer are omitted in the voids of the underlying layer. The third layer repeats the first and the fourth, the second, etc. This type of packaging is an optimal variant spatial arrangement of spheres with a coordination number of 8 and has a lower percentage of its filling volume. The pore volume is 31.98%.

If the coordination number is 6, there is simple cubic packing. A globular structure, with a coordination number of 6, is built by successive overlay balls packed in a square grid so that the centers of the balls of the second layer are located directly above the centers of the balls first, the centers of the balls of the third layer are above the centers of the second, etc. Pores in this package, enclosed between eight balls in contact, has six throats, close in shape to a quadrangular. In this case, a looser structure is formed in which the pore volume is 47.64%.

If the coordination number is 4, the packing type is of diamond and ice. The customized packaging is the most loose of all considered regular packages. It is constructed from layers of balls spaced from each other in a plane. Balls of the next layer are between the balls of the first layer. Centers of the balls of the third layer are located directly above the centers of the spheres of the second layer. The fourth layer is located exactly above the first, etc. As a result of such packing of balls, the volume of voids produced is 65.99%. In the limit, with a decrease in the coordination number, each symmetrically constructed structure with a predetermined pore system should go into a chaotic folded body with a set of pores that will differ in both shape and size. The boundary of such a transition can be considered packing with a coordination number of 3.

When the coordination number is 3, this package is like a fishnet—loose formations with a continuous three-dimensional framework. The porosity of this structure is 81.5%. Thus, relying solely on the various kinds of packing of spherical particles, it is not possible to achieve the desired porosity, and, moreover, to achieve a density less than 0.2 g/cm^3. However, if we admit the formation of mixed packing with a coordination number of 2 and 3, that is, when some particles in the considered package are present in the form of

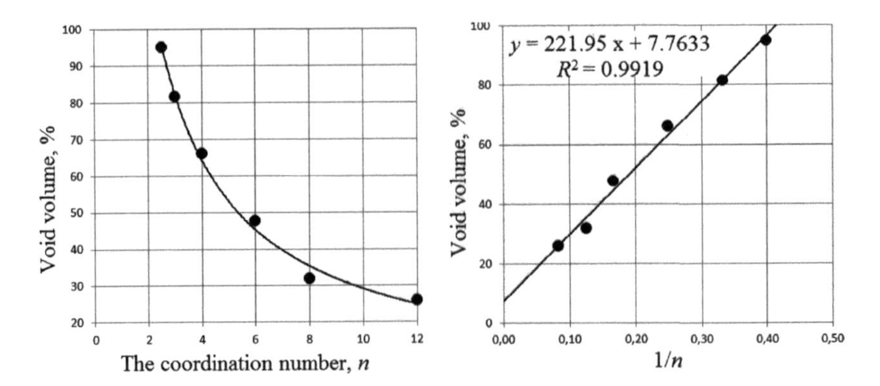

Figure 2.19 Dependence of the porosity of the packing of spherical particles on the coordination number n and $1/n$ [27].

short chains consisting of at least three particles, for such packages, porosity of 95% can be achieved.

The dependence of the porosity on the coordination number for different methods of stacking spheres is shown in Fig. 2.19 [27], which shows that the porosity is inversely proportional to the magnitude of the coordination number.

Currently, particularly among the various sorbents, such material are quite common in which a large part of their volume falls on the pores. For example, the porosity of silica gels ranges from 36 to 84%. Reference [13] describes the preparation of airgel with a porosity of 99%. The porosity of aerosils is 95–98%. Similar structures are formed from fibrous crystalline silicates such as chrysotile, during their disintegration by a surfactant. Individual fibers of the tubular type with a diameter of 18 mkm are thus formed, which due to the weave create very loose stable colloidal dispersions with a high porosity [40]. Apparently, such a net structure has liogels, aerosils, and some particularly loosely-packed xerogels, globular chains which are interconnected so that the number of contacts for the majority of the globules is 2. For stiffening the skeleton of the gel, the number of contacts between the particles and their neighbors at the nodes of this network should be equal to 3. But in any case, such structures constructed on the basis of the spherical particles do not have sufficient rigidity. Therefore, the most promising methods for the preparation of porous materials having sufficient strength is the use of different fibers or hollow microspheres.

Based on the plot of the porosity α (%) from the coordination number n of spherical particles, shown in Fig. 2.16, one can obtain the following dependence:

$$\alpha = \alpha_{\infty} + \frac{B}{n} \tag{2.12}$$

The coefficients in this equation for the dimension dim $(\alpha) = \%$ have the following meanings: $\alpha_{\infty} = 7.763 \pm 0.016$, $B = 221.95 \pm 0.45$, and $R^2 = 0.9919$. Since the equation of this curve has a singular point, it is possible to calculate the limiting value of the coordination number in which the void volume will be equal to 100%:

$$n_{100} = \frac{100 - \alpha_{\infty}}{B} \tag{2.13}$$

The value of $n_{100} = 2.406 \pm 0.010$. The physical meaning of this value is that in dispersed nanostructures with coordination numbers less than this value the more or less stiff carcass structure of the material collected from individual nanoparticles is missing. If the coordination number is below this, the particles are collected only in the individual chain structures that are not related to each other, which is realized when the coordination number is 2.

To describe the dependence of porosity on the coordination number of the packing of spherical particles, calculations were carried out for the selection of a correlation which would give a maximum correlation coefficient. The maximum correlation coefficient was obtained for the dependence of the following form:

$$\alpha = \alpha'_{\infty} + \frac{B'}{(n - n_0)} \tag{2.14}$$

The coefficients in this equation under conditions as in the previous case were $\alpha'_{\infty} = 0.14964 \pm 0.00019$, $B' = 310.10 \pm 0.40$, $n_0 = -0.7750 \pm 0.0010$, and $R^2 = 0.9975$. The resulting correlation dependence is shown in Fig. 2.20 [27]. For this relationship, you can also calculate the limit value of the coordination number at which the void volume is equal to 100%:

$$n_{100} = n_0 + \frac{B'}{(100 - \alpha'_{\infty})} \tag{2.15}$$

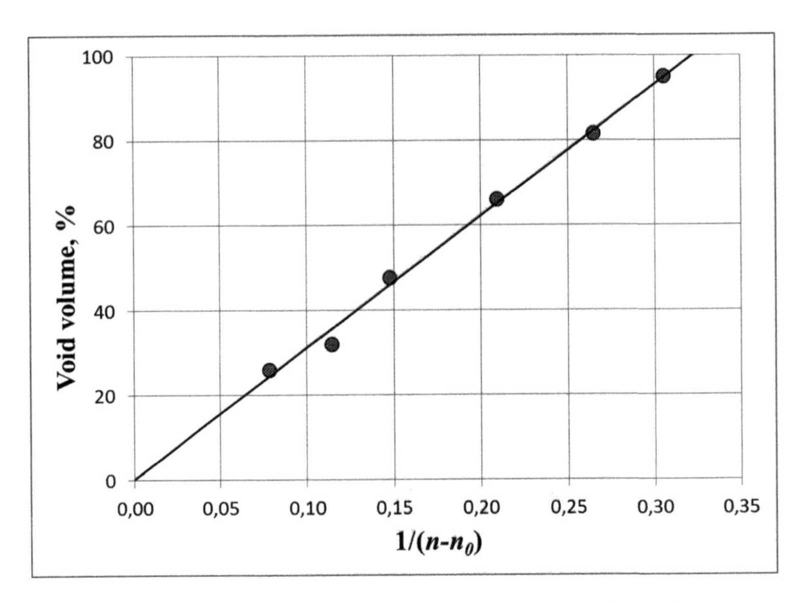

Figure 2.20 Dependence of the porosity of the packing of spherical particles from $1/(n - n_0)$ [27].

The value of $n_{100} = 2.331 \pm 0.006$. Interpretation of these results is more difficult, even though they give a more accurate result, and several are expanding the range of coordination numbers, to which still may form rigid skeleton, with openwork packing of spherical particles.

In recent decades, the rapid development of computer technology has led to the creation of fundamentally a new powerful means of investigating the physical processes—computational experiment. Numerical simulation is the only way to study these processes in conditions when theoretical methods meet with serious difficulties and practical experiments are overly laborious or fundamentally impossible. The creation of computer models has now become a recognized and rapidly developing direction of science and technology.

The application of simulation compared to traditional laboratory methods saves money and time. In studies based on computational experiments, a crucial role is played by the representation of the physical phenomenon in the form of an appropriate mathematical model, computational algorithm, or a computer program. The mathematical model in the computational experiment is simultaneously a test material and the algorithm of the experiment. In Ref. [41] it is shown that one of the most common methods of constructing mathematical models is the simulation by means of particles or particle method. It is also indicated that when used properly the particle method is able to demonstrate its clear advantages. On the basis of the accumulated theoretical and experimental data was built a computer model of disperse systems based on the particle method.

The particles method is common to the class of models in which there is a discrete description of physical phenomena, including consideration of the interacting particles. Any classical system consisting of particles can be described, knowing the law of their interaction, coordinates, and velocities. Each particle has a number of fixed characteristics (e.g., weight and size) and changing characteristics (e.g., position and velocity). As an object of numerical study of disperse systems by computer simulation aluminum powder was chosen. This is due to the fact that aluminum particles have a spherical shape.

As a result of these calculations were obtained varying degrees of the volumetric filling of system under the external pressing force. From the presented data in Fig. 2.21 [27] on the distribution of the coordination numbers for different volumetric filling of the powder follows that the limit volumetric filling of powder equal to 64% corresponds to a state where more than 40% of the particles have a coordination number of 12.

The computer model also makes it possible to study the nature of the contacts between the particles in the dispersion. To investigate the distance between "contacting particles" corresponding computational experiments were conducted. Figure 2.22 [27] shows the distribution of the number of pair contacts depending on the distance between the particles for different volumetric filling. From this figure, it follows that for all volumetric filling, the dependence of the distribution of pair contacts between the particles of the distance between them has two peaks. Reducing the volume filling of the system leads to a shift of the maxima in the region of the greater distance between the particles.

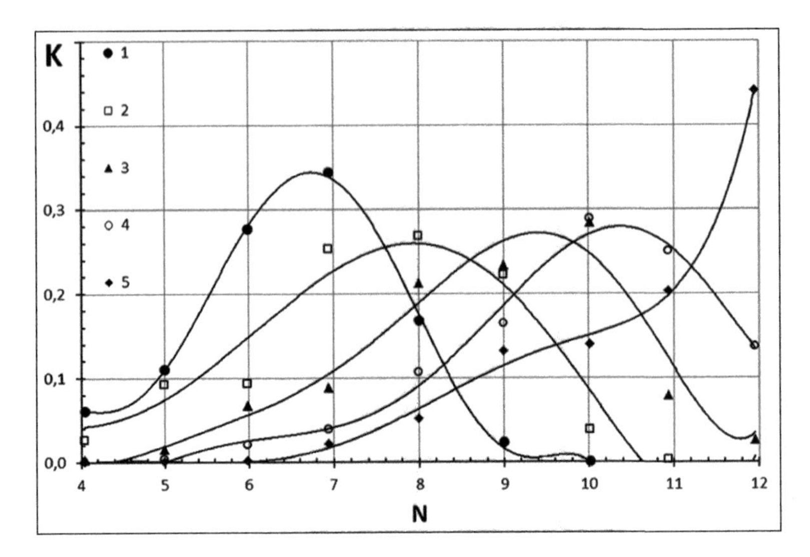

Figure 2.21 Distribution of coordination numbers for different volumetric filling of powder: **1** - 0.56; **2** - 0.58; **3** - 0.60; **4** - 0.62; **5** - 0.64 [27].

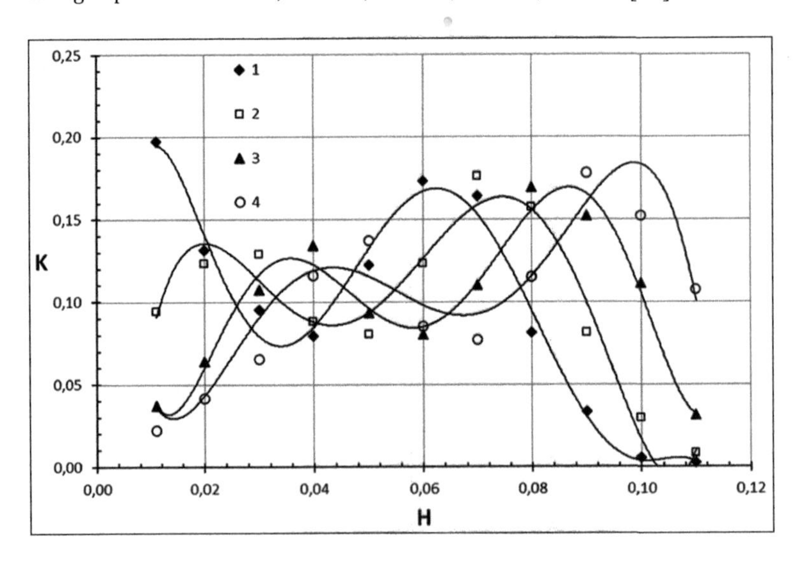

Figure 2.22 Distribution of the pair of contacts of the particles as a function of the distance between them for different volumetric filling of the system: **1** - 0.64; **2** - 0.62; **3** - 0.60; **4** - 0.56 [27].

Storage and processing of a powder in real conditions is accompanied by the absorption of moisture from the surrounding medium as a result of water vapor adsorption on the particle surface and its capillary condensation in the pores of the material. The formation of the liquid phase leads to the appearance of liquid "bridges" and capillary forces between the particles of the powdered material.

These forces have a significant impact on the behavior and properties of the powders and dispersion compositions, both in storage and when using in industrial processes. In this case, the various parameters of the particles have a significant impact on the amount of capillary forces, in particular the size and shape of the liquid "bridges," causing a type of particle "contact."

Consequently, the determination of the effect of geometrical parameters of the powder particles by an amount of the capillary forces acting between them allows better prediction of the properties and behavior of the particulate components and will also contribute to obtaining, based on them, disperse materials with specified properties.

The proposed method also allows determining the amount of liquid in a capillary "bridge" between the particles and the magnitude of the capillary pressure as a parameter, which determines one of the components of capillary forces [27]. As the computational cell, the capillary interaction of two particles connected by a liquid "bridge" and having the shape of a sphere, a cone, and a plane has been considered. In real disperse systems using the selected forms can be described as a large number of various types of "contact" between the particles.

In this case, as in the case of spherical particles of the same size, the magnitude of capillary forces is dependent on two components. The first component is determined by the surface tension of the liquid, which acts along the wetted perimeter. The second component is caused by the presence, rarefaction, or pressure arising due to the curvature of the surface of the "bridge" liquid and is described by the Laplace equation.

It is seen that in all cases, the growth of the quantity of the liquid is accompanied by the appearance of maximum of capillary forces. Growth of the capillary force occurs in accordance with

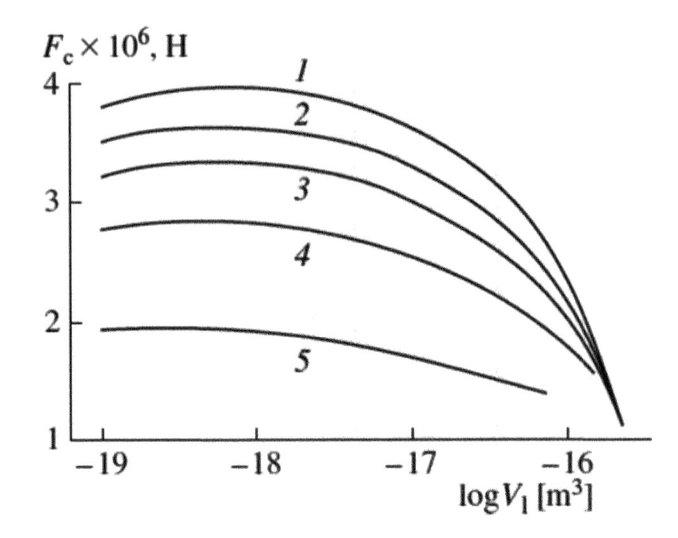

Figure 2.23 The dependence of the capillary force of the volume of liquid "bridge" in the interaction of particles of type "sphere-sphere." The radius of the particle is equal to 5 microns. The ratio of the radii of the particles is **1** - 100, **2** - 10, **3** - 5, **4** - 2.5, and **5** - 1 [27].

the increase of the ratio of particle sizes (Fig. 2.23 [27]). The magnitude of this force is committed to the values corresponding to the type of "contact" particle, such as a "sphere-plane." This effect is particularly noticeable with large volumes of liquid "bridge" when the differences in the values of force are negligible. On the basis of the obtained data, it was not only qualitatively confirmed but also quantitatively shown by the influence of differences in particle size by the amount of capillary forces acting between them.

Using these equations the dependence of capillary forces on the value of the amount of liquid in the "bridge" was determined for the considered types of "contact" of the particles. Water is considered a capillary liquid. For all types of "contact," the size of the gap between the particles is equal to 0.01 mkm and the wetting angle was 0°. From Fig. 2.24 [27] can be seen that the type of "contact" of particles has a significant effect on the magnitude of the capillary force. Its influence increases with increasing volume of the "bridge." On the basis of the proposed models have been described and tested experimentally the rheological properties of organomineral

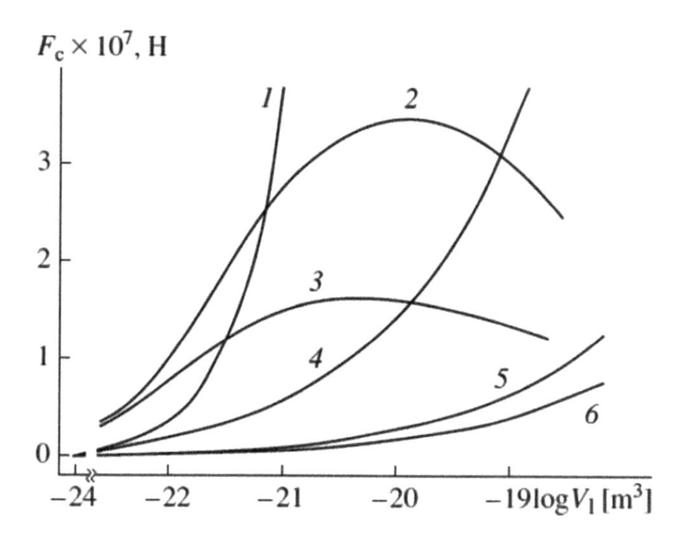

Figure 2.24 The dependence of the capillary force of the volume of liquid "bridge" for any type of "contact" between the particles: 1, "plane-plane" ($d = 0.2$ mkm); 2, "sphere-plane" ($R = 0.5$ mkm); 3, "sphere-sphere" ($R_1 = R_2 = 0.5$ mkm); 4, "cone-cone" ($\beta_1 = \beta_2 = 150°$); 5, "cone-plane" ($\beta = 90°$); and 6, "cone-cone" ($\beta_1 = \beta_2 = 90°$) [27].

nanodispersions and the possibility of obtaining on their basis the conductive polymer composites [34].

2.5.2 Packing of Fibrous Nanoparticles of the Filler

A heterogeneous structure is one of the most important factors determining the mechanical behavior of reinforced, dispersion-strengthened nanocomposites under mechanical loads. Currently, there is a need of modeling spatial patterns of nanocomposites in order to predict their effective elastic and strength characteristics. Exclusion of the possibility leads to weakly predictable results, especially relevant for the task of generating materials with a high-volume fraction of fibers, the achievement of which is made possible through the use of specially designed iterative procedures for streamlining the reinforcing elements.

Synthesis of fragments of random structures of fibrous nanocomposites, the reinforcing elements, which are fiber, round in cross section, is related to the random placement of disjoint smooth disks in the plane. In Ref. [42] are described in detail the various ways of generating fibrous and dispersion-strengthened materials. To determine the laws governing the formation of rarefied and close-packed structures of reinforced nanocomposites and estimate the influence of factors that affect their properties, the following algorithms are implemented:

Algorithm 1—the Monte Carlo method: In the studied fragment (typically a single square) are randomly placed fibers. If, in the generation of the fiber position, it is not beyond the fragment, then coordinates of the center of the fiber are recorded. In the synthesis of the following coordinates of fibers, checks crossing fibers placed previously. In the case of both conditions: the absence of intersections with all reinforcing elements and the location in the fragment generated the fiber is "turned on" in the structure of the composite. Upon reaching the necessary or maximum possible volume fraction of filling, or to fulfil the condition, number of unsuccessful attempts to accommodate reinforcing elements.

Algorithm 2—additional displacement fibers at the "hard" boundaries of the fragment: This algorithm is a modification of the method of "radial gravitational field" and features software implementation, described in detail in Ref. [42]. Additional mutual movement along a straight line connecting the centers of the cross sections is carried out, again, simulating a random structure and the previously generated fibers, at a distance, which guarantees the absence of intersections with all the structural elements. If in the process of correction of placement, any fiber comes into contact with the boundary of the fragment, the output of a portion of the cross section of the reinforcing element beyond the area is completely eliminated; fiber moves only along the boundary of the fragment. Fiber simulation and modification of their location create a random structure with the given volumetric filling or reaching a certain limiting value the amount of additional relocatable fibers

An important characteristic of the random structure of composites is the volume fraction of the reinforcing filler. For the definition of the limit of the volume fraction fibers, it is necessary to choose a parameter that is the criterion that determines the

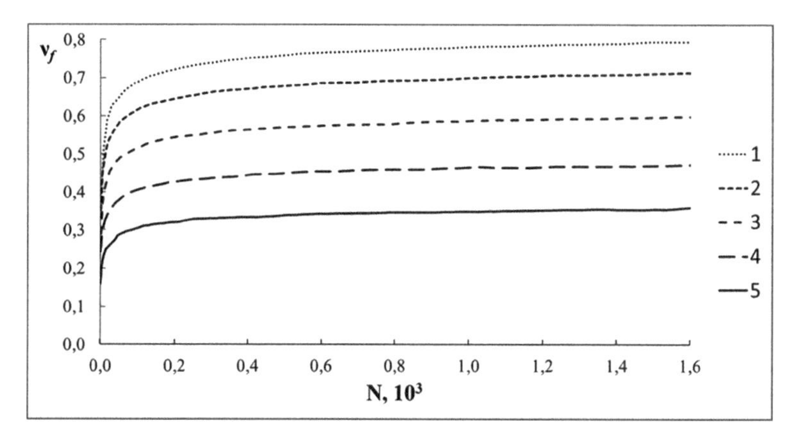

Figure 2.25 Dependence of the volumetric filling of the composite on the relative number of fibers, relocatable further. The guaranteed thickness of the layer of the matrix d/R is **1** - 0.0, **2** - 0.1, **3** - 0.3, **4** - 0.6, and **5** - 1.0 [27].

end of the iteration process. As the criterion for Algorithm 1, the total number of unsuccessful attempts at fiber placement can be used. For Algorithm 2, such a criterion is the relative number of additional relocatable fibers (the normalization is carried out to the total number of fibers) in the simulated structure of the composite. In this model, there is a parameter that determines the guaranteed thickness of the layer of the matrix d, surrounding each fiber, which is characteristic of the interparticle interaction. This parameter is given in the form of relative d/R (R is the radius of the fiber). An output dependency of the volume fraction from these parameters (Fig. 2.25 [27]) on the horizontal asymptote corresponds to the moment they reach the maximum volume filling [29].

A simulation was performed of composites formed by fibers with circular cross section of uniform diameter, surrounded by layers of the matrix material of different thicknesses. Table 2.2 [27] presents data on the limiting volume fractions of filling obtained by averaging 20 independent realizations of the random structure. The size of the guaranteed interlayer of the matrix is a parameter that depends on the nature of the components of the nanocomposite, the technology of the material composition and properties of special coatings and adhesive layers applied to the reinforcing filler in the manufacture

Table 2.2 Limiting the volume fractions v_f^{max} circular in cross-sectional fibers

The guaranteed, thickness of the layer of the matrix d/R	0.00	0.10	0.20	0.30	0.40	0.50	0.60	0.70	0.80	0.90	1.00
Algorithm 1	0.53	0.48	0.44	0.40	0.37	0.34	0.31	0.29	0.27	0.25	0.24
Algorithm 2	0.80	0.73	0.66	0.61	0.55	0.51	0.48	0.45	0.42	0.40	0.37

Source: [27]

of the composite, and the parameters of the electric double layer or the nature of the van der Waals forces. As can be seen, a guaranteed increase in the thickness of interlayers to a size corresponding to the radius of the fiber leads to a reduction in the limit of the volume fraction by more than double.

In real composites the volume fraction of the reinforcing filler is typically 0.60–0.70, and in highly materials, 0.80–0.85. To obtain fragments of random structure, with the volume, degree of filling, exceeding $v_f = 0.50$, was used by Algorithm 2, which provides an iterative procedure, the additional displacement of the fibers. It was found that the value $v_f^{max} = 0.80$ (the same cross-sectional fiber without a guaranteed layer) exceeds the highest possible degree of filling of the composite with a periodic arrangement of fibers in the tetragonal grid nodes ($v_f^{max} = 0.785$). However, it does not reach the limit of the volume fraction, typical of the material with a periodic hexagonal structure ($v_f^{max} = 0.907$). A recorded value v_f^{max} coincides with the value of the maximum filling for random dense packings of "smooth" disjoint disks of the same diameter.

In these papers were presented data, especially for materials with a high filling, which is particularly important for structural materials. However, nanocomposites are also important for a low degree of filling of the fibers. Such a calculation was presented in Ref. [29].

Comparing the packing of spheres described in the previous section, the fibers can be represented by the following analogy between openwork packing of spherical particles and fiber packing. If the chain of bonded spheres is represented in the form as described around them, or cylinders, it is possible, in this structure, to build a relationship between the average coordination number of spheres, the ratio of fiber diameter d_v, and the average distance between the points of contact between two adjacent unidirectional fibers intersecting and touching the third fiber l_{av}. For $n = 3$, $l_{av} = 2r_s = d_v$, where d_v is the fiber diameter, $n = 2$, and $l_{av} = \infty$.

The character of packaging of fibrous particles can be determined by the structural parameter having the following form:

$$p = \frac{l_{cp}}{d_v} \tag{2.16}$$

Tracing the analogy of packing fiber and spherical particles, the effective coordination number of the packing can be represented as follows:

$$n = c + \frac{1}{p}, \tag{2.17}$$

where c is a constant quantity.

From the equation of porosity (see above) we obtain:

$$\alpha = \alpha_\infty + B\frac{p}{c \cdot p + 1}, \tag{2.18}$$

where $p \to \infty$ and porosity of $\alpha \to 100\%$. From this we can estimate the value of the constant c (when $\alpha_\infty = 7.763 \pm 0.016$ and $B = 221.95 \pm 0.45$):

$$c = \frac{B}{100 - \alpha_\infty} = 2,406 \pm 0,005 \tag{2.19}$$

The given geometry package has an openwork character. Graphically the equation of such packing of fibrous particles is shown in Fig. 2.26, Curve 1.

To verify this model, we will consider other models of geometric packing of fibrous particles. The first model may be represented as alternating, mutually orthogonal layers of endless fibers. It is easy to see that the porosity in this model would have the following form:

$$\alpha = 100\left(1 - \frac{\pi}{4 \cdot p}\right) \tag{2.20}$$

Graphically, this dependence is presented in Fig. 2.26, Curve 2.

The second model is located in the space regularly as simplexes of three mutually orthogonal fibers. For this model, we can easily show the dependence of porosity on the structural parameter:

$$\alpha = 100\left(1 - \frac{3 \cdot \pi}{4 \cdot (p+1)^2}\right) \tag{2.21}$$

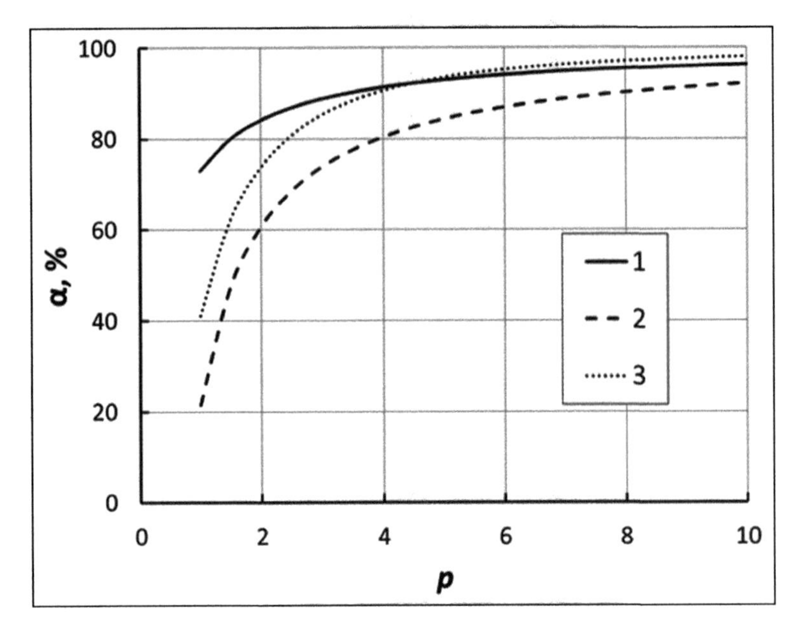

Figure 2.26 The dependence of porosity (α, %) on the structural parameter p of the fibrous packing of nanoparticles: 1, packing openwork fibers; 2, interspersed layers of fibers that are mutually perpendicular; and 3, simplexes of the three mutually orthogonal fibers arranged regularly in space [27].

Graphically, this dependence is presented in Fig. 2.26 [27], Curve 3. Comparing the dependence we can see that the densest packing has mutually orthogonal layers of fibers. Openwork packaging and packing of regular simplexes of three mutually orthogonal fibers are closest to each other, especially for $p > 4$. This effect is typical for the most rarefied structure. Different behaviors of dependencies for small values of the structural parameter are due to the fact that the model geometry openwork provides packaging of discrete fibers and other models consider packing continuous, endless fibers. Thus, we can assume that the curves obtained in the framework of openwork packaging are close enough to realistically obtain the package.

2.5.3 Nanomaterials Based on Layered Particles

Currently, these composites are widely used in industry, both as structural materials and in other uses. Adding to the polymer even small amounts of silicate nanoplates (usually 3–5 wt.%) can significantly improve the barrier diffusing material properties, thermal stability, and resistance to thermal warping. This is due to the fact that in contrast to conventional composites (single components of conventional composites having micron and submicron sizes) nanomaterials are characterized by extremely high interfacial area boundaries, resulting in their dominant role in the formation of physical properties of the material. Here it is possible to obtain significant gains in improving different physical characteristics of nanomaterials with very low concentrations of the filler. In the nanocomposites, the volume fraction of the particles is usually only a few percent points, whereas in conventional composites it is on the order above.

For the first time the use of ultrafine clay filler was proposed as early as 1974, but only recently have such materials begun to be used and are really in great demand. Currently, one of the most common types of nanocomposites are the systems based on polyolefins and layered clay minerals (smectite). In the past decade, the world experienced a rapid growth in the number of publications and patents related to the manufacturing process and the study of their mechanical properties [42]. Such systems have achieved a substantial increase in the elastic modulus, strength, fire resistance, resistance to thermal warpage, and improvement in barrier properties with respect to the diffusing substances. However, analysis of well-known publications shows that the industry of materials in terms of understanding and explanation of available evidence, data, and designed is still not enough.

For relevant research, a widespread polymer such as polyethylene is used as a binder matrix. This is a partially crystallized material and even in its pure form is a complex multilevel hierarchical structure that has a well-defined structural heterogeneity on the nano-, meso-, and microlevels.

The systems under consideration are structurally an inhomogeneous medium consisting of a polyolefin matrix implanted with ultrathin silicate flakes a few nanometers thick and with an average diameter from tens of nanometers to 1 mkm, depending on the mineral deposit and the conditions of its formation. The typical filler particles are 1–2 nm in thickness and 30 to 1000 nm in diameter.

These particles can be chaotically distributed in the volume of the material or can form individual packs, called tactoids. Tactoids consist of several (usually on the order of tens) of parallel plates, between which are located one or more molecules of the polymer matrix. In the first case, the nanocomposites are called exfoliated materials and in the second, intercalated materials.

At low thermodynamic compatibility of the filler, the matrix polymer molecules cannot penetrate into the gaps between the layers of the silicate. The resulting, in this case, material is a simple mechanical mixture of polymer and mineral—this is the usual dispersion-filled microcomposite. When forming intercalated nanocomposite polymer molecules diffuse into the space between the closely spaced parallel plates of silicate. As a result, there is a swelling of the mineral crystallite, but in this case, there is no loss order in the arrangement of its layers. In the formation of exfoliated nanocomposites, polymer molecules penetrate into the gaps between the layers of tactoids and move them apart so dramatically that there is a complete destruction of the tactoids, and silicate plates previously parallel acquire a random orientation.

When exfoliated nanocomposites are formed, polymer molecules not only penetrate into the space between the layers of tactoids, they force them to interact with each other. Thus, the structure collapses and the parallel silicate plates lose their orientation in space and are randomly distributed in the bulk material. Figure 2.27 [27] is a schematic representation of the structure of the nanocomposite in the case of intercalated and exfoliated fillers.

Mechanical properties of the nanocomposite are improved if the silicate filler particles are oriented along the direction of the external deformation. This occurs due to improved adhesion between the matrix and the inclusions due to "biting" of the matrix and also when the outer plates are bent inward and on the internal plates are absent bends. This effect can be achieved by pretreatment of

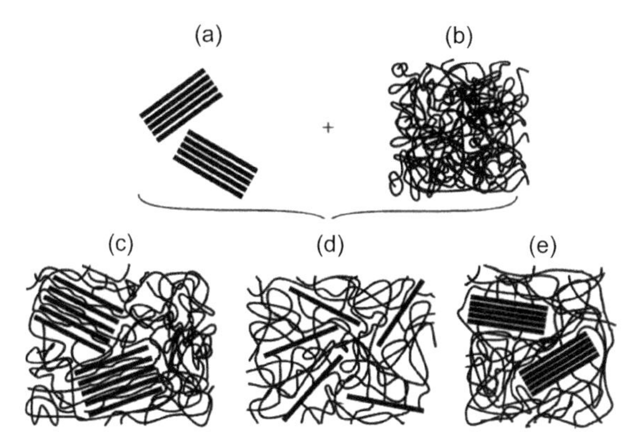

Figure 2.27 Schematic representation of the different structural types of nanocomposites polymer/clay: (a) crystallite clay (tactoid), (b) base polymer, (c) intercalated nanocomposite, (d) an exfoliated nanocomposite, and (e) phase separation nanocomposite [27].

the material, for example, by drawing through a die or other similar methods.

The concentration of the filler to a greater extent affects the increase of the macroscopic elastic modulus. When the amount of the filler is only 6–8%, its elastic modulus is increased by about 3 times. If we compare composites with granular fillers, the famous Einstein's formula gives an increase in the modulus of only 1.2 times [27].

2.5.4 Special Organosilicate Additives and Modifiers

2.5.4.1 Tetrafurfuryloxysilane as a nanostructuring agent

With the introduction of specific organic silicate additives such as tetrafurfuryloxysilane (TFS), it is possible to achieve a significant increase in the density and strength of the silicate matrix in harsh environments due to hardening of the contact between silicate globules binder gel and an alkali component because of the "inoculation" furan radical [43].

During hydration TFS forms active silicate SiO_2 nanoparticles, orthosilicic acid, and furfuryl alcohol (FA), thereby creating on the surface of the matrix grains the oligomeric silicate nanofilms. TFS is a kind of nucleus-forming, which crystallizes in the microsize, thus blocking the surface pores in the silicate matrix, which reduces the shrinkage deformation of concrete

In recent years, polymer-silicate composite materials have been used in various applications. These materials are water-soluble silicates with active additives of furan series substances. They are operated under acidic and neutral media and exposed to high temperatures. These materials are cheap and easy to produce, nontoxic, and nonflammable. The cost of the polymer-silicate materials is commensurate with the cost cement concrete and is several times lower than the polymer concrete cost. The polymer-silicate materials in the form of concrete, mortar, and putties are used for the manufacture of various structures, monolithic, and piece lining. There are some prospects for the use of composite materials based on binders of liquid glass modified by FA.

A significant increase in strength, heat, and fire resistance of a silicate matrix is achieved by introducing into the composition tetrafurfuryl esters of orthosilicic acid, which are called TFS (Fig. 2.28 [33]). The effect is achieved by strengthening the contacts between the globules of silica and modifying alkaline component because of the "inoculation" furan radical. Introduction of a TFS additive in the binder leads to the formation of nanoparticles of SiO_2 and FA, which fills the matrix and forms a cross-linked polymer. These particles act as crystallization centers and nucleation. Adding TFS increases the mechanical and chemical resistance of the binder, and this approach began to be widely used for the preparation of acid-resistant concrete and coatings [18].

This effect can be explained by the following considerations. Thermal stability of oxo-compounds can be judged by the relative strength of the interatomic bonds M–O and C–O in their crystal structure. The bond lengths of M–O and C–O within the coordination polyhedron can vary considerably, which indicates their power disparities. During the dehydration and thermal exposure, the specific part denticity of the ligands may vary. The emerging structure can at the same time perform the functions of a ligand, and the absence of a solvate. The increasing denticity of the ligands leads to distortion of the oxygen environment of the matrix element tetrakis(furan-2-ylmethyl) orthosilicate

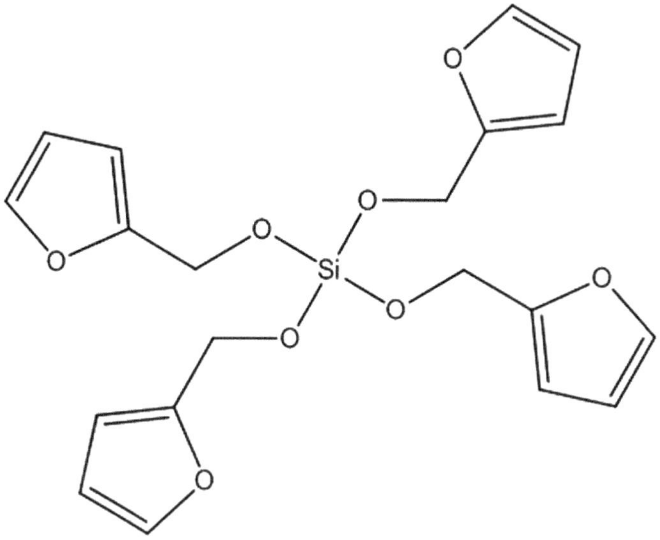

Figure 2.28 Chemical structure of tetrafurfuryloxysilane: tetrakis (furan-2-yl-methyl) orthosilicate [33].

or filler, with a corresponding change in the distance of M–O and C–O in the structure, and hence a change in their strength.

To increase the strength and acid, heat, and fire resistance of building materials and designs, they are entered into the binder tetrafurfuryl esters of orthosilicic acid (TFSs). They are synthesized by transesterification of tetra-ethoxysilane with FA.

The resulting binder comprises water glass 80–95 wt.%, TFS 2–7 wt.%, a curing agent, and sodium hexafluorosilicate 13 wt.%. Thus as part of water glass, organic alkali water glass is used, wherein 1,4-diazabicyclo[2.2.2]octane-1,4-diammonium or 1,5-diazabicyclo[3.3.3]undecane-1,5-diammonium silicate (2–4 wt.%) is taken as organic cation.

The water-soluble silicate containing organic alkali cation salts is prepared by reacting an organic quaternary ammonium derivative with amorphous silica. Soluble organic alkaline silicates, such as tetrabutylammonium silicate (TBAS), were used as the binder component for the self-extinguishing [44].

The proposed nanostructuring binder is obtained by mixing liquid laminar glass containing alkali metal cations such as Na, tetrafurfurilovogo ester of orthosilicic acid (TFS), and a water-soluble silicate containing an organic alkali cation of 1,4-diazabicyclo[2.2.2]octane-1,4-diammonium or 1,5-diaza-bicyclo[3.3.3]undecane-1,5-diammonium silicate [26]. After mixing of all components of the binder, it should be used within 2–3 h. A hardener is added in conjunction with the mill-ground mineral filler.

Introduction of a nanostructuring binder component, tetrafurfuryl ester of orthosilicic acid, leads to the formation of SiO_2 nanoparticles, which act as centers of nucleation and crystallization, and FA, which fills the silicate matrix and forms a cross-linked polymer. Adding TFS increases the mechanical and chemical resistance of the binder and is widely used for the preparation of acid-resistant concrete and fillers. Sodium silicate with a cation of 1,4-diazabicyclo[2.2.2]octane is compatible with aqueous dispersions of polyurethane and chloroprene as well as with the majority of latexes based on synthetic rubbers.

Controlling the properties of the resulting silicate polymer concrete (SPC) and preparation of nanostructured materials with optimal properties require detailed understanding of the mechanisms of processes occurring in their preparation. This requires knowledge of the details of the structure of materials used, in particular TFS. For this were conducted quantum chemical calculations of the structure, topology, and properties of TFS. The calculations were performed by standard methods. The calculated molecular structure of TFS is represented in Fig. 2.29 [33].

The structure of the molecule was calculated by minimizing the total energy of the molecule by minimizing the gradient and reaches the minimum value. Two variants were used as the basis of calculation: the original structure and after application of the procedure the molecular dynamics. The procedure of molecular dynamics included 10,000 iterations; thus these results more accurately describe the topology and energy molecules of TFS. The results of energy calculations are presented in Table 2.3 [33].

To calculate the topology and chemical properties of TFS molecules $Si(OCH_2(C_4H_3O))_4$ different computational methods were

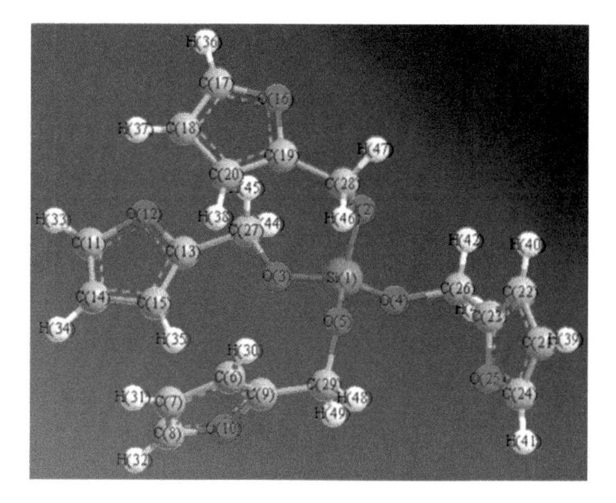

Figure 2.29 The structure of the molecule of tetrafurfuryloxysilane $Si(OCH_2(C_4H_3O))_4$ [33].

used. Computational chemistry covers a variety of mathematical techniques, which are divided into two broad categories:

- Molecular mechanics apply the laws of classical physics to the atoms in the molecule without explicit consideration of the electrons. For these calculations, we used the method MM2 and MMFF94.
- Quantum mechanics is based on the Schrödinger equation for the description of a molecule with an explicit treatment of the electronic structure. In turn, the quantum mechanical methods can be divided into two classes: ab initio (nonempirical) and semiempirical. In the ab initio method was used the process called the General Atomic and Molecular Electronic Structure System (GAMESS) Interface. In the case of semiempirical methods we used a detailed Huckel method (Huckel).

The calculation of atomic charges, which were obtained from electrostatic potentials, provide useful information about the chemical activity. The atomic point charges give a better idea of the probable locations of attack by chemical interaction of molecules with external agents.

Table 2.3 Results of minimization of the energy characteristics of a molecule of tetrafurfuryloxysilane $Si(OCH_2(C_4H_3O))_4$

Energy characteristics	The starting basis		Basis after molecular dynamics	
	Iteration 599	**Iteration 599 + 2**	**Iteration 189**	**Iteration 189 + 2**
The bonds deformation	1.0707	1.0648	1.0720	1.0738
The deformation of bond angles	48.2545	48.2542	48.7845	48.7835
The deformation of tensile bending	−0.4262	−0.4203	−0.3812	−0.3815
The deformation of the torsion angles	−11.0329	−11.0331	−11.1505	−11.1506
Non van der Waals interactions	−8.8831	−8.8831	−15.6637	−15.6636
van der Waals interactions	10.4449	10.4452	10.2485	10.2477
The dipole-dipole interaction	5.3752	5.3750	4.9159	4.9159
The total energy of the molecule, kcal/mol	44.8032	44.8028	37.8255	37.8252

Source: [33]

These results indicate the heterogeneity of furfuryloxy functional groups, which in its turn indicates the possibility of a stepwise mechanism of molecular interaction of TFSs between themselves and with the surrounding molecules. Furthermore, the molecule of tetrafurfurylhydroxysilane is not symmetrical, which also leads to the possibility of a stepwise mechanism of nanophases formation and nanostructuring of SPC in its preparation.

SPC of optimal composition was obtained, which has increased strength, durability, density, and crack resistance. We investigated the diffusion permeability of concrete and its chemical resistance in various aggressive environments.

Introduction into the SPC of 0.3% additive of TFS increases the strength and density of the material by approximately 50% in the entire investigated range of the content of water glass [18, 45]. A plastic mixture of SPC allows producing constructions having any geometric shape by enabling the placement of concrete in cramped conditions. It should be noted that the compressive strength and deformability of samples of the mixture SPC modified with TFS additive were maximal. Studies have shown that the introduction of monomeric additives leads to a drastic reduction of shrinkage strain. Shrinkage of the SPC at the age of 28 days was only 0.06%, while the content in the mixture of 3% TFS.

The formation of a concrete structure is accompanied by intense compression of the binder gel by capillary forces fluid intermicellar. To the mixture, without the presence of monomer addition, such liquid is water. Compression of the gel leads to a maximal shrinkage deformation from the start of hardening mixture. Introduction of additives of FA or TFS in the reaction mixture leads to a substantial reduction of the influence of capillary forces by reducing the surface tension of the liquid in the capillaries [18, 45].

A small change in the content of sodium silicate dramatically changes the technological characteristics of SPC. With the reduction of the content of water glass, there is an increase in the strength and density of SPC. The best composition of SPC is 11.23% water glass and 0.34% of monomeric additives (FA or TFS). The composition of the SPC modified by the addition of TFS has a high compressive strength and high deformability. Introduction into the

SPC composition of the monomeric TFS additives leads to increased stiffness of the mixture and a very significant reduction of shrinkage strain. Use of furan series of compounds as additives can be effective in reducing the diffusion penetration of aggressive environment in the SPC and increasing its resistance to corrosion.

In Ref. [27], the aim was to increase crack resistance and reduce shrinkage strain in the composite material. The initial raw material mixture contains water glass, sodium hexafluorosilicate, polymeric additive, microfiller, sand, and a placeholder. This result is achieved due to the fact that this mixture, except for the above components, has a mixture of primary condensation products of phenols with formaldehyde, a water-soluble polyamide resin, and furyl alcohol as a polymeric additive. These components are contained in a ratio of 2:1:6.5–3:2:13.5. In the present composition shungizit is used as coarse aggregate. As a result we have the following content of the main components (in mass percentage): liquid glass, 12–18; sodium hexafluorosilicate, 2–3.5; polymer additive, 0.4–1.4; a microfill agent; shungizit; sand; and the rest.

Using a mixture of the primary products of the condensation of phenols with formaldehyde, a water-soluble polyamide resin, and furyl alcohol as a polymeric additive promotes the formation of strong polymer chains, improving the uniformity of the material structure through the formation of improved chemical bonds between the contact binder and the shungizit filler. These factors provide significant crack resistance and reduced shrinkage.

2.5.4.2 Modification of an aqueous dispersion of chlorosulfonated polyethylene and other polymers compositions

Introduction of polymeric modifying addition in the composition of the silicate nanocomposite materials can solve the following problems:

- Increase in tack and curing of the adhesive layer when mounting the insulating material to the metal surfaces
- Possibility of adjusting the activities while mounting of structures

- Lowering the alkalinity of the composition to prevent the reactivity of the adhesive with respect to the primer compositions
- Decrease in the average density of the composition
- Preservation of mounting properties of compositions at temperatures up to 10°C
- Preservation of adhesion of the adhesive bonding on a high impact
- Elasticity and water resistance of the glue line

It is known that aqueous dispersions of polymer (PVD) are thermodynamically unstable systems. The polymer is in the form of particles-globules 0.1–0.3 microns in size, stabilized with surfactants. Under the action of electrolytes, stabilizers can be desorbed from the surface of dispersed particles, leading to the formation of large clumps of polymer coagulates. Such electrolytes are sols obtained from water glass and an aqueous solution of sodium or potassium silicates.

Coagulation was prevented only after directional, preliminary modification of water glass by introducing the colloidal agent in an amount from 0.5 to 1%. The resulting effect is due to the influence of additive on the structure of the colloidal particles in a hydrogel formed from sodium silicate and the possibility of condensation of the hydroxyl groups of orthosilicic acid when changing their degree of ionization. At the same time, the pH decreases slightly, by approximately 0.2–0.4 units. Modified liquid glass is combined with all studied polymeric aqueous dispersions in ratios of 99:1 to 75:25. This range is suitable for use in adhesive compositions to create a fire protection.

Liquid glass is capable of forming stable emulsions with solutions of a number of polymers. This phenomenon was observed for the compositions of liquid glass solutions with chlorosulfonated polyethylene (HSPE).

Composite materials of this type are of great interest, as they have reduced flammability and are essentially cheaper than starting polymers. The durability of such materials and the stability of aqueous emulsions depend on the properties of the starting silicate. These compositions can be used as waterproofing and fire-retardant materials due to the fact that the silicate contained in them passes into a water-resistant form during the drying of the emulsion.

Elastic coatings based on HSPE are characterized by a complex of the valuable properties. They are resistant to ozone, acid and alkaline environments, fire, oils and fuels. It allows using products from HSPE in almost all fields of engineering. When mixing the solutions of HSPE with the conventional liquid glass, emulsions are formed that are stratified faster than occurs during drying of the system. Films of such materials that are dried at the substrates are not capable of soaking silicate, which is formed by dissolving and, therefore, is peeled off when wetted. With solutions of HSPE, a liquid glass solution forms stable emulsions. These emulsions are almost transparent in appearance. If they have dried, plastic films can be prepared with a high content of sodium silicate that is uniformly distributed. When approximately equal amounts by weight of modified liquid glass and a solution of HSPE was mixed in toluene, a stable emulsion was obtained. After drying of the emulsion, films were formed that contained 65% water-insoluble silicate. HSPE concentration was 15%, and the mixing ratio by volume equaled 1:1.7. Such films are not peeled off from the metal when immersed in water or under the influence of atmospheric precipitations.

Silicates of chemical elements with multiple charged cations are insoluble in water. For example, on adding a liquid glass aqueous solution of copper sulfate, the mixture becomes turbid and after some time, on the bottom of the vessel a light-blue gelatinous precipitate starts to appear. However, cuprammonium complex sulphate solution is mixed with liquid glass to form water-soluble silicates. Spectral data show that the ammoniate of copper in contact with the liquid glass does not change its structure. This solution retains its homogeneity even during heating, although the color of the solution changes from deep blue to pale blue or aquamarine. This fact points to the restructuring of the cuprammonium complex. The dried films of liquid glass that contain copper ammoniates have less shrinkage and absence of cracking at drying. The introduction of a copper complex improves the stability of liquid glass emulsions with a solution of HSPE. When used of films formed from these emulsions, resulting from the loss of ammonia, liquid glass passes into insoluble copper silicate.

In the hydrolysis of dimethyldichlorosilane in the presence of diethyl ether, oligodimethylsiloxanes were obtained containing 98% of cyclic oligomers [15]. Introduction of oligodimethylsiloxanes or hexamethyldisilazane in the composition of the liquid glass and HSPE facilitates their emulsification, plasticizing the dried product, and the obtained film has additional water repellency. After 1 year of open-air use, the film composition containing 3.9% oligodimethylsiloxane, retaining strength and elasticity.

A combination of aqueous sodium silicate with a toluene solution of HSPE results from the formation of a water-in-oil emulsion. The polymer solution constitutes the continuous phase, and silicate is distributed are fairly evenly in it. Such emulsions can be easily diluted with a solvent for the polymer. Additives of water do not affect the viscosity. After drying, these films, the uniformity of the components stored in a fairly wide range of compositions. Mixing and emulsion formation occurs very easily, often forming almost transparent compositions.

Stable emulsions form with liquid glass and not only HSPE toluene solutions but also solutions of other polymers in nonpolar organic liquids. At present solutions of chlorinated polyvinyl chloride, copolymer VHVD-40, polystyrene, and nitrocellulose are investigated. The smaller the difference of values of surface tension and viscosity of the mixed solutions the more successful the homogenization.

NTI & Polymate have created a series of new materials based on liquid glasses, including liquid glasses with organic cations. TFS and the aqueous dispersion HSPE were used as nanostructuring additives.

To reduce the fire hazard of wooden constructions and structures, special ways and means of fire protection that are both active and passive have been developed and successfully used. So-called passive means of fire protection of building constructions are used for the prevention of fires, that is, for prophylactic purposes. Passive means, in turn, are divided into two groups: chemical and structural.

Fire protection of building constructions is done most often by treating them with various flame retardants—chemicals fire protection of a tree. There are many types of these compositions—paints, varnishes, and enamels; coatings and plasterings; and soakings. Fire-retardant compositions must possess certain characteristics in order to provide the proper level of fire protection of wood and wooden structures. When exposed to high temperature, paint and varnishes swell the surface, which prevents the penetration of heat to the material and complicates the spread of flame along the surface of the wood. With all the rich assortment of fire-retardant paints and varnishes, they all have one common drawback—they can hide the texture of wood, impairing its appearance, so they are mainly used for fire protection of internal wooden structures.

On the background of fire-resistant coatings available on the market, flame-retardant Silaguard, developed by NTI & Polymate, is unique. It includes a number of modifications for wood, for metal, and for plastic products, including high-transparency options. This allows the use of the above-mentioned composition not only for coating hidden wooden structures of buildings but also as a decorative coating.

Components in the coating composition of Silaguard are available on the world market. As the plasticizer for a given composition, the latex used is SEPOLEX CSM, which is produced by Sumitomo Seika Chemicals Co., Ltd., Japan. It has the following average parameters: particle size 1.09 ± 0.09 μm, viscosity 30 ± 7 mPa·s, and solid content 40.1 ± 0.2 mass percentage.

A distinctive feature of the flame-retardant Silaguard is the presence of antipyrenes in this composition. The transparency of the composition is not reduced during application of such layer thickness that is necessary for fire protection. It is also used as a plasticizer, increasing the resistance of the coating to weather. The advantage of the proposed coating is the absence of organic solvents in its composition. This composition has a high fire-retardant capacity, while less than its expenditure norm. It does not emit harmful substances during fire exposure, which makes this coating environmentally friendly. Production of fire-retardant coating for wood Silaguard is performed in a mixer with a stirrer intended for making a suspension containing up to 30% filler solids. The cost of coating Silaguard is 40–50% lower than the cost of fire-resistant coatings, epoxy base.

Technology has been developed to obtain a special binder on the basis of liquid glass with an organic cation and nanostructuring additive TFS, for "cold" production technology of glass and carbon composites and honeycomb structures for the aerospace industry. Such materials after heat treatment have a heat distortion temperature up to 1050°C, compared with 140–180°C for traditional composites based on epoxy resins, when they are of equal value. Compared with the more heat-resistant materials based on polyphosphazenes, their cost is 8–10 times lower. Thus, heat resistance is increased at least twice.

This invention relates to novel chemical compositions that contain water-soluble organic polymers and alkali-stabilized colloidal silica and is more particularly directed to such compositions in which the water-soluble organic polymer is an adhesive or film former and the alkali-stabilized colloidal silica is a material having an ultimate particle size not exceeding about 0.03 micron, preferably prepared by a process including the step of passing an alkali silicate solution through an acid-regenerated ion exchange resin. If an aqueous medium contains an alkali-stabilized colloidal silica sol, disperse particles with size in the range 10–30 nm and a water-soluble organic polymer dissolved in this medium, then the polymer being selected from the group consisting of polyvinyl alcohol, methyl cellulose and the silica having a molecular weight $(0.6 \div 50)10^6$ Da can be determined by light scattering in aqueous solution [17].

Production of polymer materials with predetermined properties is a very urgent problem, which is closely related to the structure of the hardened materials. That is, the structure determines the final properties of the polymer. The process of forming the space-network structure using HSPE proceeds slowly. In the presence of γ-aminopropyltriethoxysilane (AGM-9), the degree of curing is not more than 70% at 40°C. A material with stable properties is necessary to provide conditions for the most complete passing for the curing reaction.

To control the process of forming a network structure HSPE low-molecular-weight organosilicon compounds, such as alkylalkoxysilanes and octamethylcyclotetrasiloxane (MCTS), are used. The process of forming the space-network structure was studied by changing the viscosity of the HSPE solution in time. It should be noted that the curing of systems containing no AGM-9 flowed slowly, the viscosity practically unchanged during the 2 months. Curing took place only in the presence of AGM-9. Upon introduction of silane derivatives (methyltriethoxysilane [MTS], vinyltriethoxysilane [VTS], and ethyltriethoxysilane [ETS]), a partial hydrolysis of the product of tetra-alkoxysilane (ETS-40) in an amount of 1 mass percentage to 5 mass percentage resulted in a reduction of the lifetime of the system from 90 min to 60 min and increased its viscosity. Increased adhesive strength was observed in the compositions that have been modified by ETS-40. Methods have been developed of regulating the strength and technological properties of HSPE using elementorganic and organosilicon compounds of different natures. These modifiers improved the strength and technological properties of materials based on HSPE [33].

2.5.4.3 Solid alumina-silicon flocculants-coagulants-matrix-isolated nanocomposites

Aluminum-silicon flocculant-coagulant (ASFC) is one of the few binary compositions composed of only inorganic components: a coagulant (aluminum sulfate) and an anionic flocculant (active silicic acid). Step ASFC is based on the fact that the interaction of the primary components of ASFC (aluminum compound coagulant and flocculant active silicic acid) forms complex compounds with a higher flocculation ability—zeolite-like nanoscale structures develop a sorptive surface. There is a synergistic effect—increased effectiveness of the exposure as a result of the integration of individual processes into a single system. The mechanism of water purification is realized due to the volume of sorption of pollutants on self-assembled aluminum-silicon complexes.

However, existing methods to date allow you to receive such materials only in the form of solutions. Thus, their lifetime is not more than 2–3 weeks. This factor is holding back the practical use of ASFC in industrial practice for wastewater treatment.

The task was solved with the processing of aluminum-silicon raw material with sulfuric acid, separating the liquid phase from the solid and liquid phase dehydration. Processing of raw materials was carried out with concentrated sulfuric acid under conditions effective for obtaining a concentrated (by 20–30% or more) aqueous solution of flocculant-coagulant. To obtain a dry product the resulting solution was dehydrated at a temperature below the boiling point of water by evaporation under vacuum or by dispersing in a high-temperature high-speed gas stream of coolant. The resulting product is dried and separated from the coolant at a temperature below the boiling point of water.

These processing methods have allowed "freezing" and making it matrix isolation in solid phase the components of flocculant-coagulant, acid salts of aluminum sulfate and the active silicic acid, which is in nanodispersed condition. Quick translation of the active components in the solid state can dramatically reduce the rate of diffusion processes and, at the same time, preserve the activity of the material.

Introduction to the produced nanocomposite material additives of partially water soluble polymers such as polyacrylamide and polyguanidines leads to an additional stabilization of the composition and enhances the effect of matrix isolation. Furthermore, the use of the compositions modified with polymer compounds in water treatment gives additional effects—flocculating and bactericidal.

Experiments have shown that the material thus obtained can be stored for a long time. During the warranty period of storage (0.5 years), we did not observe significant changes in the properties of the material. Some samples were observed to preserve 90% of the activity for over 2 years. The aqueous solution in a concentration of 0.1–2.0% ASFC is stable during long-term storage. For effective water treatment the reagent is required in much smaller quantities. An important feature is the use of powdered ASFC in water treatment from oil.

2.6 Synthesis and Application of Hybrid Materials Based on Silica with Grafted Polymers

Organic polymer/inorganic hybrid nanocomposites are of current research interest because of their improved mechanical, electrical, and optical properties. Among the inorganic nanoparticles, silica is extensively used as a filler or reinforcement agent in polymeric matrixes. The chemical modification of the silica particles by a polymer improves the solubility, the stability, and the dispersion in various solvents. Processing techniques are based on either physicochemical routes or polymerization methods. Chemical grafting techniques include "grafting onto" and "grafting from" methods. The "grafting from" method begins by the fixation of initiating groups on the surface of the particles followed by in situ polymerization.

Organofunctional silanes such as 3-(trimethoxysilyl) propyl methacrylate (TMSPMA) are widely used for "grafting from" processes since they contain simultaneously a silicon-alcoxy function and a double-bond end group. So, free double bonds can be fixed at the surfaces of particles using a condensation step of $-Si(OMe)_3$ with $\equiv Si-OH$ groups. A variety of polymerization techniques, including conventional free radical, cationic, anionic, ring opening, and controlled radical polymerizations (CRPs), have been used for the growth of polymer chains from the solid surface of different inorganic particles. The "grafting onto" route is another effective method. It involves the reaction of functional end groups of polymers onto the silica nanoparticles bearing reactive groups. In this paper, we propose a new method to obtain fluorinated silica nanoparticles bearing free double bonds, making them reactive in a polymerization process.

Most often, fluorinated silica nanoparticles are obtained either by a sol-gel process using a perfluoroalkyl di- or tri-alkoxysilane or by condensation of perfluoroalkyl di- or tri-alkoxysilane onto preformed silica nanoparticles. In Ref. [33] processes, cotelomers bearing one unsaturation were condensed onto silica nanoparticles. So, new poly[(3-(trimethoxysilyl)propyl methacrylate)-stat-(perfluorodecylacrylate)], or P(TMSPMA-stat-PFDA), was synthesized by cotelomerization, leading to oligomers with α-functional end groups. Only perfluoroacrylate polymers with short chains can be obtained due to the insolubility of longer ones in most organic solvents.

The free-radical cotelomerization of TMSPMA with perfluorodecylacrylate (PFDA) in the presence of 2-mercaptoethanol was performed at 80°C in acetonitrile. Hydroxy end groups of the cotelomers were reacted with 2-isocyanatoethyl methacrylate (IME) to give macromonomers. The P(TMSPMA-stat-PFDA) cotelomers, containing fluoro and silanes groups, were then grafted onto silica nanoparticles. Optimal grafting conditions were found with the TMSPMA monomer alone in toluene at 110°C. The structure of the modified silica was analyzed by Fourier transform infrared spectroscopy (FTIR) and ^{29}Si solid-state by nuclear magnetic resonance spectroscopy (NMR). The amount of grafted TMSPMA or P(TMSPMA-stat-PFDA) was calculated by thermogravimetric and elemental analyses. The grafting yield increased with the copolymer/silica weight ratio until a maximum value of 2.26 $\mu mol \cdot m^{-2}$ was obtained.

Moreover, grafting density in cotelomers was quantified by thermogravimetric and elemental analyses with similar values. A maximum grafting density value has been observed while increasing the silane/silica weight ratio. As to the potential uses, such hydride nanocomposites can be considered as building blocks to fabricate more complex nanomaterials using the reactive polymer end chains [33].

Nowadays, a variety of nanomaterials with different structures have been applied in biomedicine because of the unique characteristics inherent to the nanoscale. For example, mesoporous materials have been widely investigated and effectively used in drug delivery systems; different metallic nanoclusters based on gold, iron, and silver nanodots and their hybrids have found potential application in sensing and magnetic resonance imagining.

Preparation of nanoscale systems with new functional properties is a major challenge; however, silica nanoparticles remain widely used due to their unique features, such as a high surface area and thermal and chemical stability [33]. The unique compatibility and possible size control in synthesis allow the realization of different combinations of silica particles with other available materials. One of such approaches is based on using silica as a scaffold for templating other materials such as dyes, quantum, and magnetic dots. Here, silica is used as a shell for the protection of encapsulated materials. In another way, silica nanoparticles can be considered as seeds or cores around which other materials can be grafted (Fig. 2.30 [33]). In this field, different functional polymers are being involved for surface modification of silica core. Functional groups in polymeric chain provide excellent binding properties of modified silica surface for effective captures of different small objects: magnetite or fluorescent nanoparticles, drugs, etc. A number of polymers have been already synthesized and applied for silica functionalization.

The most common studies include poly(acrylamide), polyethilenglycol, and polyacrilic acid. For this reason, a new kind of polymer should be examined for developing new surface properties of polymer-coated silica particles. Taking into account the published results about polymer science and biotechnology, we suggest considering new guanidine-containing polymers based on polydiallyl and polymethacryloyl guanidine (Fig. 2.30a) as one of the most promising candidates for silica modification due to their low hemotoxicity and high ionic properties. There are only a few studies

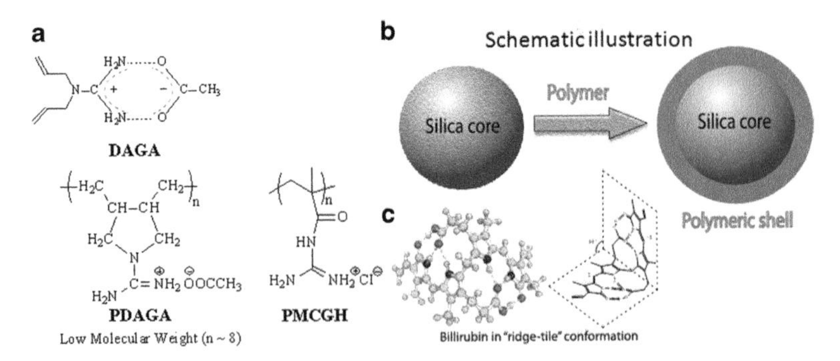

Figure 2.30 (a) Structures of diallyl guanidine acetate (DAGA), polydiallyl guanidine acetate (PDAGA), and polymethacryloyl guanidine hydrochloride (PMCGH). (b) Schematic illustration of modification of silica core. (c) Structure formula of bilirubin in "ridge-tile" 5Z, 15Z conformation stabilized by six intramolecular hydrogen bonds [33].

concerning investigation and application of guanidine-containing polymers excepting polythexamethylen guanidine and its analogues. It was reported that a guanidine fragment in a polymeric chain is more promising than its amine equivalents due to the capability of guanidine groups to bind more strongly with negative molecules compared to amine groups.

Polymer-coated silica particles have received tremendous attention as adsorbents for many organic compounds. At present, synthesis of highly effective adsorbents has become a promising technique for blood detoxification, especially for bilirubin removal. Bilirubin (Fig. 2.30c) is a dicarboxylic acid and circulates in human blood plasma, where it is bound to serum albumin to form a water-soluble complex. It is transported to the liver as a complex with albumin, where it is normally conjugated with glucuronic acid and excreted into bile. However, bilirubin can be accumulated in blood at high concentrations once a patient suffer from a liver disease (hyperbilirubinemia). Several techniques have been employed for removal of the high concentration of free bilirubin from plasma in order to prevent hepatic and brain damage [33].

However, hemoperfusion treatment based on application of hemoadsorbents is one of the most effective techniques. Many kinds of adsorbents (carbon nanotubes, mesoporous silica, graphene oxide, etc.) have been exploited. All of them have some drawbacks related to low adsorption capacity, biocompatibility, or high toxicity. Therefore, the search and synthesis of new adsorbents for efficient bilirubin removal is very demanding.

In the work in Ref. [33], polymer-coated silica particles were synthesized via the sol-gel method in the presence of a guanidine-containing polymer. A subsequent treatment of silica particles with positively charged guanidine-containing polymers leads to a great increase in the adsorption capacity for bilirubin compared to unmodified silica. The authors demonstrated successful modification of the silica core by guanidine-containing polymers using electron microscope, FTIR spectroscopy, and thermal analysis.

The development and design of adsorbents with high adsorption performance is the actual aim of successful toxin removal. In this work, we report the performance of new guanidine-containing polymers as shells orientated around silica particles in effective bilirubin removal. To evaluate polymer-shell modification, FTIR spectroscopy and high-resolution electron microscopy were performed. Changes in the surface morphology of polymer-coated silica were detected.

It was shown that the polymeric shell completely covers silica surface. The grafting amount was evaluated by thermogravimetric and was in the 18–34% range. Incorporation of guanidine-containing polymers into a silica matrix leads to a great increase in adsorption capacity. According to the Langmuir model, the maximum adsorption capacity for bilirubin was 43.47 mg/g. We also demonstrated that prepared polymer-coated silicas could well adsorb bilirubin from the bilirubin-albumin solution at low concentrations of albumin.

Therefore, this study revealed that guanidine-containing polymers have great potential to serve as effective agents for improving of the adsorbent for bilirubin removal [33].

2.7 Conclusion

Synthesis and study of the properties of new organomineral hybrid nanocomposites is of great importance. On the basis of these studies can be found solutions to many specific technological problems. Promising areas in the field of organomineral hybrid nanocomposite materials will be methods of synthesis of multicomponent materials, as well as materials such as "net in net" and "host-guest". There are promising ways of using alkoxides of various metals, including complex, polynuclear, and multifunctional, into the sol-gel processes. The fundamental problem with the chemistry and physics of nanocomposites remains the dependent "structure-property." The solution of this problem is in shifting from research of materials to their purposeful designing. Various experts around the world are motivated to actively work on a solution.

Only some of the features and applications of sol-gel and nanotechnology to produce organomineral-hybrid nanocomposites are reflected in this chapter. Of course, this is an incomplete review and one or another aspect or example may be missing.

We have tried to capture some of the principles of synthesis and application of hybrid nanomaterials in various fields of technology, discussing concrete examples. Objects of the sol-gel and nanotechnology are fractal systems. Therefore, even small changes in the parameters of the processes for their production can have a significant effect on the final product. On the one hand, it increases the complexity of the system and, on the other hand, offers you a great opportunity to develop your own, customized solutions of practical problems, often with minimal changes in the composition and manufacturing technology.

References

1. Kerber M. L. (2008). *Polymer Composite Materials: Structure, Properties, Technology*, Professiya Publishing House: St. Petersburg, 560 p.
2. Friedrich K., Fakirov S., Zhang Z. (2005). *Polymer Composites: From Nano- to Macro-Scale*, Springer: New York.
3. Kobayashi N. (2005). *Introduction to Nanotechnology*, BINOM: Moscow, 134 p.
4. Kickelbick, G. (2007). Introduction to hybrid materials, in: *Hybrid Materials: Synthesis, Characterization, and Applications* (G. Kickelbick, ed.), Wiley-VCH Verlag GmbH & Co. KGaA: Weinheim, Germany, 516 p.
5. Pomogailo A. D. (2000). Hybrid polymer-inorganic nanocomposites, *Russian Chemical Reviews*, **69**(1), 53–80.
6. Guglielmi M., Kickelbick G., Martucci A. (eds.), (2014). *Sol-Gel Nanocomposites*, Series: Advances in Sol-Gel Derived Materials and Technologies, Springer: New York, IX, 227 p.
7. Dimitriev Y., Ivanova Y., Iordanova R. (2008). History of sol-gel science and technology (review). *Journal of the University of Chemical Technology and Metallurgy*, **43**, 181–192.
8. Livage J., Henry M., Sanchez C. (1988). Sol-gel chemistry of transition metal oxides. *Progress in Solid State Chemistry*, **18**(4), 259–341.
9. Hench L. L., West J. K. (1990). The sol-gel process, *Chemical Reviews*, **90**(1), 33–72,
10. Jonschker G. (2014). *Sol-Gel-Technology in Praxis*, Vincentz Network: Hanover, European Coatings Library.
11. Brinker C. J., Scherer G. W. (1990). *Sol-Gel Science: The Physics and Chemistry of Sol-Gel Processing*, access online via Elsevier.
12. Mandelbrot B. (2002). *Fractal Geometry of Nature*, SRI: Moscow, 656 p.

13. Aegerter M. A., Leventis N., Koebel M. M. (eds.) (2011). *Aerogels Handbook*, Springer: New York, 965 p.
14. Hay J. N., Raval H. M. (2001). Synthesis of organic-inorganic hybrids via the nonhydrolytic sol-gel process, *Chemistry of Materials*, **13**(10), 3396–3403.
15. Andrianov K. A. (1955). *Organosilicon Compounds*, State Scientific, Technical Publishing the Chemical Literature: Moscow.
16. Park M., Komarneni S., Choi J. (1998). Effect of substituted alkyl groups on textural properties of ORMOSILs. *Journal of Materials Science*, **33**(15), 3817–3821.
17. Iler R. (1982). *Chemistry of Silica*, vol. 1–2, MIR: Moscow, 416 p.
18. Figovsky O., Beilin D. (2013). *Advanced Polymer Concretes and Compounds*, CRC Press, Tailor & Francis Group: Boca Raton, 245 p.
19. Korneev V. I., Danilov V. V. (1996). *Soluble and Waterglass*, Stroyizdat Publ.: St. Petersburg.
20. Kudryavtsev P. G. (2014). Alkoxides of chemical elements - Promising class of chemical compounds which are raw materials for HI-TECH industries, *Journal Scientific Israel - Technological Advantages*, **16**(2), 147–170.
21. Grigoriev P. N., Matveev M. A. (1956). *Water-Glass (Production, Properties and Application)*, Moscow.
22. Vail J. G. (1952). *Soluble Silicates*, vol. 1,2, Reinhold: New York.
23. National Primary Drinking Water Regulations-Description of Corrosion Control Treatment Requirements, 40 CFR 141.82, revised July 1, 1991.
24. Figovskiy O. L., Beilin D. A., Ponomarev A. N. (2012). Successful application of nanotechnology in construction materials, *Nanotechnology in Construction*, **3**, 6–21.
25. Kudryavtsev P., Figovsky O. (2014). Soluble silicates such as a promising basis for obtaining hybrid nanocomposite materials, *Book of Abstracts IV International Conference, "Technical Chemistry, from Theory to Practice"*, 20-24.10.2014., p. 71.
26. Kudryavtsev P., Figovsky O. (2016). Nanocomposite organomineral hybrid materials, Part I and Part II, *Nanotechnologies in Construction, A Scientific Internet-Journal*, **8**(1, 2). http://nanobuild.ru/en_EN/journal/Nanobuild-1-2016/16-56.pdf; http://nanobuild.ru/en_EN/journal/Nanobuild-2-2016/20-44.pdf

27. Kudryavtsev P., Figovsky O. (2015). Nanocomposite organomineral hybrid materials, *Journal Scientific Israel - Technological Advantages*, **17**(3), 7–60.
28. Kudryavtsev P., Figovsky O. (2014). Nanostructured materials, production and use in construction, *Nanotechnology in Construction*, **6**(6), 27–45. http://nanobuild.ru/en_EN/journal/Nanobuild-6-2014/27-45.pdf (in Russian).
29. Figovsky O., Kudryavtsev P. (2014). Advanced nanomaterials based on soluble silicates, *Journal Scientific Israel - Technological Advantages*, **16**(3), 38–76.
30. Cooper P. W. (1996). *Explosives Engineering*, Wiley-VCH: New York.
31. Kudryavtsev P., Figovsky O. (2015). *The Sol-Gel Technology of Porous Composites*, Monograph, LAP Lambert Academic Publishing, 466 p. (in Russian).
32. Anderson M. L., Rolison D. R., Merzbacher C. I. (1999). Composite aerogels for sensing applications, *Proc. SPIE*, **3790**, 38–42.
33. Kudryavtsev P. G., Figovsky O. L., Kudryavtsev N. P. (2016). Nanocomposites based on hybrid organo-silicate matrix, *Journal Scientific Israel - Technological Advantages*, **18**(1), 112–133.
34. Val'tsifer V. A., Gubina N. A. (2003). Rheological and electrical properties of an oligomeric formulation as influenced by fractional composition of conducting filler, *Russian Journal of Applied Chemistry*, **76**(10), 1659–1661.
35. Tretyakov Yu. D., Goodilin E. A. (2009). Key trends in basic and application-oriented research on nanomaterials, *Russian Chemical Reviews*, **78**(9), 801–820.
36. Feynmann R. P. (1961). *Miniaturization*, Reinhold Publishing Corp.: New York.
37. Shabanova N. A., Sarkisov P. D. (2004). *Fundamentals of Sol-Gel Technology Nanosized Silica*, Akademkniga: Moscow, 208 p.
38. Figovsky O., Beilin D., Blank N. (2008). Advanced environment friendly nanotecnologies, *Proceeding of the NATO Advanced Research Workshop on Environmental and Biological Risks of Hybrid Organo-Silicon Nanodevices*, St. Petersburg, 18–30 June 2008, pp. 19–29.
39. Pomogailo A. D., Dzhardimalieva G. I. (2009). *Monomeric and Polymeric Metal Carboxylates*, Fizmatlit: Moscow, 400 p.
40. Figovsky O. L., Kudryavtsev P. G. (2014). Liquid glass and aqueous solutions of silicates as a promising basis of technological processes of new nanocomposite materials, *Engineering Journal Don*, 2, 2448.

41. Zvereva N. A., Val'tsifer V. A. (2002). Computer simulation of the structure of disperse systems by the particle method, *Journal of Engineering Physics and Thermophysics*, **75**(2), 317–323.
42. Garishin O. K. (1997). Geometric synthesis and study of random structures // structural mechanisms of mechanical properties of granular polymer composites, Ural Branch of Russian Academy of Sciences: Ekaterinburg, pp. 48–81.
43. Kudryavtsev P., Figovsky O. (2014). *Nanomaterials Based on Soluble Silicates*, LAP Lambert Academic Publishing, 241 p.
44. Kudryavtsev P., Figovsky O. (2014). *Nanomaterials Based on Soluble Silicates*, LAP Lambert Academic Publishing, 165 p. (in Russian).
45. Kudryavtsev P., Figovsky O. (2014). Nanocomposite organomineral hybrid materials, *Injenernyi Vestnik Dona*, 2 (in Russian).

Chapter 3

Polymer Nanocomposites with a High Resistance to an Aggressive Environment

3.1 Introduction

Coating application is one of the traditional methods of protecting materials against corrosion. The main function of a coating is to isolate the protected material from an aggressive medium. Coating and lining on the bases of chemically resistant polymeric, silicate, and other inorganic nonmetal corrosion-proof materials are widely adopted for the protection of metals, concrete, and other structural materials. At present, polymeric mastics, reinforced, and film coatings are the most prevalent protective materials for coating and lining of structural materials.

In the general case, the service life of a coating, τ, is made up of the following components:

$$\tau = \tau_d + \tau_a + \tau_e + \tau_c, \quad (3.1)$$

where τ_d is the time for the diffusion front of the components of the medium to reach the surface of the substrate; τ_a is the time of the fall in the strength of the adhesive linkage between the coating and the substrate, including the induction period of corrosion; τ_e is the time to reach the critical level of corrosion losses or the local

Green Nanotechnology
Oleg Figovsky and Dmitry Beilin

ISBN 978-981-4774-10-9 (Hardcover), 978-1-315-22928-7 (eBook)
www.panstanford.com

destruction of the coating by corrosion products from the beginning of the contact of the substrate with the medium as a phase; and τ_c is the time to reach the critical level of corrosion losses at the bottom of a pore or a fissure.

An increase in τ at each stage gives an increase in the protective properties as a whole.

The most common cause of the failure of coatings is their inadequate corrosion durability due not only to the occurrence of chemical degradation of the material of the coating and to a lowering of its strength during use but also to the appearance in it defects, such as fissures, pores, and craters, caused by the structural inhomogeneity of the coatings and their cracking under the action of internal stresses.

Consequently, improved protective properties of nonmetallic materials and coatings can be achieved by:

- Reducing of the penetration rate of the aggressive medium's components to substrate surface
- Increasing of the adhesion between coating and the substrate or reducing of the underfilm corrosion rate
- Reducing of the material's defectiveness and rate of chemical degradation
- Reducing of internal stresses in the coatings

Recently, it was assumed that contact between medium and material can only occur when the coating contains defects, which form during processing or exploitation. However, it has been shown that for macro-molecular coatings, several elements of aggressive media may penetrate protected surfaces even without coating defects. This is because of the diffusional permeability of macromolecular membranes toward some substances. Their permeability through gases and vapors is their principal feature, and no transformation in their composition or structure could change this. In most polymers, chemical destruction takes place in diffuse-kinetic fields, and its rate depends on the penetrability and the stress-strain state of the coating materials. Only numerical changes in their diffusional characteristics are attainable.

Polymeric materials for structures and coatings are increasingly dominating corrosion-protection technology. As was shown in works in Refs. [1, 2], the most effective method of improving protective properties is to reduce the permeability of coatings due to the change in the molecular and meta-molecular structures of the material and use additional components reducing the rate of diffusion of electrolytes in polymers and anticorrosive silicate compounds. We were the first to propose the set of inorganic substances of composite polymeric materials that selectively interact with the water or water solutions of acids, salts, and alkalis in order to decrease their penetrability and increase their chemical resistance simultaneously. As a result of the said reactions, high-strength hydrate complexes and inorganic glue cements are formed. As a consequence of this, not only protective features are retained during penetration in aggressive medium, but also the coatings become stronger and less penetrable (Figs. 3.1 and 3.2 [4]). Such additives are metal oxides—the pulverulent part of acid- or alkali-hardening adhesive cements.

Permeability of the coating decreases because of the following processes in the polymeric composite materials:

- While the active additive compound (the powderlike part of the glue cement of the aqueous, acidic, basic, or salt formation of solid) interacts with the medium in macropores, microdislocations and other material contact-surface defects form hydrate complexes and, as a result, defects are "repaired," filled with the stable cement, and the material is reinforced.
- The crystallohydrates formation leads to an increase in the volume and the surface of the contact of the active additive in the constant volume of the polymeric matrix. This results in the appearance of compressive stress, which compacts the material structure, increases the density of the material's structure, and reduces its free volume and decreases its free volume. Due to increased specific surface of the filler and formation of a material with binding features, the adhesion on the polymer-additive boundary becomes higher. Therefore, the relative surface

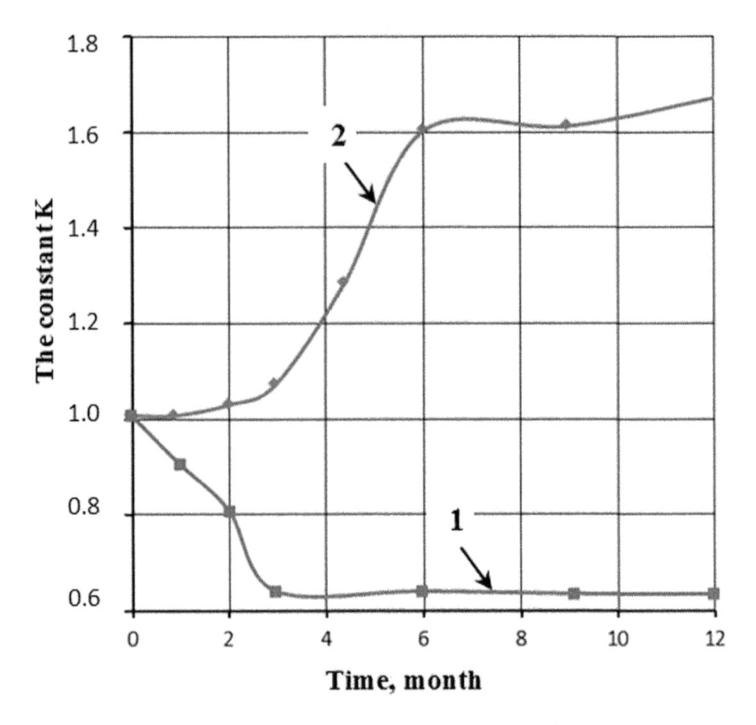

Figure 3.1 Correlation between the tensile strength of liquid ebonite covering materials and the time of exposure (10% HNO_3, 293 K): 1, without additive; and 2, with additive (7% by weight) based on oxide of bismuth [1, 4].

area of the active filling material increases and forms a new crystallohydrate phase exhibiting gluing properties, which creates adhesion on the polymer-filler interface.

- The penetrating fluid aggressive medium is used in the upper layers of the coatings to form a new mineral phase, and this decreases the rate of medium flux and its sorption by the next layers.

In the case of chemically stable hydrophobous thermoplastics (fluorine plastics, polyolefins, etc.), the decrease of the rate of the penetration of aggressive media, especially aqueous solutions of acids, becomes the main technical aim.

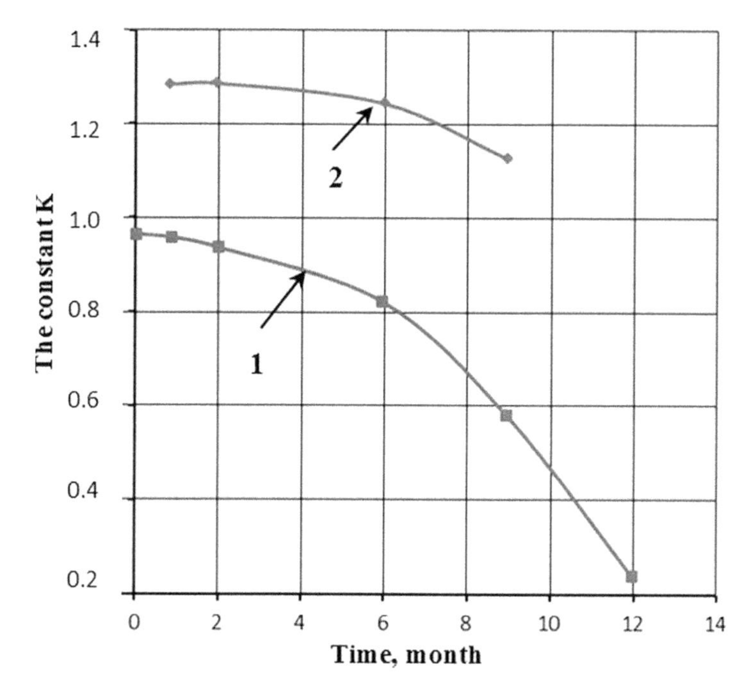

Figure 3.2 Correlation between the compressive strength of epoxy-phenolic composite materials and the time of exposure (30% H_2SO_4, 293 K): 1, without additive; and 2, with additive (20% by weight) based on copper oxide [1, 4].

Developments in civil engineering and industrial growth have created a continual demand for building materials with new and improved performance attributes. Polymer concretes (PCs) appear to offer possibilities for meeting these new requirements. By PC is meant a polymer composite with a polymer matrix and sand and rocks, like those used in Portland cement concrete, as inclusions. Service conditions often dictate specific material requirements that may be met by PC when several composite properties are considered simultaneously.

Advancements in PC materials have slowed over the past 25 years compared to the rate of advancements in the 1970s and 1980s. The knowledge base in concrete polymer materials has matured as many products have been made commercially available. There are now many polymer-based construction materials that have been shown to perform very well for their intended. purposes: concrete spall repair, crack repair, concrete overlays, and precast concrete components. The cost of polymer-based systems is high relative to conventional Portland cement concrete materials, and it is necessary to demonstrate the improved durability, reduced thickness/size, ability to be placed in difficult environmental conditions, and/or the fact that other nonpolymer materials will not work. There are many situations for which concrete-polymer materials prove to be the most appropriate materials for the intended application.

Understanding of the nature of PC is necessary for the design of the most cost-effective PC composites and to produce materials with desired properties.

PC is usually used in severe conditions in industrial and public buildings as well as in transportation and hydraulic structures. The main uses are repair, strengthening, and corrosion protection of concrete structures. The main advantages of PC over ordinary concrete are improved mechanical strength, low permeability, and improved chemical resistance. The main limitation is their relatively high material cost. This is why it is important to find the optimum technical/economic compromise. To solve this problem, it is necessary to formulate a reliable predictive mathematical model of PC material properties.

The way of creation of chemically resistant PCs by use of the effect of so-called positive corrosion is offered [1–5].

The essence of the proposal is the use of special additives that by interaction with some aggressive medium form insoluble compounds capable of condensing and strengthening the structure of a PC.

3.2 Modeling of Diffusion in Polymeric Materials

In the light of the above-mentioned points, it is reasonable to describe the process of diffusion in such systems using models with the ability to capture particles (capturance), especially those of diffusion with irreversible sorption. The main feature of the mechanism of diffusion in such models is the diffusional transfer appearance in the zone between the sample boundary and the diffusional boundary. In the diffusional boundary, the diffusant is irreversibly captured; as a result, the sorbed molecule cannot participate in the diffusion process.

The mathematical description of the diffusion process using the above-mentioned model allows a freedom because it is applicable in principle to various mechanisms of capturance of the diffusion on condition that the process of capturance is faster than that of diffusion.

If the coating material is chemically stable, the protection effectiveness can be lost because of the adhesion lost or under-film corrosion but only after delays needed for the diffusion of the aggressive agent through the thickness of the coating—the so-called time till passage, τ_0. If the coating thickness is specified, τ_0 is the time duration till passage.

Mathematically, the described process can be described by the following system of diffusional kinetic Eq. 3.2:

$$\left.\begin{aligned} \frac{\partial C}{\partial t} &= -D_1 \frac{\partial^2 C}{\partial x^2} \\ C &= C_0 \text{ at } x = 0 \\ C &= 0 \text{ at } x = X \\ -D\left(\frac{dC}{dx}\right)_{x=X} &= E\frac{dX}{dt}, \end{aligned}\right\} \tag{3.2}$$

where C is the concentration of the diffusive substance dissolved in a material with "inactive" functional groups, E is the concentration of chemically active functional groups of the polymer, and X is the coordinate of the reaction zone, changing with time.

The solution of this system of equations against the concentration of the noncaptured acid is given by:

$$\left.\begin{aligned} C = C_0 \left[1 - \frac{\operatorname{erf}\left(x\sqrt{4Dt}\right)}{\operatorname{erf}(\beta)}\right] \\ \text{for } 0 < x < X \text{ and } 0 < t < t^*, \\ \text{and } C = 0 \text{ for } x > X \text{ and } t > 0, \end{aligned}\right\} \tag{3.3}$$

whereas the motion of the reaction zone is described by the following equation:

$$X = \beta\sqrt{4Dt}, \tag{3.4}$$

where t^* is the duration of the walking of the moving boundary along the membrane and β is the numerical parameter found from the following equation:

$$\sqrt{\pi\left[\frac{\exp\left(-\beta^2\right)}{\operatorname{erf}(\beta)}\right]} = \frac{E}{C_0}\beta \tag{3.5}$$

where erf(β) is the error function.

The distribution of the total concentration of the acid (free and captured) can be presented as:

$$\left.\begin{aligned} C_{\text{tot}} &= E + C_0\left[1 - \frac{\operatorname{erf}\left(x\sqrt{4Dt}\right)}{\operatorname{erf}(\beta)}\right] \text{ at } 0 < x < X \\ \text{and } C_{\text{tot}} &= 0 \text{ at } x > X \end{aligned}\right\} \tag{3.6}$$

According to the experimental data on profiles of the distribution, the gradient of the concentration of acid in the diffusion zone is negligible. In this case $C_0/E \ll 1$ and Eq. 3.5 can be rewritten as follows:

$$\beta = \sqrt{\frac{C_0}{2E}} \tag{3.7}$$

Substitution of Eq. 3.6 into Eq. 3.3 gives:

$$X^2 = \frac{2C_0}{E}Dt \tag{3.8}$$

or

$$\lambda^2 = \frac{2C_0}{E}D \tag{3.9}$$

Thus, the model of diffusion with capturance allows the distribution of the diffusant with the strike diffusional boundary, Fick-like motion of the boundary, and the estimation of the relationship of the constant of diffusion with the principal diffusion-sorption characteristics (D, C_0, E).

3.3 Modeling of Diffusion in a Random Porous System

The process of the transport of acid in pores and defects of the coating material can be described by the model of random fluxes

[1, 4]. This model considers mass-transfer processes in complex systems (like micropores) as the evolution of nonequilibrium systems consisting of numerous subsystems possessing excess energy (negentropy).

In the case of micropores, these subsystems are distributed in energy according to Gibbs:

$$f(U) = \mathrm{EXP}(-U/\alpha), \tag{3.10}$$

where U is the value of the excess energy of the subsystem and α is a parameter characterizing the specific (nonequilibrium) properties of the system. The total excess energy of the system is given by:

$$E_\Sigma = Q_0 \int_{U_{\min}}^{U_{\max}} U f(U)\, dU, \tag{3.11}$$

where Q_0 is the normalization coefficient. The value of U is determined by the thermodynamic forces of diffusion, and the mass transfer causes the dissipation of total excess energy, changes the energy distribution of subsystems, and reduces the value of E_Σ. The dissipation is given by the following balance equation:

$$\frac{dE_\Sigma}{dt} = -\sum_{k=1}^{M} g_k (E_\Sigma)^k, \tag{3.12}$$

where g_k are coefficients and M is the number of terms taken into account. On the basis of specified values of M and g_k, one finds the change of E_Σ and α with time, then one estimates all required characteristics of the system.

It can be noted that the use of effective anticorrosion protection leads to a serious decrease in values of g_k.

It can be noted that the considered method of increase in delays of coating lifetime is not only very effective but also technological, because the inhibitor involved only changes the composition of the coating, not its preparation and application processes.

3.4 Decrease of the Rate of Diffusion

In the case of chemically stable hydrophobous thermoplastics (fluorine plastics, polyolefins, etc.), the decrease in the rate of penetration of aggressive media, especially aqueous solutions of acids, becomes the main technical aim.

The diffusion of acids in hydrophobous thermoplastics is characterized by the strike front of diffusion, the displacement of which with time is described by Eq. 3.13:

$$X = \lambda\sqrt{\tau}, \tag{3.13}$$

where X is the deepness of the penetration of the acid into the polymer during the time duration τ and λ is the "the penetration constant" determined by the nature of the polymer, the nature and concentration of the acid in the exterior medium, and the temperature.

If the thickness of the coating is 1, according to Eq. 3.13, the value of τ_0 is found from:

$$\tau_0 = \left(\frac{l}{\lambda}\right)^2 \tag{3.14}$$

Methods of estimating the extent of acid diffusion in filled polymers are the same as those used in studies of acid diffusion in nonfilled polymers, that is, indicator and trace methods. The involvement of inhibitors does not change the tendencies of the diffusional process in most cases. This one is described by Eq. 3.13, while the corresponding value of λ_f is shorter than its analog in the nonfilled polymer λ_0. In some cases, the slow motion of the diffusion front compared with the process described in Eq. 3.13 appeared. Stopping the diffusion front altogether was even suggested.

3.5 Influence of Additives on the Operating Characteristics of Polymeric Materials

Metal oxides, which are powderlike parts of acid and alkali glue cements, are suited for such kind of additives. These complexes were investigated and their hardening, strength, and volume characteristics were considered. This allowed one to choose purposefully the compositions of the powders to be penetrated into polymeric materials.

We investigated the influence of specific inorganic additives on the operating characteristics of polymeric materials after the penetration of the additives into the polymeric matrix. The influence of additives on the diffusion permeability of a polymeric material was evaluated based on the mass change of the samples after exposure in aggressive media, as well as on the penetration depth of the aggressive medium into the polymeric material. The volume-ohmmeter and coulomb-metric methods were used to investigate the modification influence on protective features.

In accordance with the aims of the experiments, inhibitor concentrations have been presented in grams or moles per unit of mass or volume. Inhibitors have been involved into polyolefins, fluorized copolymers of polyolefins, and poly-3,3-bis(chlormethyl) oxacyclobutane. The diffusion of HCl, HF, and HNO_3 from solutions of various concentrations into polymers has been studied

The effectiveness of the inhibitors can be characterized by the parameter γ showing on how many occasions the time for the acid to appear from the filled polymer τ_{0f} is longer than it is from the nonfilled polymer τ_0:

$$\gamma = \frac{\tau_{0f}}{\tau_0} \tag{3.15}$$

For equal-thickness coatings, according to Eq. 3.10:

$$\gamma = \left(\frac{\lambda}{\lambda_f}\right)^2 \tag{3.16}$$

The dependence of γ of various factors on various "polymer-inhibitor-acid" system combinations is the same, and only numerical characteristics change. The following factors confirm that an acid's concentration influences the efficiency of its diffusion inhibition: for HCl, the divergence in the coefficients of diffusion into filled and nonfilled polymers rapidly rises with the increase in acid concentration (Fig. 3.3 [1, 4]). For example, when the concentration changes from 10 to 30%, the parameter λ_0/λ_f increases from 1.7 to 5, which corresponds to the increase of λ from 3 to 25—more than 8 times.

The influence of a polymer on parameter β is not simple but depends on the type of acid (Table 3.1 [1, 4]). The data in Table 3.1 show that for HCl and HF, β decreases in the range of polymers F-30. However, for HNO_3, F-40 changes the last position for the first one (F-30 = the copolymer trifluorinechloride-ethylene; F-40 = the

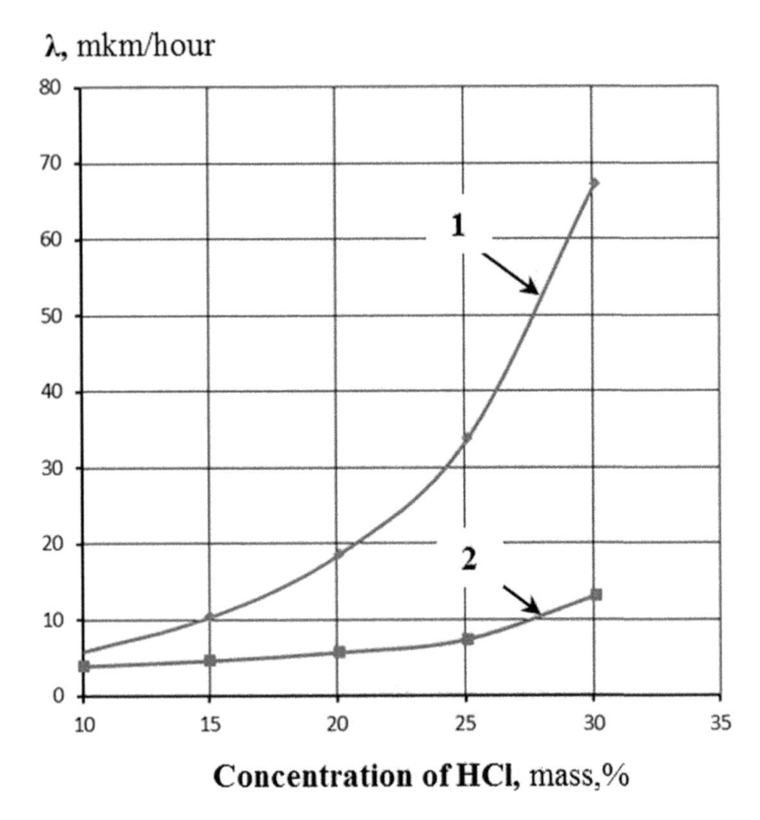

Figure 3.3 Permeability coefficient of HCl in penton λ depending on the acid concentration: 1, nonfilled polymer; 2, polymer containing NaOH, 0.72 mole/kg, and $T = 343$ K [1, 4].

copolymer tetrafluorine-ethylene). Theoretical considerations allow the assumption that the influence of the nature of the polymer on the efficiency of the inhibition is related to its sorptive capacity against the acid: more the sorption constant, striker the inhibition effect.

The permeability coefficients (λ) and the time of acid penetration through protective layers of low-pressure polyethylene (LPPE) with CaO additives are illustrated in Table 3.2 [1, 4].

Introducing 3% on mass of the additive $A1_2O_3$·CuO-CdO in the latex composition "elasticit" decreases its permeability into one of the main media for production of mineral fertilizer-hydrate extractive phosphoric acid by more than 300 times, while keeping

Table 3.1 Effectiveness of CaO β (kg/mole) for inhibition of the diffusion of various acids in various polymers

Acid (mass %)	Type of polymer			
	F-30[a]	PELP[b]	F-2M[c]	F-40[d]
HCl (25)	41	26	15	11
HF (27)	110	19	6	3
HNO_3 (27)	11	9	9	21

Source: [1, 4]
[a]F-30: Trifluorinechlorinethylene-ethylene, copolymer
[b]PELP: Polyethylene of low pressure
[c]F-2M: Modified polyfluorevinylidene
[d]F-40: Tetrafluorine-ethylene, copolymer

Table 3.2 Influence of calcium oxide on the diffusion parameters of LPPE

Aggressive medium at 70°C	Amount of CaO (parts by weight)	λ (mkm/hr)	Time of acid penetration for unfilled τ_0 and filled polymer matrix τ_{0f} Thickness of LPPE layers (mm)			
			1	3	5	τ_0/τ_{0f}
20% HCl	0	125	65	595	1623	—
	1	41	596	5355	14,875	9.1
	5	21	2265	20,400	56,650	34.8
30% HF	0	60	62	2340	6500	—
	1	50	50	3600	10,000	1.5
	5	33	33	8280	23,000	3.5

Source: [1]

its tensile strength at the same value. By comparison the same latex composition loses 30% of its tensile strength under the same condition if such an additive is not introduced.

Electron-microscopic research and x-ray analysis of the latex surface coating modified by active oxide fillers after exposure in hydrate-extracted phosphoric acid enabled the detection on the surface of a dense film made from hard crystal-hydrates of mixed alumo- phosphates serving as a barrier layer that prevents a deeper penetration of the aggressive medium.

Of interest is increasing of chemical resistance of the ecological friendly hybrid nonisocyanate polyurethane matrix (see Chapter 1) by introduction of the special additives.

Table 3.3 Coefficient of chemical resistance to an aqueous 30% sulfuric acid solution

Indexes		Sample no. 21	22	23	24	25	26
Copper aluminate content[a]		0	10	20	40	100	160
Exposure time	6 months	0.87	1.12	1.14	1.06	0.98	0.98
	9 months	0.62	0.98	1.06	1.08	0.99	0.92
	12 months	failed	0.96	1.05	1.07	0.99	0.95

Source: [1, 6]
[a] In parts by weight based on 100 parts of nonisocyanate polyurethane network

Table 3.3 [1, 6] summarizes the test results of composite samples with hybrid nonisocyanate polyurethane network matrix and with different amounts of copper aluminate filler. These samples were tested after 6, 9, and 12 months of exposure in 30% sulfuric acid at 20°C. The coefficient of chemical resistance K_{CR} was defined as:

$$C_1 = \sigma_\tau / \sigma_{\tau 0}, \tag{3.17}$$

where σ_τ and $\sigma_{\tau 0}$ are tensile strengths of the samples after exposure and in an unexposed state (control samples), respectively.

It is apparent that the control sample without an active filler (sample 21), under prolonged exposure to an aggressive medium, for example, after 6 months has a relatively high K_{CR} of 0.87. However, all of the above nonisocyanate polyurethane network samples comprising the active filler retained even more of their initial tensile strength than did the control after 6 months of exposure. Even after 12 months of exposure in 30% sulfuric acid, by which time the control sample had failed, samples 22–26 all had excellent tensile strength retention. In fact, the tensile strength of samples 23 and 24 increased substantially over their initial tensile strength even after 12 months of exposure.

Let us consider the effect of the active additives on the chemical resistance of the liquid ebonite mixture (LEM), which is an advanced abrasive resistant material for monolithic thick-layer coverings (see Chapter 1).

Nitric acid resistant coating compositions were prepared by using powdered bismuth oxide additive with a mean particle diameter of 20–40 μm. Table 3.4 [1, 7] illustrates the chemical

Table 3.4 Coefficient of chemical resistance (Eq. 3.14) of LEM covering for the protection of a steel substrate in a 10% aqueous nitric acid solution (thickness of the vulcanized covering = 100 μm)

Index	Samples				
	19	**36**	**37**	**38**	**39**
Bismuth oxide content[a]	0	2	5	7	10
Exposure time and temperature:					
360 days at 60°C	0.47	0.61	0.79	0.98	0.76

Source: [1, 7]
[a]In parts by weight based on 100 parts of LEM

resistance of LEM samples with the coating containing this additive in a 10% aqueous nitric acid solution.

Additional physical property evaluations were carried out for sample number 38; the following properties were obtained: tensile strength, 28 MPa; elongation at break, 7%; and adhesion strength to steel, 10.3 MPa.

The influence of additive modification on the strength characteristics of polymeric materials and the protective features of composition after exposure in aggressive media were studied as well.

The analysis of the obtained results allows one to state the following:

- Inorganic modification additives, which interact selectively with diffusant aggressive media, improve the operating characteristics of the polymeric compositions of all classes that were investigated.
- Additives of the said kind not only prevent the decrease in strength characteristics resulting from the exposure of polymer in aggressive medium but also improve the strength characteristics of polymeric material during the operation. The modification has an influence on the polymeric material both in simple and complex tension conditions (Fig. 3.4 [1, 5]).

The dependence of the modifying effect on the content of aggressive inorganic additives is extreme (Fig. 3.5 [1, 5]). Increase

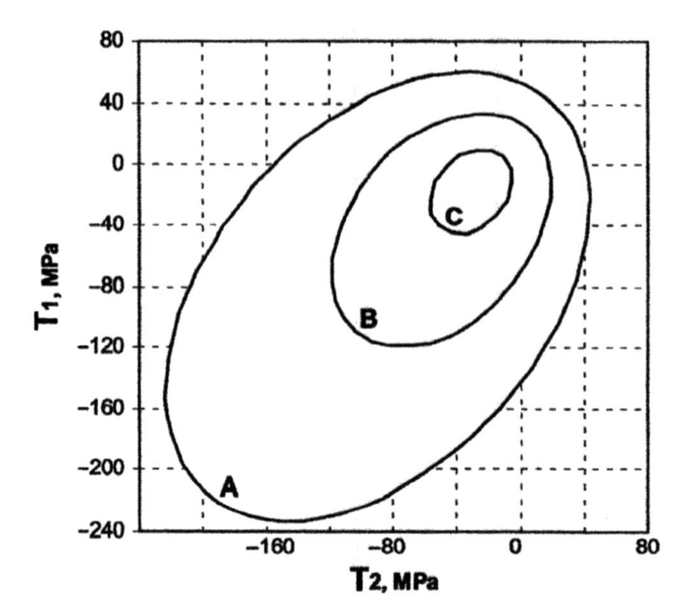

Figure 3.4 Destruction curves for a flat stress state: A, samples before exposure in the aggressive medium; B, samples with an active filler after exposure in 30% solution of SO_4 for 500 h; and C, the same samples as in B but without an additive [1, 5].

in brittleness of the polymeric matrix results in the narrowing of the optimal field, and the right side of the curve lowers more rapidly. The optimal additive quantity decreases with the decrease in the porosity of the polymeric material. The extremity of this dependence is caused by the formation of aqua-components in the polymeric matrix.

The left side of the curve demonstrates the field, where the forming binder fills up the pores of the material. By exceeding the optimal additive quantity, the porosity of the material is not sufficient to compensate for the increase of the additive volume, and this results in the appearance of tensions, which cause the destruction of the material in spite of the further increase in the share of inorganic binding agent.

Electronic, microscopic, x-ray, and structural investigations showed that the crystal-hydrates of metals are formed inside

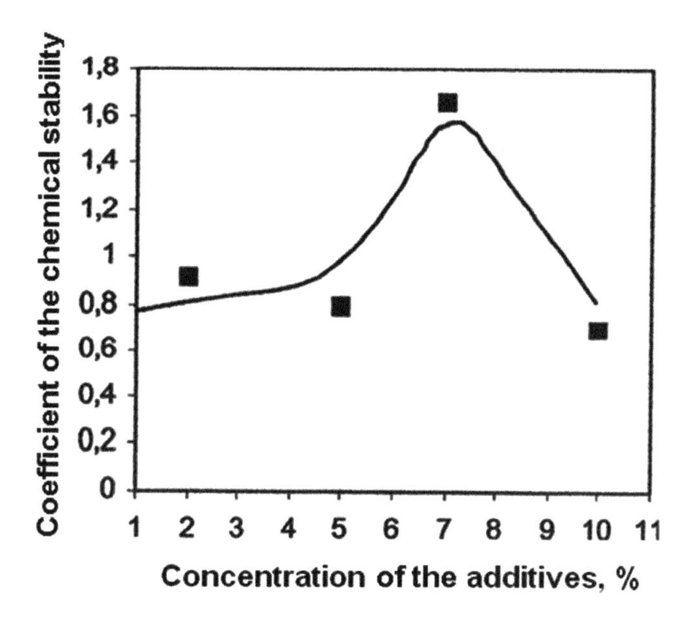

Figure 3.5 Dependence of the tensile strength index of a modified ebonite compound on the modification component content after exposure in a 10% solution of HNO_3 for 1 year [1, 5].

the polymeric matrix, but they are different from those that are formed outside of it. It is caused by the membrane mechanism of mass transit in polymeric materials, as the penetration depth of phosphate and sulfate ions is lower than that of water.

As is known the chemical destruction of a polymer matrix of a composite material in a liquid aggressive medium occurs in the diffusion kinetic region. Therefore, the additives form a high-strength inorganic cement that not only promotes to a drastic (about 1.5–2 times) reduction in permeability but also in some cases leads to an increase in operational strength of the material.

For example, introduction of the complex additive $nTiO_2 \cdot mCuO \cdot pCdCO_3$ in epoxy-rubber mastic that is for a long time in the 85% phosphoric acid medium increases the strength by 1.5–1.6 times. Without the additive the material is destroyed during the first month.

By chemical destruction of the polymeric matrix some radicals can be introduced in the aqua-complexes. Therefore, the nature of the effect of additive fillers on the protective features of composite materials is based on the formation of aqua-complexes. Transformation of the active filler into the polymeric matrix is based on the hardening mechanism of inorganic glue cements.

In the process of the production of modified material, technological defects appear as a result of thermal tensions in the composition. Introducing the filler increases the share and size of defects. Locally, near the filler particles, a higher porosity level can be registered. In the process of diffusion of the medium into the polymeric composition, the active filler contacts the acid solution. Therefore, in the hydration process the filler is transformed into an insoluble glue cement that changes significantly the physical and chemical features of the composition. The relative free-space volumes of the matrix (V^{m}) and the filler (V^{f}) can be defined as some functions of time (τ) and of a medium concentration in the polymer (c), that is,

$$V^{\mathrm{m}} = f_1(c, \tau); \quad V^{\mathrm{f}} = f_2(c, \tau). \tag{3.18}$$

The transforming rate of the filler into cementing glue depends on the rate of chemical reaction (V_x) and mass transfer (V_{m}). This process can occur in kinetic or diffuse fields. If the diffuse field takes place, that is, $V_{\mathrm{m}} \ll V_x$ the transformation stage depends on the quantity of fluid penetrating in this point of the composition during the time (τ). When the rate of chemical reaction is significantly slower than the rate of diffusion, $V_x \ll V_{\mathrm{m}}$, the transformation stage of the active filler into a glue cement can be described as:

$$V^{\mathrm{m}} = V_0^{\mathrm{f}} e^{-kc\tau}, \tag{3.19}$$

where k is the efficient constant of the rate of chemical reaction depending on the concentration of water solution in the fluid and V_0^{f} is the initial relative free-space volume of the filler.

In the process of the transformation, the active filler increases the initial absolute volume and this results in the appearance of local compressing tension in the polymeric matrix and partial filling up of material defects.

The diffusion process, where the diffusion coefficient depends on the medium concentration and time and is accompanied with mass absorption, can be described in cylindrical coordinates as follows:

$$\frac{\partial c}{\partial \tau} = \frac{1}{r} \cdot \frac{\partial}{\partial r}\left(D \cdot r \cdot \frac{\partial c}{\partial r}\right) - q, \tag{3.20}$$

where D is the diffusion coefficient, q is the volumetric absorption intensity of the aggressive medium due to the chemical reaction, and r is the current radius of the hollow cylindrical sample.

The diffusion coefficient in quasi-stationary conditions ($\frac{\partial c}{\partial \tau} = 0$) is:

$$D = \frac{\partial \theta}{\partial \tau}\left[\frac{\ln(r_{\text{out}} - r_{\text{inn}})}{2\pi l\,(c_{\text{inn}} - c_{\text{out}})}\right], \tag{3.21}$$

where $\frac{\partial \theta}{\partial \tau}$ is the velocity of the acid solution flow and r_{out}, c_{out}, r_{inn}, and c_{inn} are, respectively, the radius and saturation concentration of the outer surface and the radius and saturation concentration of the inner surface of the hollow cylindrical sample with length l.

The data about the constants of sorption of acids by polymers are very poor; therefore, it is impossible to correctly check this assumption. The mass change of the nonfilled polymer after 24 h in the aqueous HCl medium is compared with the value of γ. These results are shown in Fig. 3.6 [1, 4], plotted against double-logarithm coordinates. The results do not contradict the assumption made. However, the change in qualitative characteristics needs emphasizing. At low degrees of filling, γ increases linearly with the inhibitor concentration φ. For most of the systems, this correlation remains until the inhibitor concentration reaches 1 mole/kg:

$$\gamma = 1 + \beta\varphi \tag{3.22}$$

The value of β can be employed to compare inhibitor efficiency. Following on from Eq. 3.22, this one is equal to the relative change in time duration of nonpermeability $\Delta\tau_{0f}/\tau_0$ when the inhibitor concentration is changed to 1:

$$\beta = \frac{\Delta\tau_{0f}}{\tau_0} \cdot \frac{1}{\Delta\varphi}, \tag{3.23}$$

where β can be taken as the specific relative efficiency of the inhibitor.

Table 3.5 [1, 4] contains values of β for various inhibitors involved in penton. It is obvious that β changes in the large spectrum: from 2 for $Ba(OH)_2$ to 86 for KOH. No clear relation

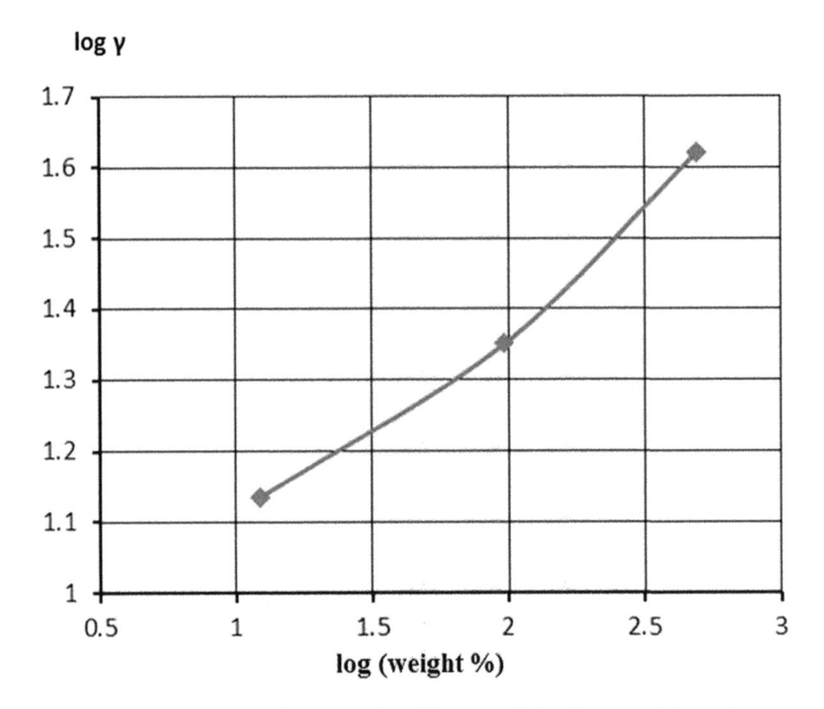

Figure 3.6 Correlation between the value γ and the change of the polymer mass (% of the initial) after being exposed to HCl [1, 4].

between β and the chemical nature of the inhibitor was found except in such obvious cases as chromium and silica oxides, which cannot slow down the diffusion of HCl as they are inert. On the other hand, some forms of silica oxide are effective inhibitors of HF diffusion. It is difficult to explain the large value of β of some inhibitors by the size and surface area per mass of their particles.

For example, for both "good" and "bad" inhibitors, the particle size of bivalent metal oxides is roughly the same: 1.5–2 mm. The surface area per mass of one of the best inhibitors CdO is close to zero, whereas that of BeO is 56.5 sq.m/g; yet the efficiency of this one does not differ much from other oxides whose surface area is about 10 times less. Probably, β is the complex function both of the diffusion parameters of acid in polymer and of the kinetic parameters of the interactive acid inhibitor.

Table 3.5 Inhibitor effectiveness β (kg/mole) for various inhibitors of the diffusion of HCl into penton

	Inhibitor and valence						
Additive	**Oxide**		**Hydroxide**			**Carbonate**	
element	**2**	**3**	**1**	**2**	**3**	**1**	**2**
Li			29				
Be	29–14						
Na			63			11	3[a]
Mg	24–14					16	
Al		24			5		
K			86				
Ca	20–14						29–24
Mn							24
Co	16						
Cu	2.5			16			7[b]
Zn	8						
Sr							29–24
Cd	47						
Ba	8			2			47
Bi		7					

Source: [1, 4]
[a]Sodium bicarbonate
[b]$(CuOH)_2CO_3$

The former is the product of the permeability constant of the acid in the polymer λ_0 and the acid vapor pressure in the exterior medium. The latter comprise not only the rate of inhibitor reaction with the acid but also complications in the reaction process itself, determined by the formation of reaction products. The formation of a core layer on the inhibitor particle can hinder the diffusion of the acid into the reaction zone. The direct result of complications in the reaction process should be the incomplete conversion of the inhibitor into the reaction product.

Indeed, the analysis of a sample filled with CaO and exposed for 150 days in 33% HCl showed that only 32 mass percent of compounds of Ca contain calcium chloride, the rest being nonreacting CaO. Taking into account the extent of HCl penetration into the sample, the average degree of CaO converted to chloride in zones where HCl has penetrated is 35%. Hence, the operating concentration of the inhibitor is not that of gross concentration. Only a part of it is determined by the correlation of the process's kinetic and diffusional parameters and represents the effective concentration of the inhibitor—its "activity" determining the inhibiting action.

As mentioned above, the linear dependence of γ on φ remains at small φ only. A further increase in φ leads to a reduction in $\Delta\gamma/\Delta\varphi$. The "penton-aluminum powder" system stemmed the increase in γ with an increase in φ at $\varphi > 5$ mass percent when inhibiting the diffusion of both HCl and HF. More degrees of filling cause the opposite effect: an increase in φ does not lead to a decrease but rather an increase in the value of λ_f and even more so for the nonfilled polymer. These phenomena are probably related to an increase in the permeability of the polymeric matrix in zones where the acid reacts with the filler. This makes it easy for the acid to move to the diffusion front. The increasing permeability can be caused by empty spots and dislocations formed as a result of the interaction of the acid with the inhibitor.

These assumptions are indirectly confirmed by the following facts: the acid reaction products were removed with inhibitors from high- degree filling polymers. This means that a permeability phase occurred, which is possible only in systems with communicating pores. In some special experiments, the rate of diffusion of the inert dye-stuff into the high-degree filled polymer before its treatment by an acid was compared with that after treatment.

It was found that the diffusion rate is much greater in the second case. With small degrees of filling, neither removal nor divergence in the dye-stuff's rate of diffusion was observed. It was supposed that the communicating dislocations' system formation is due to local tensions in the polymer at the conversion of the inhibitors to the reaction products. In some cases, the equivalent volume of the reaction products is significantly more than that of the inhibitors. This local increase in volume is the cause of local tensions in the polymer zones that interface with the inhibitor particle. The most serious changes in volume are realized when the reaction products are hydrated. For example, the transformation of MgO into $MgCl_2 \cdot 6H_2O$ should cause more than an 11-fold increase in the volume of empty space. Exposing the inhibitor containing polymers to an acid led to a mass change much more than that for nonfilled polymers: to 1.5% instead of 0.1–0.2%, which is the norm. This is because the equivalent weight of the reaction products is more than that of the inhibitor.

If the relation of weights is contrary, as in the inhibition of HF diffusion with sodium carbonate, the mass defect appears. The mass defect was also found in high-filled samples exposed to acids—because of the capturance of the filling substance by the acid. Sometimes, the mass change crisis with time can be extreme, that is, an initial increase in the sample mass until the maximum gained and then a decrease. Such curves can be explained by the superimposition of two simultaneous conversion processes and the substance capturance by acid under conditions of the matrix's progressive defectiveness.

It is interesting to study the influence of acid on the effectiveness of inhibition, although a definite answer is impossible. The influence of acid is shown by its diffusion capability in a given polymer and its diffusion coefficient, the rate of interaction with an inhibitor, and its thermodynamic activity in an aqueous solution. It seems that no experiment can ensure the equal values of all the above-mentioned parameters except one, the influence of which is being studied. Thus the data cited below only illustrate the complexity of the problem. The effectiveness of different acids' diffusion inhibition in modified polyfluorvinylidene was compared with MgO and CaO.

It was found that MgO is more effective as an inhibitor of HCl than CaO. For the inhibition of HNO_3 diffusion, both oxides are equivalent, whereas in HF, calcium oxide is more effective than MgO. The acid concentrations were chosen at random in this series of experiments. Perhaps other concentrations of acids would lead to other correlations.

The mass changes are caused by the sorption not only of acid but also of water.

The temperature dependence of the permeability coefficients of HCl and HF into pentan containing NaOH has been studied and compared with the temperature dependence of the permeability coefficients of these acids in the nonfilled polymer. It was found that the inhibitor involved did not significantly influence the temperature behavior of λ. Hence, also the coefficient γ does not significantly depend on the temperature of the inhibitor considered.

Inhibitors involved in the polymer do not significantly influence the strength of the adhesive bond of the penton coating with aluminum. Prolonged tests of such coatings exposed to HCl have been carried out. It is assumed that the involvement of inhibitors in the composition of coatings significantly strengthens the coating's adhesion to the metal. For filled and nonfilled coatings, the deterioration in adhesion caused by a penetrating acid is the same.

The time until adhesion deterioration sets in is estimated from Eq. 3.14 with probably a little more, because the diffusion front motion characterized by the value A in deep layers does not take place quite as fast as it does near the surface. Another reason for this phenomenon may be the rough estimation of the exact moment when adhesion deterioration sets in. One may assume that evaluation of this moment from the time till passage is allowed and sure because it means a reserve. In experiments with coatings of 1 mm thickness, for nonfilled coatings adhesion, deterioration sets in after 40 days, whereas for coatings containing 0.5 and 1.0 mass part of MgO per 100 mass parts of the polymer after 160 and 240 days, respectively.

Table 3.6 [1, 4] presents the time duration till passage for "model" coatings with thicknesses of 1,000 mkm containing 1 mole CaO per 1 kg of the polymer evaluated from the following equation:

$$\tau_{0f} = \frac{l^2}{\lambda_0^2}(1 + \beta\varphi) \tag{3.24}$$

Table 3.6 Calculated the time till passage for various polymer-acid systems (months) with CaO as an inhibitor

Acid (mass %)	Type of polymer				
	F-30	**F-40**	**F-2M**	**PEPL**	**Penton**
HCL (25)	0.1 (5)	0.8 (9)	0.1 (1)	0.04 (1)	0.8 (13)
HF (27)	0.01 (1)	0.1 (0.6)	0.03 (0.2)	0.4 (7)	
HNO_3 (27)	4 (51)	4 (94)	0.1 (1)	0.7 (7)	

Source: [1, 4]
Figures in parentheses are CaO%.
Polymer layer thickness is 1,000 mkm.
CaO concentration in the polymer: 1 mole/kg.
Temperature: 343 K.

This equation is obtained by combining Eqs. 3.14, 3.15, and 3.22. The calculations employ experimental data on λ, for various acids and polymers, in comparison with analogous results for nonfilled polymers.

The analysis of data given in Tables 3.1–3.6 allows some conclusions:

- For the choice of the coating polymer, the value of λ is important. Polymers with a small λ are preferable even at moderate β (of course, other factors are also taken into account). An example is the "F-40 HCl" system, in which the value of β is the smallest (see Table 3.1), but due to the small value of λ_0 it is over all other polymers in the value of λ_{0f} except penton, for which β is larger and λ_0 slightly smaller.
- The most complicated problem is protection from HF penetration. For nonfilled coatings this was known beforehand (from results of estimation of λ). Also the use of polyethylene is perspective. As the analysis of Eq. 3.18 shows, the increase of the thickness of the coating, the use of more effective inhibitors, and an increase in its concentration allow a gain in the time duration till passage value of quite a few years. Although the values of β in the inhibition of HNO_3 diffusion are often smaller than those for HCl and HF (see Table 3.1), the time duration till passage—when inhibitors are used—can be significant because of small values of λ_0.

3.6 Development of Chemical Resistant Polymer Concretes

Resistance of PCs in inorganic acids can be improved by creating specific barriers (physical and chemical) on the path of motion of corrosive reagents.

At the first stage of aggressive media attacks, the free water contained in a solution of acid diffuses into PC. The acid in itself starts its penetration into the body of a polymer composite as if on "walkway" of the free water. To eliminate this walkway and to create a barrier to water by means of hydrophobic additives such as graphite, it is possible to obtain acid resistance compositions. However, graphite does not allow obtaining a solid polymer concrete (PC).

It is known that the most widespread and cheapest filler is quartz sand, with good resistance to acids, so increasing the resistance of PC using this filler would be most effective. It can be achieved by introducing special additive substances such as strontium carbonate, oxides or peroxides of alkaline earth metals, bismuth nitrate, activated carbon, fluorides, and triaryl-imidazoles and their dimers.

These additive substances bind acids, form insoluble compounds, and adsorb on their surface chemical reagents.

Table 3.7 [1, 8] illustrates the effect of the proposed additives on the strength and deformability of polyester PCs in 30% sulfuric acid solution.

One can see that an aggressive environment had a dramatic impact on strength and deformation of the polyester PC. So, after 30 months of exposure, the strength of the polyester polymer without an additive decreased by 2.5 times. More significantly, its deformability reduces. Introduction of additives significantly alters the course of this process.

The most effective additives are strontium carbonate and lophine + piperidine (Fig. 3.7 [1, 8]).

Introduction of additives into the polyester polymer composition reduces its compressive strength by 30% and the modulus of deformation by 30–40% during 30 months of exposure in a 30% solution of sulfuric acid.

Let us take a brief look at the course of chemical attack of the corrosive environment on the PC.

Intense penetration and absorption of water from the sulfuric acid solution into body of the PC is observed in the first 2–3 months of the exposure. The acid concentration over the time of exposure increases, which leads to destructive processes in the material. These processes are accompanied by a decrease in the strength characteristics of the PC. It should be noted that reduction of these characteristics of the compositions containing strontium carbonate

Table 3.7 The effect of a 30% solution of sulfuric acid on the strength and deformability of polyester polymer concretes

Kind of additive in composition of polyester polymer concrete	Strength (numerator) and modules of deformation (denominator), MPa at exposure during (month)							
	1	2	3	6	12	18	24	30
Without additive	76/17,000	72/16,000	68/15,000	61/14,000	53/11,000	45/8000	38/7000	36/5000
Strontium carbonate	77/17,000	74/16,000	71/15,000	68/14,000	62/14,000	58/14,000	56/14,000	54/13,000
Barium oxide and peroxides	71/17,000	64/16,000	60/15,000	55/13,000	51/12,000	45/11,000	43/10,000	42/9000
Bismuth nitrate	75/12,000	70/15,000	64/12,000	58/11,000	52/12,000	50/12,000	49/13,000	48/12,000
Activated carbon	63/14,000	62/13,000	60/12,000	56/10,000	52/8000	48/8000	44/8000	42/7000
Lophine	76/17,000	75/16,000	72/14,000	69/12,000	61/11,000	57/11,000	54/10,000	52/10,000
Lophine and piperidine	79/17,000	78/17,000	77/16,000	74/16,000	70/15,000	66/15,000	62/14,000	59/13,000

Source: [1, 8]

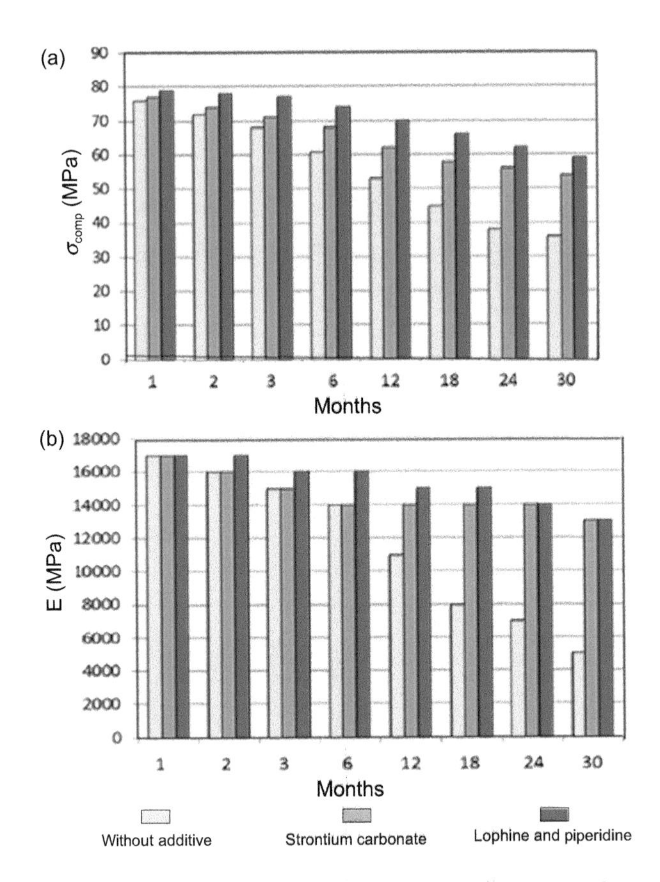

Figure 3.7 The influence of additives on compression strength σ_{comp} (a) and modules of deformation E (b) of polyester polymer concrete during 30 months of exposure in a 30% solution of sulfuric acid [1, 8].

and with bismuth nitrate occurs at a high speed in the first 6 months of exposure and then the degradation process is retarded.

It must be stressed that the strength of the all polyester PC samples containing additives is stabilized after 30 months of exposure in a solution of sulfuric acid. This phenomenon is associated with the formation and accumulation of the insoluble compounds as a consequence of sulfuric acid and additive reaction resulting in sealing of the PC, thus increasing its strength and stiffness. The strengthening effect is also connected with a change of its structure owing to penetration of sulfuric acid, which polymerizes the unreacted portion of the oligomer and thereby "clogs" the pores of the composite by molecules of the acid.

Let's consider the behavior of the polyester and furfural-acetone PCs in 20% hydrochloric acid solution.

The additives were used:

- For the polyester PC:
  - Lophine
  - Lophine and piperidine
  - Strontium carbonate
  - Barium oxide
  - Bismuth nitrate
  - Activated carbon
- For the furfural-acetone PC:
  - Sodium fluosilicate (10%) and quartz sand (90%)

The behavior of these composites is similar to the above-described behavior of the polyester PC in a 30% sulfuric acid solution that is evidently due to predominantly process of water diffusion, which is a general for these materials.

The test results of the PCs in a 20% hydrochloric acid solution after 30 months of exposure are shown in Table 3.8 [1, 8]. Chemical

Table 3.8 Test results of the polymer concretes in a 20% hydrochloric acid solution

Kind of polymer concrete	Kind of additive	Relative chemical resistance after 30 months of exposure
Furfural- acetone polymer concrete	Without additive	**1.00**
	Sodium fluosilicate and quartz sand	1.56
Polyester polymer concrete	Without additive	**1.00**
	Strontium carbonate	1.48
	Barium oxide	1.12
	Bismuth nitrate	1.45
	Activated carbon	1.48
	Lophine	1.48
	Lophine and piperidine	1.38

Source: [1, 8]

resistance of the PCs was obtained by the compression strength tests.

The most effective additive is sodium fluosilicate + quartz sand for furfural-acetone PC and the number of additives are strontium carbonate, bismuth nitrate, activated carbon, and lophine for the polyester PC (Fig. 3.8 [1, 8]).

Let's go over to next issue on the problem of chemical resistance of PCs. We mean an alkali resistance.

Increasing physical and mechanical properties of the polyester PC in an alkaline environment is possible by decreasing the rate of hydrolysis of ester bonds at the expense of application of special additives. The interaction of these additives with alkalis leads to partial or complete neutralization of alkalis. These additives include aluminosilicates, especially zeolites and micas.

Zeolites are crystalline substances with developed porosity; their activity is condition by an exchange of sodium cation on multiple-valent cations of magnesium, calcium, manganese, aluminum, etc. A zeolite additive in the polyester PC composition changes the nature of penetration of an aggressive environment as a result of exchange of the cations. Moreover, calcium and magnesium cations create in the composite's body additional bonds through free carboxyl groups. Finely divided zeolites are "molecular forces" that arrest the ions of an aggressive environment as well.

The authors used the zeolite with cations of calcium and magnesium, which formed calcium and magnesium hydroxides less aggressive to the polyesters than sodium hydroxide or potassium hydroxide (Table 3.9 [1, 8]).

Alumino silicate–dispersed micas in the polyester PC exploited in alkalis have the property of cation exchange. So, muscovite mica replaces aluminum cations with the alkali metal cations. The vermiculite replaces magnesium cations with cations of alkalis with a consequent significant improvement in the corrosion resistance of the polyester PC. It has been found experimentally that replacement of magnesium cations with potassium cations occurs at a higher rate, and therefore the vermiculite additive is more effective in a medium of potassium hydroxide (Table 3.9). Increase in corrosion resistance of the polyester PC containing aluminum silicate micas plate is also favored by the laminar form of their particles, which in turn reduces

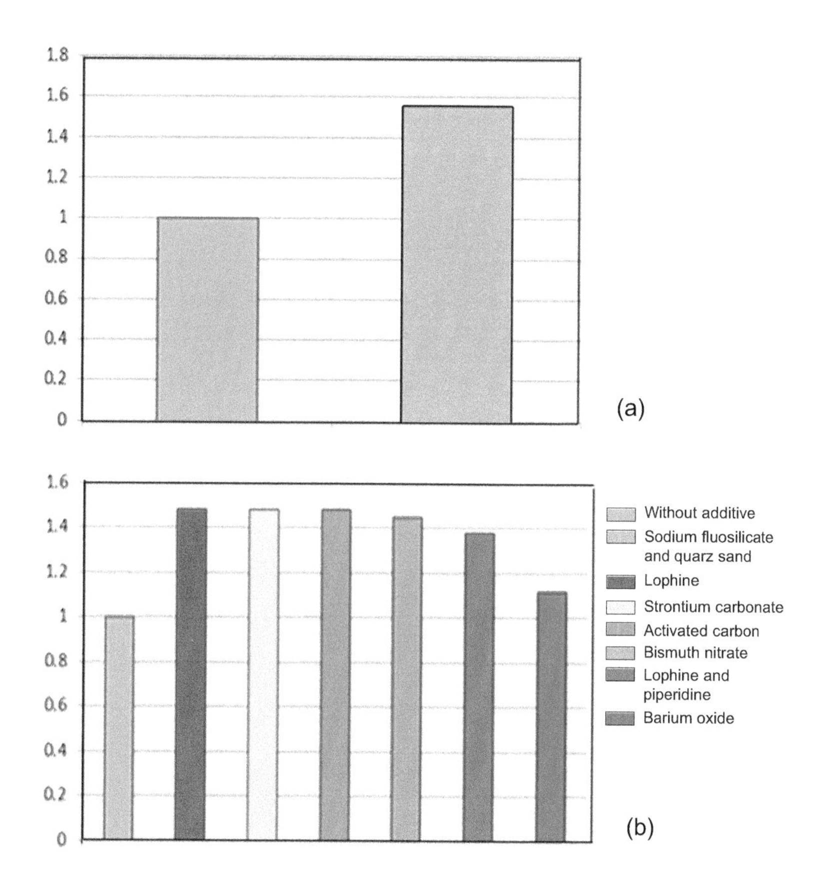

Figure 3.8 The effect of the additives on the relative chemical resistance of the polymer concretes after 30 months of exposure in a 20% hydrochloric acid solution: (a) furfural-acetone polymer concrete; and (b) polyester polymer concrete [1, 8].

the permeability of the polymer composites. All the investigated additives have promoted the improvement of alkali resistance of the polyester PC. Cobalt salts and lophine with piperidine were the most effective additives (Fig. 3.9 [1, 8]).

Additives promote the formation of large amounts of spatial bonds "cross-linking" of the polymer, thereby reducing its reactivity. The additives allow the formation of insoluble compounds, which seal the polymer structure and prevent penetration of corrosive reagents.

Table 3.9 Chemical resistance of polyester polymer concrete in a 15% solution of alkalis

Kind of additive	Relative chemical resistance after 7 months of exposure	
	Sodium hydroxide	**Potassium hydroxide**
Without additive	**1.00**	**1.00**
Zeolite with calcium cations	1.55	1.49
Zeolite with magnesium cations	1.65	1.68
Vermiculite	1.59	1.76
Muscovite mica	1.51	1.56
Biotite	1.69	1.66
Aluminum oxide or hydroxide	1.62	1.50
Bismuth nitrate and activated carbon	1.63	1.59
Zirconium phosphate	1.50	1.52
Cobalt acetate	1.76	1.68
Cobaltous chloride	1.62	1.74
Selenium and zinc	1.59	1.65
Aluminum and cuprum	1.57	1.66
Metol	1.44	1.50
Lophine	1.58	1.65
Lophine and piperidine	1.74	1.77

Source: [1, 8]

Increased corrosion resistance of the polyester PC containing aluminum oxide or hydroxide is attributable to the ability of these additives to collect the cations or anions on surface of the polymer depending on the pH of the solution. Aluminum hydroxide as an additive partially neutralizes the alkali by the interaction $Al(OH)_3 + NaOH \rightarrow NaAlO_2 + 2H_2O$, resulting in the improved chemical resistance of the composite.

This effect increases with an increase in the content of the additives on the composite's surface, since its corrosion resistance is directly related to the strengthening of surface layers.

During development of the chemically resistant polyester PC formulations, the method used involved the so-called sacrificial additives, that is, powder mixtures of dissimilar metals. For example, the mixture of aluminum and copper gives the effect of electrolytic corrosion of metals. Corrosion of aluminum reduces the influence of the alkali environment on the polymeric binder. The

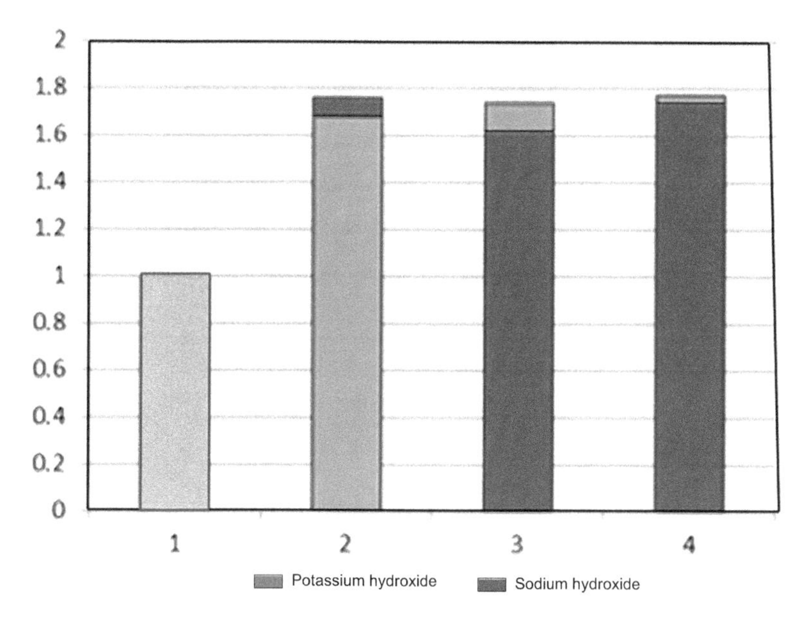

Figure 3.9 Influence of additives on the relative chemical resistance of the polyester polymer concrete in an alkali environment: 1, without additive; 2, cobalt acetate; 3, cobaltous chloride; and 4, lophine and piperidine [1, 8].

reaction product is alumina oxide, which exchanges with corrosive environment ions, resulting in the reorganization of the interaction mechanism of the additive with alkalis.

Alkali resistance of the polyester PC with this metal addition is much higher than the control samples. Similarly, it is observed when using active pairs of metal as additives (copper and zinc; zinc and selenium).

A special mention should be made of the important role of additives from the family of triaryl-imidazoles and their dimers, namely lophine and piperidine.

Even a small amount of these additives significantly increases the corrosion resistance of the polymer composites. Triaryl-imidazoles and their dimers "capture" the active radicals arising from polymer degradation, react with them, and thus form the reaction products, which retard or prevent the penetration of the aggressive media in the PCs. Additives zirconium phosphate, cobalt acetate, and

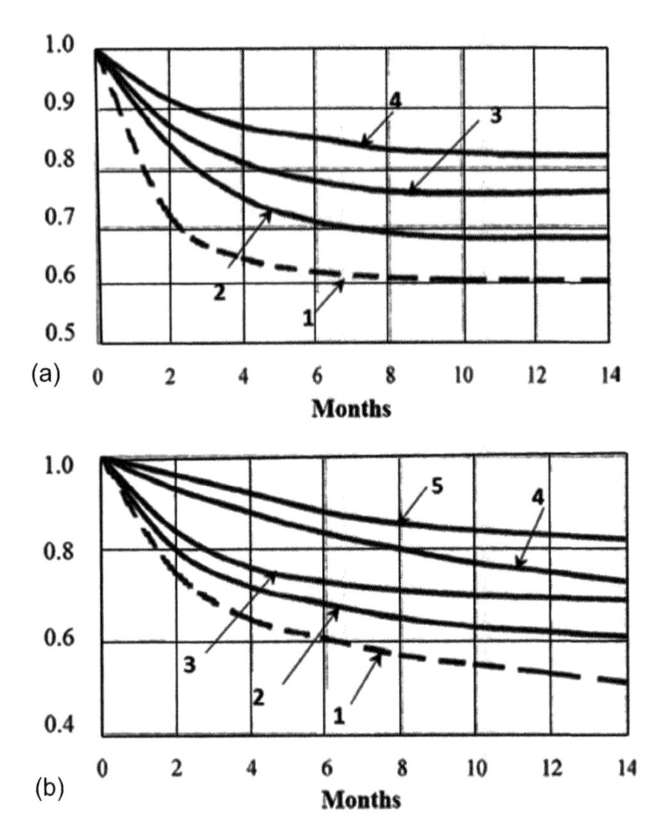

Figure 3.10 Water resistance of the two polymer concrete compositions. (a) Furfural-acetone concrete: 1, without additive; 2, with acid resisting cement; 3, with tetrahydrofuran; and 4, with lophine and piperidine. (b) Polyester polymer concrete: 1, without additive; 2, with acid resisting cement; 3, with barium oxide; 4, with lophine and piperidine; and 5. with cobalt acetate [1, 8].

cobaltous chloride and also metol act to increase the chemical resistance of polyester PC (Table 3.9).

Figure 3.10 [1, 8] shows the water resistance of the two PC compositions: furfural-acetone and polyester PCs.

One can be seen that the PCs without special additives have poor water resistance. So water resistance of furfural-acetone concrete without additives goes down to 38% after 6 months of exposure (Fig. 3.10a). The sharp decrease in water resistance is due diffusion of water molecules in the polymer's body, separation and rupture of the polymeric binder ties, and elution of the free components as well.

Increased water resistance of the furfural-acetone concrete is achieved by the formation of a structure with more branched bonds. This structure is created by the introduction into the PC mixture of special additives—acid resisting cement, tetrahydrofuran, and also lophine and lophine + piperidine—capable of forming with furfural-acetone monomer and benzene sulphonic acid the sustainable spatial bonds and insoluble compounds, sealing the structure and reducing the permeability of the composite. The greatest effect is achieved when triaryl-imidazoles and their dimers are used as additives (Fig. 3.10). It is worthy of note that lophine plays the part of "a protective belt"; it closes the pores of the composite from penetration by an aggressive environment.

Introduction of acid resisting cement into a polyester PC slightly increases its water resistance, as metal oxides and hydroxides of calcium, magnesium, and barium interact with polyethers containing COOH groups and form insoluble salts. As a result the amount of hydrophilic end groups decreases and the density of the cross-link of copolymers increases.

The introduction of lophine and piperidine and also cobalt acetate leads to a very significant increase in the water resistance of the polyester PC because these additives with polyester form the strong and reliable three-dimensional cross-linking. These sort of additives provide a high water resistance of the PC, which, even after a year of exposure, was not less than 0.8 (Fig. 3.10).

3.7 Application

The additives described above have been found in large industrial applications. For instance, in the US they are employed to increase the stability and decrease the permeability of the polyethylene coatings of domestic water boilers' interior surfaces and the semiebonite coatings of chemical installations (aggressive media: mineral acids and salts). They are tested in Germany for use in vynylesther armory coatings in urethane gas cleaning.

Polymate Ltd., International Nanotechnology Research Center (Israel), has developed an extensive product range of additives for upgrading the most common polymers against a wide variety of aggressive media, including acids, seawater, fluorine, and alkalis. Anticorrosive additives (AAdd) are an effective solution for many applications. More than 80 products have been tested for use in the chemical industry.

In Table 3.10 [1, 4], the data on the elaborated technical additives and the fields of their use are presented. The amount of the modifying additive should be chosen proceeding from the fact that an increment of the inorganic component volume must be equal to or slightly higher as compared with the pores volume of the polymeric composition in the process of chemical interaction and hydration.

Its largest application is in Russia, with the industrial production of fluoroplastic and pentaplast plates with additives. These plates

Table 3.10 Elaborated additives for different aggressive mediums

Polymer	Aggressive medium									
	1	2	3	4	5	6	7	8	9	10
Polyurethane					+	+				
Epoxide (amine hardening)	+	+	+	+	+	+	+	+	+	+
Epoxide (anhydride hardening)	+	+	+	+	+	+	+	+	+	+
Unsaturated maleunate polyester	+	+	+	+	+	+	+	+	+	+
Vinyl polyester	+	+	+	+	+	+	+	+	+	+
Phenolic or furanic								+	+	+
Liquid ebonite	+	+	+	+	+	+	+	+	+	+
Soft rubber (sheet)	+	+	+	+	+	+	+	+	+	+
Natural rubber latex					+	+				
Polyethylene	+	+	+	+						
Polypropylene	+	+	+	+	+	+				
PVC soft		+	+	+	+	+			+	+
PVC hard	+	+		+					+	+
Polycarbonate		+	+		+	+	+			
Polyamide					+				+	+
Polytetrafluorin-ethylene	+	+		+				+		
Polyvinyludene-fluorine	+	+		+				+		
Copolymer (eth/other)	+	+	+							
Polysulphide (tiokol)				+	+	+		+		

Table 3.10 (*Continued*)

Polymer	Aggressive medium										
	10	**11**	**12**	**13**	**14**	**15**	**16**	**17**	**18**	**19**	**20**
Polyurethane		+	+							+	
Epoxide (amine hardening)	+	+	+	+	+	+	+	+	+	+	+
Epoxide (anhydride hardening)	+	+	+	+	+	+	+	+	+	+	+
Unsaturated maleunate polyester	+	+	+	+	+	+	+	+	+	+	
Vinyl polyester	+	+	+	+	+	+	+	+	+	+	
Phenolic or furanic	+	+	+					+	+	+	
Liquid ebonite	+	+	+	+	+	+	+	+	+	+	
Soft rubber (sheet)	+	+	+	+	+	+	+	+	+	+	
Natural rubber latex		+	+		+	+				+	+
Polyethylene		+	+				+			+	+
Polypropylene		+	+				+			+	+
PVC soft	+		+				+	+	+		
PVC hard	+	+			+	+	+				
Polycarbonate				+						+	+
Polyamide	+	+	+	+							
Polytetrafluorin -Ethylene			+			+	+			+	+
Polyvinyludene -fluorine			+	+		+	+	+	+	+	+
Copolymer (eth/other)		+	+							+	
Polysulphide (tiokol)	+	+	+	+	+	+	+	+	+	+	+

Source: [1, 4]

Mediums' numbers: 1, HF; 2, HCl; 3, HBr; 4, HNO_3; 5, H_2SO_4; 6, H_3PO_4; 7, H_3BF_6; 8, CH_3COOH; 9, KOH; 10, NaOH; 11, liquid fertilizer; 12, seawater; 13, lactonic acid; 14, mixed H_3PO_4-H_2SO_4; 15, mixed H_3PO_4-HNO_3; 16, mixed HCl- HBr; 17, KOH and HCl turn; 18, NaOH and H_2SO_4; 19, hot water; 20, drinking water

have also been employed in the preparation of biplastic constructions of chemically stable technological installations in mineral fertilizer and cryolyte production. Nonisocyanite polyurethane acid-resistant coatings, containing various additives depending on the exploitation conditions, were successively used in the treatment of floors, whereas phenol mortars with additives increase alkali stability—for cladding [4].

More complete results data of testing of the various additives for typical aggressive media are illustrated in Table 3.11 and Figs. 3.11–3.19 [1].

The coefficient of chemical resistance was calculated using the Eq. 3.17 or

$$C_2 = \sigma_t \cdot \varepsilon_t / \sigma_{t0} \cdot \varepsilon_{t0}, \tag{3.25}$$

Table 3.11 Data of testing

Polymer	Aggressive media	Brand of additive	Without additive		With additive		Permeability
			C_1	C_2	C_1	C_2	τ_0/τ_{0f}
Polyurethane	H_2SO_4	38U	–	0.67	–	1.02	42.7
	H_3PO_4	41U		0.69		1.04	44.1
	seawater	211U		0.83		0.98	37.5
	hot water	241U		0.77		1.09	36.2
Epoxide (amine hardened)	HNO_3	17E	0.32	–	0.94	–	30.8
	H_2SO_4	23E	0.55		1.03		46.8
	H_3PO_4	34E	0.78		1.29		47.3
	KOH	114E	0.88		1.01		37.4
	NaOH	126E	0.87		1.03		38.2
	seawater	218E	0.81		1		35.6
	$H_3PO_4 + H_2SO_4$	319E	0.48		0.97		39
	hot water	320E	0.75		0.98		40.3
Epoxide (anhydride hardened)	HNO_3	14Ep	0.51	–	0.98	–	34.2
	H_2SO_4	27Ep	0.72		1.08		39.2
	H_3P04	37Ep	0.7		1.17		27.8
	KOH	117Ep	0.68		0.93		39.7
	NaOH	123Ep	0.7		0.94		40.2
	seawater	214Ep	0.81		0.98		40.4
	$H_3PO_4 + H_2SO_4$	314Ep	0.63		0.06		37.9
PVC (hard)	HNO_3	18P	0.48	–	0.91	–	33.2
	KOH	119P	0.62		0.93		38.3
	NaOH	125P	0.65		0.92		38.7
	$H_3PO_4 + H_2SO_4$	311P	0.78		1.07		37.8

Table 3.11 (*Continued*)

Polymer	Aggressive media	Brand of additive	Without additive		With additive		Permeability
			C_1	C_2	C_1	C_2	τ_0/τ_{0f}
Phenolic	KOH	116Ph	0.67	–	0.94	–	28.2
	NaOH	121Ph	0.69		0.97		28.7
	seawater	213Ph	0.82		0.08		27.6
Ebonite mixture	HNO_3	12Eb	–	0.88	–	0.93	17.4
	H_2SO_4	26Eb		0.92		0.98	24.9
Ebonite mixture	H_3PO_4	31Eb	–	0.98	–	1	27.7
	KOH	112Eb		1.01		1.03	22
	NaOH	127Eb		1		0.98	23.1
	seawater	215Eb		1.04		1	28.2
	$H_3PO_4 + H_2SO_4$	317Eb		0.81		0.98	27.3
	hot water	534Eb		1.12		1.03	30.8
Polycarbonate	H_2SO_4	25Pol	0.84	–	1.07	–	29.2
	H_3PO_4	32Pol	0.89		1.01		34.3
Polyethylene	seawater	522Et	–	1.02	–	1.04	35.7
	hot water	533Et		1.07		1.03	34.9
Polypropylene	seawater	215Pr	–	1.06	–	1.05	33.8
	hot water	501Pr		1.09		1.1	34.5
PVC (soft)	HNO_3	14PVC	–	0.81	–	0.9	37.3
	H_2SO_4	23PVC		0.92		0.94	34.8
	H_3PO_4	35PVC		0.97		0.94	32
	KOH	115PVC		1.03		1	31.2
	NaOH	122PVC		1.04		1.06	31.7
	seawater	212PVC		1.17		1.15	36.5

Table 3.11 *(Continued)*

Polymer	Aggressive media	Brand of additive	Without additive		With additive		Permeability
			C_1	C_2	C_1	C_2	τ_0/τ_{0f}
	HNO_3	16Es	0.87		0.93		40.2
	H_2SO_4	26Es	0.97		0.97		38.5
	H_3PO_4	38Es	0.97		0.09		37.2
Vinyl polyester	KOH	118Es	0.81	–	0.98	–	35.4
	NaOH	124Es	0.83		0.96		35.4
	seawater	211Es	1.02		1.18		39.7
	$H_3PO_4 + H_2SO_4$	318Es	1		1.04		35.9
	seawater	402Ir		1.08		1.05	40.5
Isoprene rubber	$H_3PO_4 + H_2SO_4$	406Ir	–	0.9	–	0.92	29.8
	hot water	405Ir		1.17		1.17	40.2
	H_2SO_4	21Nr		1.02		1.01	28.7
	H_3PO_4	35Nr		0.97		0.97	27.5
Natural rubber	seawater	214Nr	–	0.08	–	1.06	38.4
	$H_3PO_4 + H_2SO_4$	314Nr		0.96		0.98	28.6
	hot water	299Nr		1.14		1.1	38.3
	HNO_3	13Am		0.47		0.82	35.5
	KOH	117Am		0.92		0.98	24.7
Polyamide	NaOH	124Am	–	0.92	–	0.97	24.5
	seawater	217Am		1.07		1.11	38.3
Copolymer Eth/Oth	seawater	701Cp	–	1.03	–	1.04	37.2
	hot water	702Cp		1.08		1.05	35.6
	HNO_3	804F	1				48.2
PTFE	seawater	805F	1.02	–	"1	–	49.5
	hot water	806F	1.07				47.6

Source: [1]

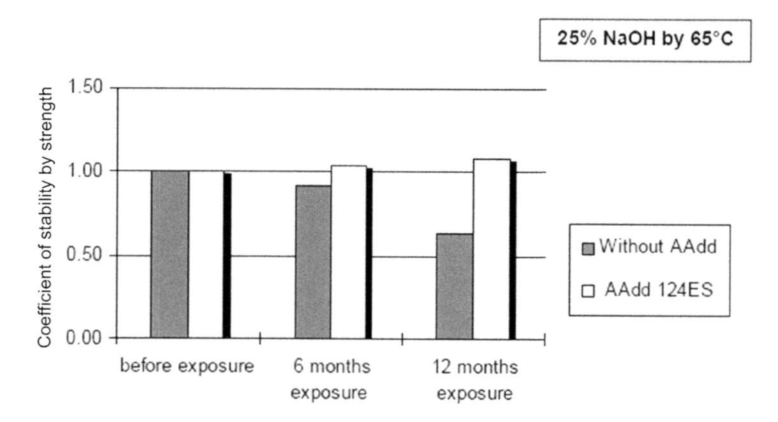

Figure 3.11 Effect of anticorrosive additives on the tensile strength of polyester reinforced plastics based on vinyl-ester resin [1].

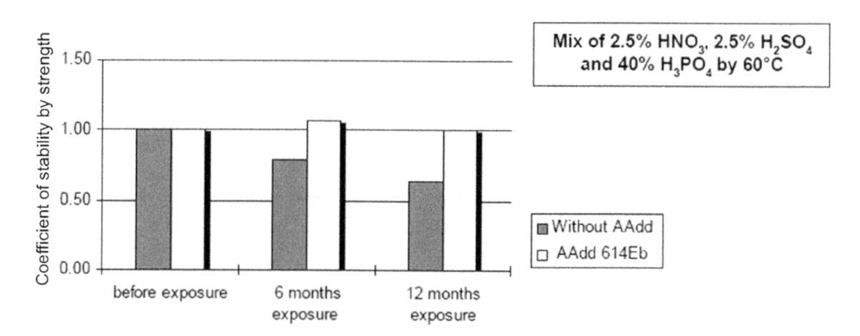

Figure 3.12 Effect of anticorrosive additives on the tensile strength of liquid ebonite mixture for covering [1, 7].

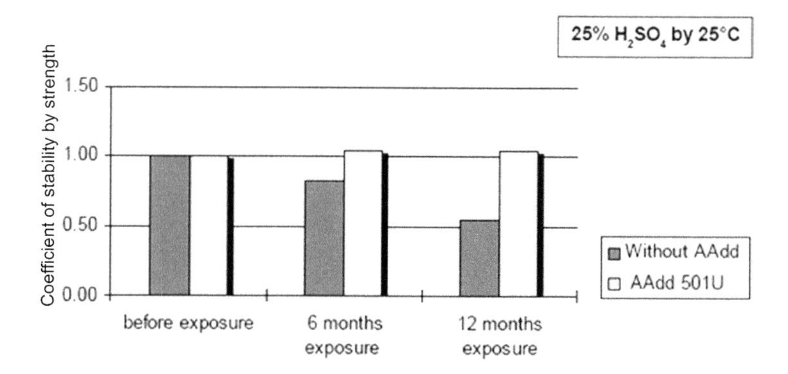

Figure 3.13 Effect of anticorrosive additives on the tensile strength of nonisocyanate polyurethane–filled composition for coverings [1, 6].

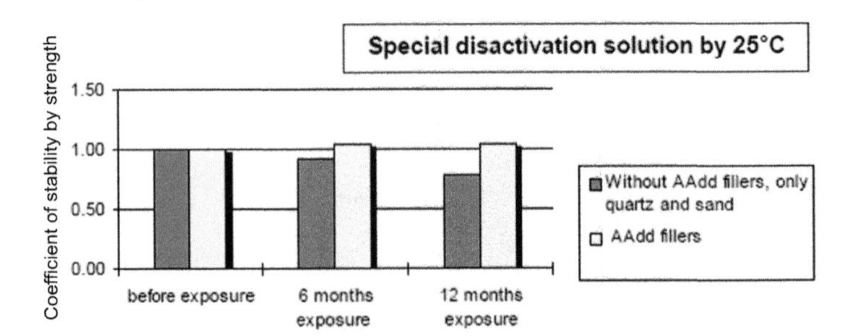

Figure 3.14 Effect of anticorrosive additives on the tensile strength of chlorine-contained epoxy mastics for flooring application in nuclear power stations and various chemical plants [1]. *Note*: AAdd in nonburning monolithic flooring provides upgraded chemical resistance against special decontamination solutions. Materials with AAdd are able to pass the IAEC method test for "natural fire" burning where flooring materials without the additive cannot.

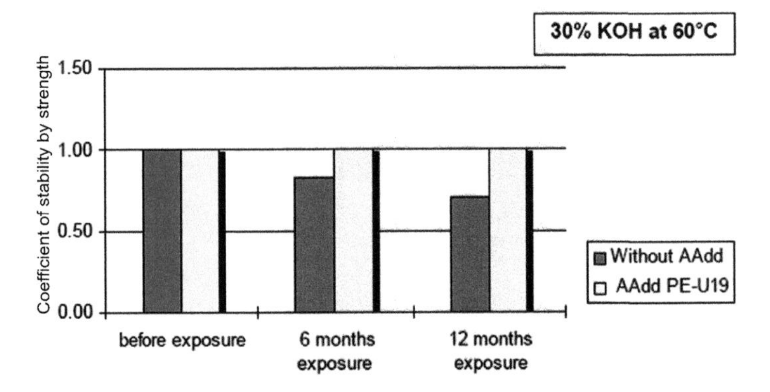

Figure 3.15 Effect of anticorrosive additives on the tensile strength of epoxy reinforced coatings [1].

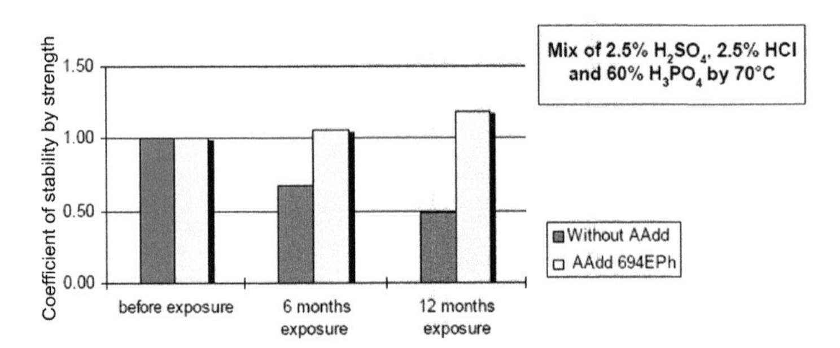

Figure 3.16 Effect of anticorrosive additives on the tensile strength of epoxy-phenolic putty for cladding [1].

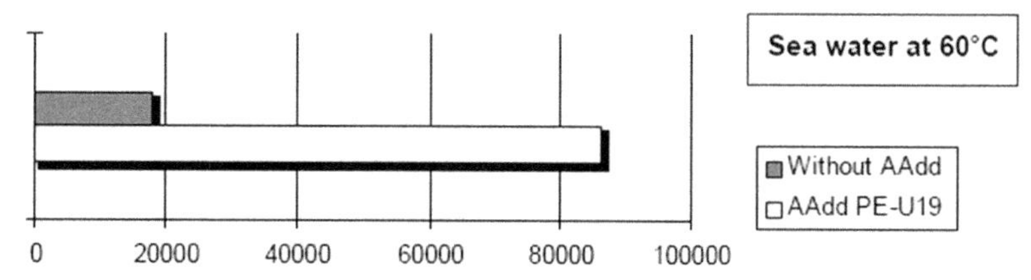

Figure 3.17 Effect of anticorrosive additives on the time of aggressive medium penetration through the protective layer from polyethylene coatings for pipes [1].

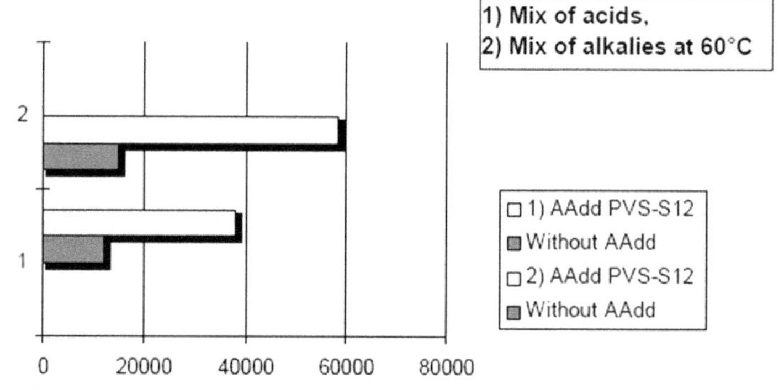

Figure 3.18 Effect of anticorrosive additives on the time of aggressive medium penetration through the protective layer from PVC films for covering [1].

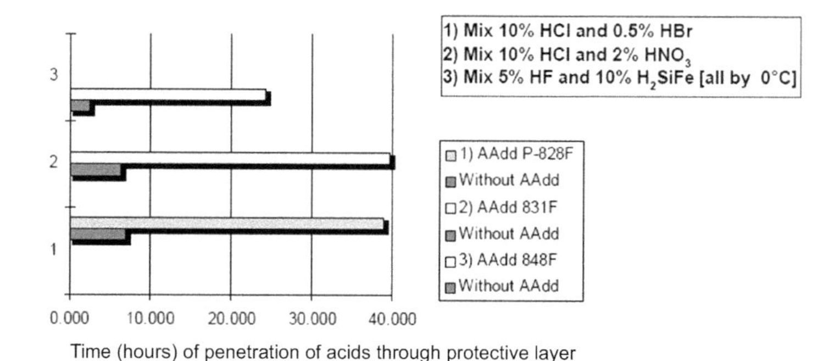

Figure 3.19 Effect of anticorrosive additives on the time of aggressive medium penetration through the protective layer from PVC films for covering [1].

where σ_t and σ_{t0} are tension strength after and before exposure in aggressive medium, respectively; and ε_t and ε_{t0} are relative elongation by tension fracture after and before exposure in aggressive medium, respectively.

3.8 Conclusions

The advanced method for increasing protective properties of polymer materials has been proposed based on the theory of "positive corrosion" for nonmetallic materials. The method implies adding of powdery inorganic active fillers interacting into the polymer matrix with penetrated aggressive media. The mechanism of formation of new phases as a result of such interacting was observed. A lot of elaborated additives were successfully tested for many kinds of polymer protective materials and different types of matrix polymers. Elaborated materials have shown significant increasing protective properties.

AAdd are an innovative approach to creating highly chemical resistant polymer materials. AAdd are specially designed to upgrade the chemical resistance characteristics of base polymers to achieve optimal performance capabilities of materials operating in aggressive environments. AAdd can be mixed into a wide range of polymer materials, offering a significant increase in product life and reducing product permeability.

These custom-made specialty formulations are designed to meet specific client requirements. When cured with polymer-based materials, AAdd can dramatically improve the capabilities of poly-based materials by upgrading their chemical resistance properties. The additives are inorganic powders that react with aggressive environments into which they are introduced, forming a new phase of high-strength hydrate complexes. This enhanced bonding occurs upon the penetration of aggressive media into the AAdd-containing polymer material. The chemical resistant properties of AAdd are activated by harsh environmental conditions where polymer systems without additives remain defenseless to chemical corrosion.

AAdd can be mixed into a wide range of polymer materials, such as epoxies, polyurethanes, glues, nylons, polyolefins, synthetic rubbers, and PVC, offering performance-enhancing attributes that increase the value of the end product.

The way of creation of chemically resistant PCs by use of so-called "positive" corrosion effect is offered. The most effective additives are from the family of triaryl-imidazoles and their dimers, namely lophine and piperidine.

AAdd provide a number of distinct enhancements for polymers offering manufacturers products with stronger, corrosion resistant products. Chemical resistance tests were conducted on polymer systems over a period of 1 year. The results revealed that AAdd-containing polymer systems significantly outperform those systems without the additive. Furthermore, extensive testing has shown that AAdd can increase product life by some 20 times. This extended life offers substantial savings for users who can extend the life of their polymer-based products, whether it is pipes, flooring, or other materials that are exposed to specific corrosive environments.

Products that have been enhanced with AAdd yield a higher impact strength than products without the additive. In addition, material permeability is reduced significantly, by 15–20 times. The percentage of AAdd mixed with a polymer matrix is relatively low, requiring only a small amount to obtain upgraded resistance characteristics of polymer materials.

Polymate Ltd., has developed an extensive product range of additives for upgrading the most common polymers against a wide variety of aggressive media, including acids, seawater, fluorine, and alkalis. AAdd are an effective solution for many applications. More than 80 products have been tested for use in the chemical industry.

Novel chemically resistant polymer coatings were elaborated by adding nanosize inorganic active fillers that react with an aggressive medium into which they are introduced, forming a new phase of high-strength hydrate complexes. This enhanced bonding occurs upon the penetration of aggressive media into active nanofillers containing polymer coating material. The chemical resistant properties of the forming polymer coatings are activated by environmental conditions where polymer systems without additives remain defenseless to chemical corrosion.

These custom-made nanofillers formulations are designed to meet specific client requirements. When cured with polymer-based materials, such nanofillers can dramatically improve the capabilities of polymer-based materials by upgrading their chemical resistance properties. The additives are inorganic nanopowders that react with aggressive environments into which they are introduced, forming a new phase of high-strength hydrate complexes.

Investigation of the protective mechanism of composite materials with additives that selectively interact with aggressive media, as well as the influence of modifying additives, provides a hint about choosing the modifying additives taking into account the kind of composite polymeric materials and their operating conditions.

The modifying additive should form with an aggressive medium influencing the polymeric material a high-strength glue cement. Introduction of ions, which were formed as a result of the destruction of the polymeric matrix, as well as the interaction of the forming glue composition, is also expedient. The most optimal case is when the concentration of an aggressive medium is slightly higher than the optimal concentration of the medium of the glue cement forming.

In the case of significant differences between the concentrations of aggressive media influencing the modified polymeric material and optimal concentrations of the same media that dictate quantities of the modifying additives, it is advisable to introduce a complex additive designed to have optimal concentration for a given media, that is, one designed for both higher and lower concentrations of the aggressive medium influencing the polymeric material.

The amount of the modifying additive should be chosen on the basis of the criterion that an increment of the inorganic component volume must be equal to or slightly higher than that of the pores volume of the polymeric composition in the process of chemical interaction and hydration.

After considering the problem of preventing the penetration of corrosion-active components into various protective coating materials, it was concluded that inhibitors involved into micropores and coating defects seriously reduce the corrosive agents' penetration rates.

Further, after analyzing the mass-transfer aspects of the inhibition of acid penetration, it was shown that the effectiveness of inhibition cannot be easily related to the chemical compositions of the inhibitors.

References

1. Figovsky O., Beilin D. (2016). Polymer nanocompoisites with high resistance to aggressive environment, *Journal Scientific Israel - Technological Advantages*, **18**(1).
2. Figovsky O. (1988). Improving protective properties of nonmetallic corrosion resistant materials, *Journal of Mendeleev Chemical Society*, **33**(3), 31–36.
3. Chalusch A. (1987). *Diffusion into Polymeric Systems*, Moscow.
4. Figovsky O., Romm F. (1988). Improvement of anti-corrosion protection properties of polymeric materials, *Anti-Corrosion Methods and Materials*, **45**(3), 167–175.
5. Figovsky O. (2006). Active fillers for composite materials: interaction with penetrated media, *Encyclopedia of Surface and Colloid Science*, Taylor & Francis, CRC Press: Boca Raton.
6. Hybrid nonisocyanate polyurethane network polymers and composites formed therefrom, Patent US 6,120,905.
7. Liquid ebonite mixtures and coatings and concretes formed therefrom, Patent US 6,303,683.
8. Potapov Yu., Figovsky O., Beilin D. (2015). Development of chemically resistant polymer concretes, *Journal Scientific Israel - Technological Advantages*, **17**(4), 189–195.

Chapter 4

Environmental Friendly Method of Production of Nanocomposites and Nanomembranes

4.1 Introduction

It was established in the middle of the past century that under any conditions of an impact, the depth of a crater could not exceed 6 calibers of a striker. A ratio of a crater depth to a striker size (caliber) was taken as a universal criterion of impact evaluation. However, it should be taken into account that a physical limitation exists for an impact process. Numerous investigations made in various countries on barriers punching allowed us to solve the problems of increase in the relative depth of a crater only in the range of 6 calibers.

As a result it was found that the experimental results, being out of the limit of 6 calibers, cannot be explained with theoretical models of the impact created during the last two centuries. Nevertheless, more and more frequently the data on abnormalities of relative depth penetrations of the calibers <100 have been published since the twentieth century. However, the experimental results, inexplicable from the point of view of mechanical and hydrodynamical models,

Green Nanotechnology
Oleg Figovsky and Dmitry Beilin

ISBN 978-981-4774-10-9 (Hardcover), 978-1-315-22928-7 (eBook)
www.panstanford.com

were not given. For example, balls of mercury inside thick steel samples obtained in experiments could not be explained, so these samples were thrown out to a dump. There is a problem of reproducibility of such abnormal results. In the planet's conditions, natural processes causing such abnormalities are not known.

When a static strength of the barrier material decreases, the resistance to a striker motion in the barrier sharply decreases too and considerable increase in the depth of striker penetration is observed. It is known that for a shot in sand, it is possible to gain the punching depths of tens of calibers. For a shot in water, what is used in gun expertise, the maximum depth of penetration can attain 100 calibers.

In the 1970s S. Usherenko [2] analyzed the known abnormalities of crater formation. It turned out that the so-called anomalous results have been obtained in a region of micro-objects interaction. Special attention was paid to the fact that the experimentally established physical limit of a relative depth of craters formation is explained by the existence of the known constants of mass and heat transfer of the barrier material. Therefore, increase in collision velocity, relative density of the striker material, and increase in the angle of impact cannot lead to an increase in the relative depth of cratering. All these parameters of impact cause a change of magnitude of the kinetic energy. However, the energy excess (in an open system) cannot be stored in the barrier material. The increase in impact energy leads to the increase in velocity of reverse emission of the striker and the barrier materials, melting of walls and of the bottom of a crater, and in extreme regimes intense radiation. In the studied variant of interactions at impact, which can be named "macroimpact," there is no opportunity to realize the phenomena of abnormal and superdeep penetration (SDP). The excess of energy, in one or another way, will be removed from the open system of bodies. The experimental result showed that the limiting value of energy density during a macroimpact does not exceed 10^9 J/m^3.

In natural conditions, a SDP process can be observed in the stratosphere and in the free space, as there are the clots of cosmic dust colliding at high velocity. A principal reason of making difficult detection of SDP on space stations is the absence of reach-through holes and, accordingly, absence of depressurization of the module with the equipment and people. Available information allowed us to assume that in the conditions of SDP implementation, the existing protection of space modules is not effective. Apparently, because of it, on operating space stations, there was a problem with stability of work of computer elements and control systems. The same new factors will represent additional danger for the modules moving with the equipment and people through the regions of the space with high concentration of dust objects.

The set of experimental conditions was determined for which the penetration on relative depths of 100–10,000 calibers proceeds stably [2, 3]. After reception of the evidence that the phenomenon of SDP exists and that there is a necessity to use physical effects that are observed in SDP conditions, there was a requirement to comprehend a fundamental result. Special attention, for more than 30 years, has been paid to the modeling of a mechanism of effective utilization of the kinetic energy of the SDP process [3–5]. However, the presented concept of the phenomenon of SDP has not appeared to be a successful one. With the new experimental results obtained, the models of SDP process were rejected because they could not be used for explanations.

4.2 Brief Review of the Known Models of Superdeep Penetration

In the first cycle of experimental investigations (1974–1978), the proofs of the existence of SDP of a clot of particles into the barrier have been obtained. It has been confirmed that it is impossible to explain the results of SDP on the basis of the known models of macroimpact. The known models can be divided into four types. Basically, these models describe the SDP process as interaction of an individual striker with a zone of its movement in the barrier.

The models, in which the energy was expended due to the elastic reaction from the barrier material, are included in the first type of the models. The hypothesis that freezing of plastic deformation occurs at SDP has been given [6, 7]. The authors have assumed that SDP is realized through a system of cracks. The energy during penetration is spent only on elastic deformation. To explain the experiment results that contradicted the described model, additional assumptions have been made [8]. A new hypothesis saying that nonplanar cracks appear is offered. On the basis of this hypothesis it was assumed that after the collapse of hundreds of thousands of unusual cracks, the matrix material can be strengthened and the tops of cracks are melted and deformed. At present, the hypothesis about cracks is used for qualitative explanation of unusual results [9]. For example, this hypothesis explains the striker's penetration into the steel at the temperature of 196°C.

It is possible to include the models in which the particles transfer into the barrier occurs in the special transport elements "solitons" (whirlwinds) to the second type of hypotheses. The hypothesis says that unusual transport of the elements and strikers proceeds with no energy expenditure. In 1998, the hypothesis was offered that SDP of the particles is a process of exchange in mass and energy due to the developed instabilities in material [10]. The degree of nonequilibrium at SDP is described by the dependence of entropy change on the deformation of a matrix material. The hypothesis of hydrodynamical instability in local areas of the barrier surface, loaded by a stream of microstrikers, is known [11]. At microcumulating, transport whirlwinds are created at the front of a shockwave, initiated by the background shockwave. The striker material is carried by the transport whirlwinds into the barrier. According to the author, recrystallization and amorphization of the matrix material, as well as the traces of microparticle material transfer, can be explained by high pressures and temperatures in the zones of microcumulating. A hypothesis on the creation of volumetric "soliton," in which the striker at SDP moves without the expenditure of energy, has been offered [12].

The third type of hypothesis is based on the assumption that there are special mechanisms of flow and crushing and a loss of strength that were not known earlier. At implementation of special mechanisms, the strength of the barrier material is reduced. Thus, as a rule, the static and dynamic strength, macrodeformations of the barrier, crushing of grains, and emission are neglected. For example, the hypothesis on the existence of a specific flow of the microstriker by a stream from a matrix material is offered [13]. Material destruction in the penetration zones is assumed under various conditions of the stress produced in the elastic-plastic medium [14]. In a regime of free oscillations, in the target, the standing wave is formed with the invariable phase and amplitude varying with time. The kinetic energy is thus expended only for crushing of the barrier material. The analogous hypothesis is presented [15, 16]. The authors assume that in a volume of the matrix material at SDP, long and narrow zones of tensile stress appear, into which the microstrikers are propagated. The SDP model, based on a typical mechanism for a cumulative jet, is offered [17]. According to the authors, new mechanisms of penetration reduce by 90% the known expenditures of kinetic energy.

The authors have developed the concept based on the fourth type of hypothesis. The hypothesis is based on a conception that at the barrier hit by a clot, variable pressure fields are created. In the barrier, the pressure attains the magnitude that is necessary for the dynamic phase transition. In Ref. [18], for the first time, the hypothesis has been presented that penetration of the particles from a clot into the metal barrier at SDP occurs during the period when the dynamic phase transition is not completed. Special features of such a hypothesis were: the barrier material has no long-distance connections, the process duration is limited, and the size and striker velocity and boundaries between various phases are limited too [19]. To confirm this hypothesis, special experiments have been carried out and the time of a dynamic phase transition in the barrier material has been defined. The dynamic losses of the kinetic energy of the strikers were considered. Experimental demonstrations of the implementation of the dynamic phase transition at SDP have been presented much later [20, 21]. The mechanism of the additional energy supplied to the striker [21] has been offered, which increases the efficiency of the kinetic energy used. In the works in Refs. [2, 3, 18, 22], the deficiency of the kinetic energy of a clot for SDP and the possibility of existence of an additional source for energy generation were observed. The possibility of generation of additional energy from the chemical reaction between the introduced substance and the matrix material is reported [23]. The reviews on various attempts to modernize the SDP mechanisms are considered and can be found in the works in Refs. [3, 4, 18].

The new concept of the physics of the SDP phenomenon [24, 25] is based on consecutive implementations of a set of physical effects. Growth of the energy density (energy accumulation) in local zones of the barrier material results from the system closing and the creation of the dynamically stable local zones of high pressure, the level of which is sufficient for a dynamic phase change.

Intensive plastic deformation in the channel elements (high-pressure zones) during the movement of the striker in the barrier material leads to the rupture of long-range connections (loss of static strength) in the material of these zones at the stage of incomplete dynamic changes of a phase. Dense plasma conditions exist in a volume of the channel elements (superplasticity conditions), and microcavities are formed in the channel elements, the collapse of which leads to "hot" points' production and energy generation. The direct and return microjets of the material of the channel elements (dense plasma) are formed, and the strikers are accelerated. A stream of high-energy ions ($\geq$100 MeV) is generated from a hot point of radiation, and a high-pressure field is produced from a hot point, the action of which leads to the expansion of the channel element and extrusion with acceleration of the central zone material of the channel element. Thus, the energy is transferred and the striker is accelerated.

The offered concept predicts the formation of a closed system of the channel elements in the barrier material volume. These elements, at the closing process, absorb the energy of the high-pressure field and in a hot point, the plasma cavity (bubble) collapses with high velocity (at pressure 10^{11}–10^{18} N/m^2). During the cavitation of the plasma cavity (bubble), the energy is generated, for example, in the form of nuclear fusion. The energy is released from a hot point in time due to the emission (radiation). The barrier material generates a wide spectrum of electromagnetic radiation. A hot point is also a source of high-energy ions (energy of an individual ion $\geq$100 MeV). Simultaneously a hot point is a source of high pressure. Expansion of a hot point is impeded by inertia of the barrier material in which microexplosion occurs. When the pressure increases, the dynamic perturbation in the barrier material is absorbed by the adjacent channel elements that are in the opposite phase. In a volume of a primary (initial) channel element, the pressure reduction occurs at the hot point expansion. The pressure drops below the critical value and the process of energy generation stops. The channel element is disclosed when the return jet, originating at the successive cycles of the collapse of the channel element along the depth, passes through the channel zone. When high turbulence appears, additional centers of cavitation can be formed in the material of the channel element (dense plasma). Exchange of pressure occurs between the channel elements. Dynamically stable high-pressure zones (steady-state oscillations of the medium) are important for the energy accumulation in the closed system "barrier—a clot of discrete microstrikers" [26]. In the system, the strikers' material interacts with the material of channel elements. The mass losses of the strikers at their penetration into the barrier material are accompanied by the origins of the moving charges, strong electromagnetic fields, controlling streams of the charged particles, and microjets of dense plasma. Coincidence of the processes of intensive deformations and strains of highly energetic ions in time and space causes an amorphous state of the channel elements' material.

4.3 Investigations of Superdeep Penetration

The first experimental condition of SDP process was formulated. The craters with a depth-to-striker size ratio of above 6–10 are recorded at the collision of the barrier with a stream of strikers having sizes less than 500 microns [3, 4, 27].

The second experimental condition of SDP is the presence of a band of impact velocities. The impact velocity cannot be lower than the velocity of superficial perturbations at the barrier's surface. For the impact velocities higher than the velocity of a shockwave passing in the barrier material, the strikers, at first, are broken according to the known mechanism and only then they are in the SDP regime.

The third experimental condition of SDP is the existence of a stage of preliminary formation of a pressure in the barrier material.

Consideration of qualitative, semiquantitative, and quantitative aspects of SDP allows us to analyze this unusual process. Figure 4.1, [1, 24] shows the schemes of various craters.

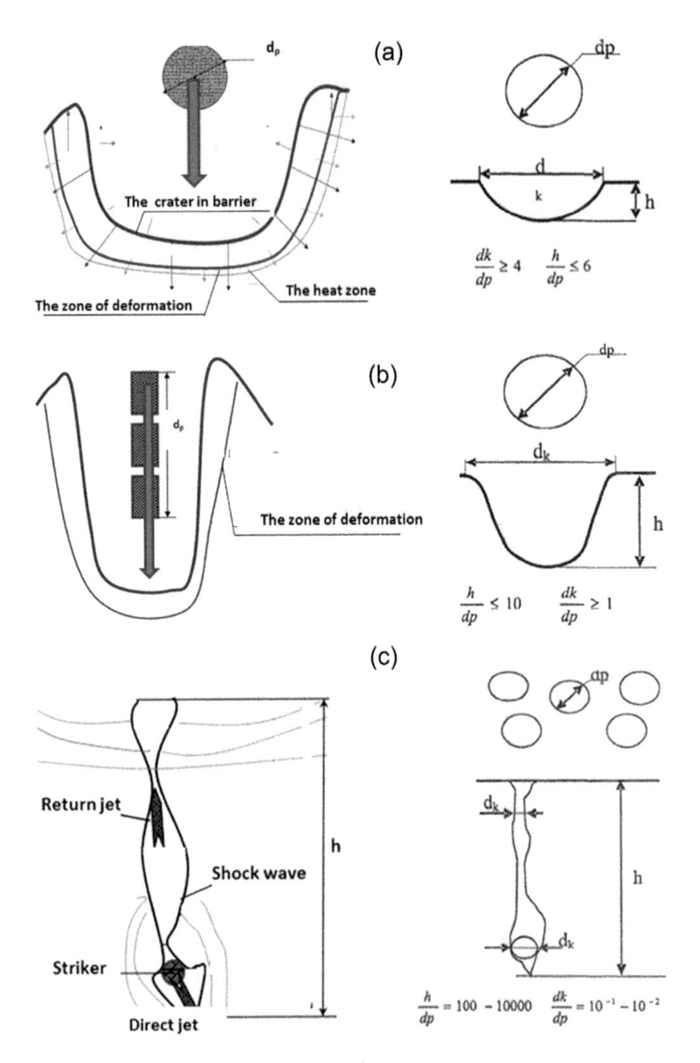

Figure 4.1 Comparison of characteristic features of various impacts with a barrier: (a) usual, (b) anomalous, and (c) SDP (d_p, diameter of a striker; d_k, diameter of a crater; and h, crater depth) [1, 24].

For SDP, the hardness of the striker material does not affect essentially the penetration depth. In Fig. 4.2 [1, 24], the zone of retardation of a striker, made of highly plastic material (lead), is shown. By comparison with usual and abnormal craters it becomes

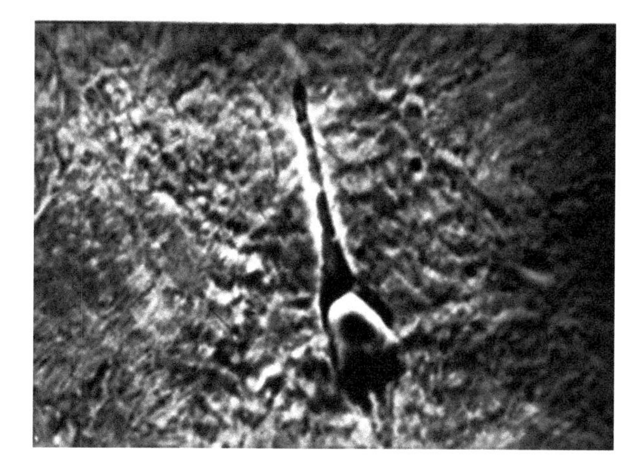

Figure 4.2 Rest of a lead particle in a steel barrier, ×5000 [1, 24].

obvious that for SDP, the visible diameter of the channel is always smaller than the initial size of the striker (Figs. 4.1 and 4.2).

The diameter of the channel is called the transverse dimension of a penetration trajectory (penetration zone), which can be seen after polishing and etching processes of a metallographic specimen.

By using, as a striker, very hard ceramic (VC) particles, the same symptoms of superdeep channel formation were observed (Fig. 4.3 [1, 24]).

A similar character of channels formation (zones) at macroimpact is observed for a shot into an elastic material, for example, into rubber. In this case, the zone of a puncture (channel) completely collapses under the action of pressure forces. The fragment we can see as the longitudinal cavity (Fig. 4.3) has appeared as a result of etching solution influence on the activated iron zone. Using the solutions of acids and alkalis of different concentrations, it is possible to gain various diameters of a puncture zone (channel).

Traces of a bullet (copper or iron) in rubber after its penetration are displayed only on the axis of its movement. On the basis of this similarity, it is possible to explain experimental results of microalloying of the barriers at SDP.

At SDP, the channel collapses (closes). It is possible to confirm this assertion in the experiments with lead dust. Lead (Pb), injected

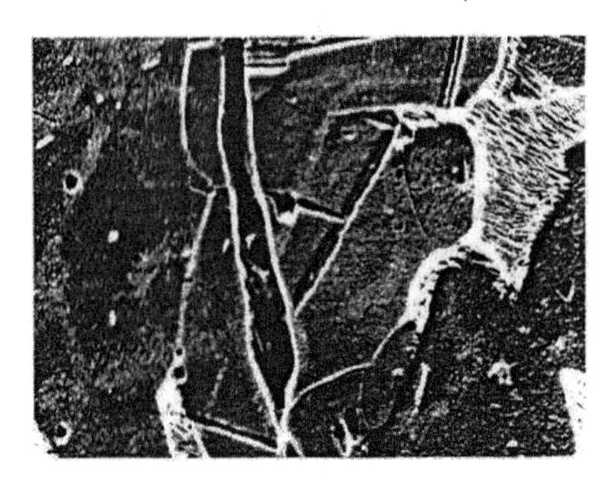

Figure 4.3 Channel structure in an iron barrier, generated by VC particles, ×5000 [1, 24].

into the channel zones, at SDP, during preparation of a specimen for investigation, reacts with the etching solution. However, the products of a chemical reaction do not leave the surface.

The trajectory of motion of lead particles in a barrier in the form of white strings can be seen at a metal surface of a barrier (Fig. 4.4 [1, 24]). Only those sections of a trajectory that are at the plane of a longitudinal cross section of a barrier are visible.

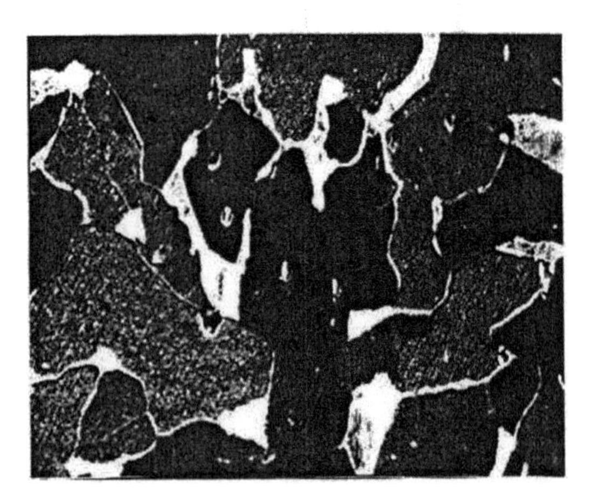

Figure 4.4 Trajectory of the motion of lead particles in a steel barrier, ×1000 [1, 24].

We treat similarly the process of punching a rubber barrier by bullets. On this basis, we try to answer the question, what causes the collapse of a channel at SDP? Strong compressive stresses appear in the volume of the barrier's material at SDP. What is the reason for these stresses? At usual impact, compression forces (pressure) are not significant and the crater diameter exceeds the striker caliber more than four times. An individual striker cannot give such results.

It is obvious that when a clot of dust particles affects a barrier, the pressure fields appear. The pressure fields in a barrier originate due to the kinetic energy of the impact. Emergence of a variable pressure field explains the presence of velocity gradients and density in the volume of a clot (stream) of separate strikers. After the striker starts to create a channel element, just in this zone the reduction of a pressure field occurs. The channel element closes. Therefore, the presence of a pressure field in the volume of a barrier is the necessary condition of SDP [2, 3].

Studying these questions has led to a formulation of the second condition of SDP implementation. SDP of strikers into a barrier can occur only at the impact of a clot (stream) of microstrikers. The SDP process never occurs for a single particle.

4.4 Effects Appearing at Superdeep Penetration

When a clot of strikers hits the barrier, a variable background pressure is generated in the barrier's material. During the interaction between the strikers and the barrier, separate strikers penetrate into the medium, loaded preliminarily by the hit of other strikers that did not penetrate the barrier but that transferred their energy (pressure) into it. Thus, when the depth of striker penetration is bigger than the striker size, the crater cavity, behind a striker, collapses under influence of the generated background pressure. Since this moment, a qualitative change of the energy reduction has started. During the interaction process, lasting 10^{-8}–10^{-5} s, the power consumption of the closed system sharply increases. As a result, the unusual effects that are characteristic for the SDP conditions should occur.

4.4.1 Interaction Conditions Determining the Cratering Type

Typical cratering for $h < 6d_p$ is observed under the following conditions:

- The crater cavity (vessel) is open.
- The compression of crater walls does not lead to crater collapse and to jet formation.
- The expansion of crater walls is significant.

Anomalous cratering for $6d_p < h10d_p$ takes place under the following conditions:

- The crater cavity (vessel) is opened.
- Collapses and formation of both direct and reverse cumulative jets (mainly reverse ones) occur.
- The expansion of walls is insignificant.

SDP for $10^2d_p < h < 10^4d_p$ occurs when the following conditions are fulfilled:

- The crater cavity (vessel) is closed.
- Collapse and formation of both direct and reverse cumulative jets (mainly direct ones) occur.
- The expansion of walls is insignificant.
- The background pressure P_b operates in the barrier and the material strength σ_m in crater's region is low ($\sigma_m < P_b$).
- The duration of the background pressure τ_b is equal to, or larger than, the crater (channel) formation time $\tau_c(\tau_b \geq \tau_c)$.

As a result of severe deformation and superposition of variable loads of the crater in the cratering process, the crater's material along the crater walls and its bottom loses its crystalline structure and has no static strength [2, 3]. Thus, to change typical cratering into an anomalous one, the following conditions should be met:

- The penetration depth into the barrier $h > d_p$.

- At the wall's compression, the material strength σ_m should be lower than the wall compression pressure P_c (elastic aftereffect).
- The compression time τ_e (elastic aftereffect) should be longer than the time of wall's compression.

An analysis of the experimental conditions, necessary for anomalous cratering, has shown different possible variants of these conditions' fulfilling.

The use of single microstrikers [3] causes shortening of the time of the wall's compression. When a stream of separate strikers is used, the background pressure is generated and the compression time τ_e is longer than the time of the wall's compression.

To change anomalous cratering into the SDP, it is necessary to close a cavity and to ensure $P_b > \sigma_m$ at $\tau_b \geq \tau_c$.

- The first effect of penetration at the SDP, due to the presence of a background pressure, is closure of the crater (channel cavity) on the whole distance of the striker movement in the barrier. As a result, the formation of a superdeep channel does not cause any loss of air tightness of the metal barrier.
- The second effect is no direct dependence of the initial hardness of the striker material on the penetration depth. It means that the hardness of the barrier material, in the SDP zone, is always lower than that of the striker hardness made of any known material. Such an effect proves that SDP transforms the barrier's material, in a local zone (during interaction $\Delta\tau < \tau_c$), into a dense plasma state. The striker's material, which is affected during penetration for the long time τ_c, is in a solid or a liquid state [2, 3, 5, 27].
- The third effect is nonuniform distribution of the pressure fields in the barrier's material. Penetration of the strikers occurs in these zones of the material where the pressure is at least of the order higher than the background pressure in the barrier. The pressure fields are mainly in long and narrow zones on the whole barrier thickness. At the beginning of the SDP investigations, this effect has been

formulated as a modeling assumption [2, 3, 27], and then it has been confirmed experimentally [18].

- The fourth effect, which directly results from the third effect, is strong local deformation in the zones of the barrier material [2, 5, 27].
- The fifth effect is the loss of the striker's mass at the increasing depth of the striker penetration into the barrier. During the striker passage through the barrier, its initial size decreases by hundreds of times. Decrease in the size and mass of the striker is nonuniform along the penetration depth, which proves the changes of SDP conditions in the barrier [3, 4, 27].

Additional effects arising at SDP, as a result of the collective action of all the before-determined conditions are appearance of electric charges at the interaction of materials of the striker and the barrier and appearance of a wide spectrum of electromagnetic fields during the movement of a clot of strikers in the metal barrier [19, 20].

Simultaneously, in the volume of the barrier material, in the point sources the flows of massive charged particles, apparently ions, occur. The energy of such particles is so high that they can pass (starting from "a hot point") through the barrier material. The experiments employing additional filters (protection screens situated at the particles' exit from the barrier) have shown that the energy of the particle at the barrier's surface (after its passage through the barrier material) was 100 MeV.

On the back surface of the barrier, at the strikers' exit, microjets of dense plasma appear. These jets possess high penetrating power. Their velocities attain hundreds of meters per second. The diameter of such a jet is not larger than 1–2 μm, and its length does not exceed some millimeters. The jets have the charges and interact with the electromagnetic fields.

Interaction between the striker and the barrier materials, in local zones (channel elements), leads to the appearance of chemical elements [8] that were absent early from this material, and it causes a synthesis of metastable compositions [27], which are not shown in the known constitution diagrams. It can be explained by simultaneous action of high pressure, intensive deformations, and radiation in the interaction zone [5, 8]. Neither mechanical nor hydrodynamical factors of SDP have essential influence on a situation inside the metal barrier.

4.5 Expenditure of Energy in a Process of Superdeep Penetration

The energy expended in the process of SDP was estimated using a principle of minimization of energy expenditures. Within this approach, at each analysis stage, the minimal possible estimation of the energy expenditures was accepted. When the energy expenditures were too high, even using a minimization principle, the assumptions for reuse of the spent energy were offered. For the calculations we have used the results obtained experimentally. To compare the introduced and the spent energy, the following approach has been accepted. For calculation of the kinetic energy, injected by a clot of high-speed particles, the overestimated assumptions were taken, and for the calculation of the expenditures of the energy, implemented during the loading, only the assumptions obviously underrating the energy values were accepted. Such an approach allows us to focus on qualitative aspects of the SDP process (Fig. 4.5).

From calculations [24] it follows that 90–98% from the general expenses of energy for crater formation is spent for overcoming of static durability. It is possible to admit that the barrier material at a combination of some parameters loses static durability. By estimations only dynamic losses of energy on the formation of channel zones and channels shutting after penetration of microparticles exceed kinetic energy of impact. At SDP other high-energy effects are realized also.

4.5.1 Clot of High-Speed Microparticles: Estimation of the Kinetic Energy

For the calculation of a quantity of the energy gained by the metal barrier as a result of the shockwave loading with a high-velocity stream of microparticles, the following values of parameters were

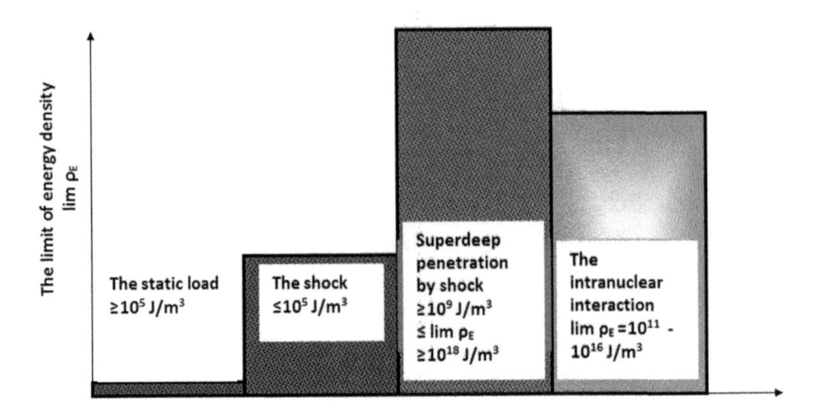

Figure 4.5 The diagram of the density of energy of various processes (depending on the speed of interacting [1]).

used: the mass of the ejected (driven) material $m_1 = 0.1$ kg, the mass of the explosive charge $m_2 = 0.2$ kg, and the velocity of the microparticles' clot = 1000 m/s. For these parameters, the kinetic energy of the ejected (driven) material was $1.5 \cdot 10^5$ J.

At the macroimpact, restriction of a relative depth of the penetration is caused by the fact that the impact energy is extended for overcoming the static impact resistance (90–98%) and the dynamic component of the resistance causes the energy loss of 2–10%. The dynamic expenditures of the energy originate at the transfer of the striker and barrier materials with some velocity, during ejection of the barrier material from the crater and during the movement of the material of the cavity walls and its bottom [3, 18]. In addition, the dynamic expenditures of the energy at usual impact include the energy expenditures for the dynamic settlement and macrocrater formation.

In Ref. [18], the energy expenditures on macrochanges of the barrier are estimated. The observable changes of the barrier's geometry are shown in Fig. 4.6 [1, 18].

As a barrier material, the cylindrical samples, made from the alloy of iron with 0.4% of carbon, having the diameter of 50 mm and the height of 100 mm, have been taken. As the striker materials, microparticles of SiC powder of the fraction of 63–70 μm were used. For this case, the energy expenditure for the macrocrater

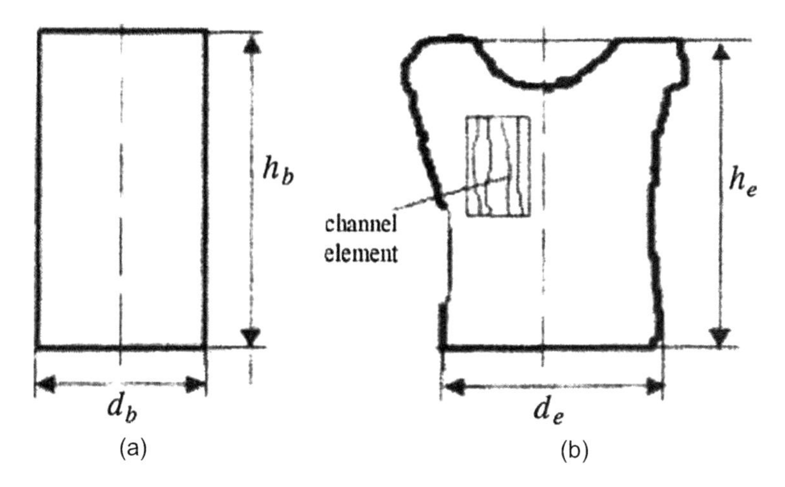

Figure 4.6 Changes in the metal barrier, arising as a result of its impact with a clot of particles: (a) initial cylindrical barrier (h_b, d_b) and (b) the barrier after the dynamic loading (h_e, d_e) [1, 18].

formation was $E_k = 2548$ J and the energy expenditure for the dynamic settlement was of the minimum value $E_d = 401{,}821$ J.

4.5.2 Formation of a Channel Structure during the Superdeep Penetration Process

In a steel sample, the zones (channel elements) with the changed structure have appeared. Figure 4.7 shows the scheme of such a channel element (zone). In the channel element zone, the number of defects in the material sharply increases. The maximum number of the structure defects is observed in the central zone of the channel element (Fig. 4.7) [1, 24].

When the special alloying additions are not used, the activated central zone completely disappears (is etched) during metallographic examinations. However, with the decreasing intensity of chemical or electrochemical effects, this zone is filled with a specific material. This specific material is the product of interaction between the injected substance and the matrix material of the barrier.

Two kinds of channel elements can be observed. The first kind represents an element of a crack type with the striker fragments

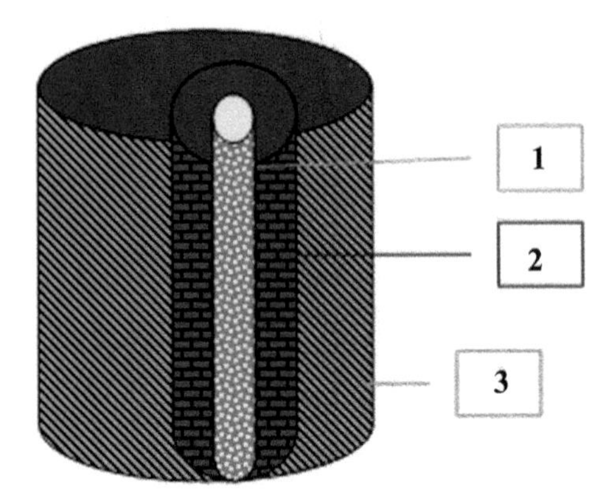

Figure 4.7 Scheme of a channel zone: 1, amorphous material; 2, microcrystalline material; and 3, defective material [1, 24].

fixed in it. These fragments can be also observed far from the axis of the channel element. As a rule, such kind of channel element is observed when the striker's substance is a fragile ceramic material. For example, silicon carbide (SiC) undergoes fragmentation in a pulse regime when it is heated to high temperature. Penetration of the fragments from the central zone of the channel element to other zones of the barrier material occurs in a pulse regime, that is, it has an explosive character.

The second kind of channel element is characterized by the maximum number of defects in the central zone. As a rule, the material injected by a clot of separate strikers has higher plasticity when compared with the ceramic's plasticity and it intensively interacts with the matrix material. In this case, the material of the central zone of the channel element was a metastable composite of various concentrations of both injected and matrix substances. For such kind of channel element, traces of intensive plastic deformation can be observed. The amorphous state of a thin structure of the central zone was stated due to the study of this structure by means of a transmission electron microscope.

In a lot of experimental investigations, nano- and microstructures were recorded in the central zone of the channel element. It is logical to assume that a higher number of defects appear in the narrow zone of the channel material. In our opinion, the existence of various channel elements testifies to various regimes of SDP.

In such an approach, the difference between two kinds of channel elements appears because of various regimes of the removal of the central zone material from the channel element. On the basis of the study of the first kind of channel element, it is possible to observe that owing to the relaxation processes in a macrovolume of the barrier material, the number of defects decreases. However, it is necessary to consider the real time of implementation of the process of the channel element formation (10^{-8}–10^{-5} s) and intensity of the process of energy reduction in a volume of the barrier material (metal hardening). In our opinion, more logical is the assumption that the part of the defective material being formed in the narrow central zone of the channel element is ejected with the direct and return jets from a solid macrobody. Thus, the difference between the first and second kind of channel element formation results from the mass and velocity of the jets of the ejected defective material. On this basis, it can be stated that the closing (collapse) process of the channel elements is of significant meaning.

For estimation of the energy expenditures on the formation of a zone with the channel microstructure, the following standard assumptions are used: the static resistance of the barrier material is equal to null, and the velocity of the moving walls of the channels is equal to the velocity of an individual microstriker.

As a result of the channel element formation, energy is spent on disclosing of the channels, in the direction perpendicular to the axis of the microstrikers' movement, and on their collapsing (collapse of cavitation bubbles) after the particles' passage.

Let us estimate the SDP process by employing the energy of the dynamic phase changes. We will consider a case of injection of a clot of TiB_2 particles in the iron barrier. Also, we will estimate the minimal volume V_{iz} of the channel element of the barrier, in which a change of the phase proceeds. It results from the experimental data that the residual part of the striker (striker rest) has the size smaller than or equal to 0.05 d_p.

The experimentally obtained limiting depth is $h = 0.3$ m. We will assume that the minimal channel volume of the barrier has the cone form. The cone diameter equals $66.5 \cdot 10^{-6}$ m (a cross section of the striker), and its height is 0.3 m. In this case, the mass of the barrier material that moves in a dynamic regime will be also underrated because, in a deformation process, the collapse of nearby layers is not taken into account. Then:

$$V_{iz} = \pi h d_p^2 \frac{1.0525}{12}, \tag{4.1}$$

where h is the barrier (zone) thickness and d_p is the striker diameter.

The cone volume is $V_{iz} = 0.347 \cdot 10^{-9}\ \mathrm{m}^3$. The material volume V_f, in which a dynamic change of a phase proceeds, will be determined by a ratio of the kinetic energy E_p and the pressures of initiation of the dynamic change of the phase P_i:

$$V_f = \frac{E_p}{P_i} = 12.5 \cdot 10^{-6}\ \mathrm{m}^3 \tag{4.2}$$

Essential nonuniform distribution of the pressure fields in local zones of the metal barriers and behavior of the phase changes occurring in them have been proved. In particular, $\alpha \rightarrow \varepsilon \rightarrow \alpha$ transformations are shown using low-alloy steel as an example. The high-pressure phase in them is the ε phase. The pressure for initiation of the $\alpha \rightarrow \varepsilon$ transition is 12 GPa [19]. The phase ε is not preserved after the loading removal, but the traces of $\alpha \rightarrow \varepsilon \rightarrow \alpha$ cycle are registered after the microstructure change. The transverse size of the zones, being tested in a cycle of transformation, changes in a wide range, from a fraction to 1–2 μm [24].

The quantity of the calculated high-pressure zones, formed due to the impact energy of a strikers' clot, is:

$$N = \frac{V_f}{V_{iz}} = 36{,}000\ \text{pcs}$$

However, the channel elements are not distributed uniformly along the crater depth.

For the determined conditions, the penetration velocity v_p is:

$$v_p = v_s \frac{\lambda}{1+\lambda},$$

where

$$\lambda = \sqrt{\frac{\rho_s}{\rho_m}} \text{ and } \lambda = 0.73535.$$

Then, $v_p = 127$ m/s. The average penetration velocity is $v_{pm} =$ 423 m/s.

The minimal movement velocity in the metal barrier is $v_p =$ 127 m/s, and the length of a clot of discrete strikers (distance for acceleration) is 0.15 m. The strikers' clot moves to the barrier with the velocity of 300–3000 m/s. Due to it, the loading by a jet is realized in the time interval from $5.0 \cdot 10^{-5}$ to $5.0 \cdot 10^{-4}$ s, that is, during $\tau_l \approx 450$ μs. As a clot material we use TiB_2 (titanium boride) particles of the density of $4.38 \cdot 10^3$ kg/m^3 at the melting temperature of 2790°C and the microhardness $H\mu = 3370 \pm 60$ kg/mm^2 [18].

The striker moving with the maximum velocity penetrates into the cylindrical barrier during $\tau_{p1} = 2.36 \cdot 10^{-4}$ s. The striker with the minimum velocity could penetrate (but it is impossible) during $\tau_{p2} = 23.60 \cdot 10^{-4}$ s. The striker, driven with the average velocity ($v_{pm} = 423$ m/s), could penetrate into the barrier only during the time $\tau_{pm} = 7.0 \cdot 10^{-4}$ s. Hence, the strikers having average and lower velocities during the loading can penetrate the barrier only to the depth of 0.2 m. It corresponds to the experimental results (Table 4.1 [1, 24]) obtained for the tested steel barrier high strength steel (HSS) with TiB_2 particles. With the increasing barrier depth, a part of structure defectiveness has decreased by 246 times. The density of the channel elements has decreased most intensively (2.8 times), and the average (visible) diameter of the channel has increased.

The quantity of channel elements on the penetration depths from 0 up to 0.2 m is $N_{0-2} = 4.8 \cdot 10^5$ pcs and $N_{2-3} = 1.7 \cdot 10^5$ pcs. The volume of the calculated individual channel element on the penetration depths from 0 up to 0.2 m is $V_{z(0-3)} = 347 \cdot 10^{-9}$ m^3, and the volume of the calculated channel element on the depths from 0.2 up to 0.3 m is:

$$\Delta V = V_{z(0-3)} - V_{z(2-3)} = 0.192 \cdot 10^{-9}\ \text{m}^3$$

The volume of the high-pressure zones is:

$$V_{zc} = N_{0-3} V_{iz} + (N_{0-2} - N_{0-3})\,\Delta V = 0.119 \cdot 10^{-3}\ \text{m}^3$$

The number of cycles of the high-pressure zones' appearance is:

$$N_{cf} = \frac{V_{zc}}{V_f} = 9.5.$$

Then, a part of the kinetic energy injected during one cycle into the barrier material amounts to ~10% of the energy that is required for

Table 4.1 Parameters of a structure of the barrier material (HSS steel)

Barrier zone	Striker size d_p	Structure defectiveness	Average (visible) diameter of a channel	Density of channel elements	Volume of defective structure V_d	Mass of defective structure M_d
m	10^{-6} m	10^{-3}%	10^{-6} m	mm^{-2}	10^{-8} m^3	10^{-3} kg
Depth 0–0.2 m	0–60	6.38	0.576	245	2.504	0.2028
Depth 0.2–0.3 m	0–60	0.0259	0.616	87	0.508	0.0411

Source: [1, 24]

the loss of static strength (for creation of the high-pressure zone):

$$E_{\Sigma f} = \Sigma E_p = 1.42 \cdot 10^6 \text{ J}$$

This unusual result was obtained without taking into account the energy expenditures on the channel's collapse and in other processes.

Prom the principle of minimization, it is accepted that the high-pressure regions originate 9.5 times in a pulsing regime of the metal cylinder volume.

However, the time of the barrier loading at the impact with a clot of separate strikers $\tau_l \approx 4.5 \cdot 10^{-4}$ s cannot be neglected. The SDP process can be realized only during the barrier loading with a particles' clot. Therefore, the particles with the minimal velocity cannot penetrate through the whole barrier thickness during the loading process. Then, the time of an individual cycle of a high-pressure action can be estimated as:

$$\tau_{imp} = \frac{\tau_{loading}}{N_{cf}} = 47.2 \cdot 10^{-6} \text{ s}$$

For this penetration time, the penetration depth is $H_{i1} = v_p$ and $\tau_{imp} = 6.0 \cdot 10^{-3} \div 60.0 \cdot 10^{-3}$ m. The average penetration depth for one pulse is $H_{medium} = 20.0$ mm.

During the first pulse, it is impossible to create the channel elements on the whole depth of a metal cylinder. Thus, on the basis of the gained estimations of an average velocity, the number of channel elements on the depths up to H_{medium}, formed during the time τ_{imp} can be determined. A volume of an individual channel's element, from a surface up to the depth H_{medium} is defined from the difference of two cone volumes. The first cone has the base diameter $D_1 = 66.5$ μm and the height $h_1 = 0.3$ m. The second cone has the base diameter $D_2 = 62.0$ μm and the height $h_2 = 0.280$ m. Then, if $V_f = 12.5 \cdot 10^{-6}$ m^3, the exact quantity of the channel elements of the first cycle is $N_{amend} = \frac{E_f}{\Delta V_{1-2}}$ =190,840. The quantity of the channel elements, registered experimentally is $N_{0-2} = 480{,}812$. During the first cycle, the high-pressure zones, corresponding to the channel elements number N_{amend} are formed on the depth of a metal cylinder. Other channel elements are formed in the next cycles. During ~2.5 cycles, all the channel elements are formed on the distance between the surface and the 20 mm crater depth. What will happen next? Will the channel elements exist (pulsate) after $\tau_{imp} = 47.2 \cdot 10^{-6}$ s.

If the channel elements are closed after the striker passage and the further pulsation stops, the return jet is sharply broken. Correspondingly, the blocking of the outlet holes cannot affect the mass transfer process in the SDP regime. Such an assumption contradicts the known experimental results [4].

We accept the underrated assumption that opening of the channel elements proceeds with the velocity equal to the penetration velocity of the strikers into the barrier. In this case, the energy is spent only on the movement of the mass of high-pressure zones with the penetration velocity. Static resistance of the striker material equals null. Then, energy expenditure for the channel elements' opening E_{op} can be defined from Eq. 4.3:

$$E_{\mathrm{op}} = \frac{M_{\mathrm{f}} V_{\mathrm{pm}}^2}{2} \tag{4.3}$$

where $M_{\mathrm{f}} = N_{\mathrm{exp}} V_{\mathrm{iz}} \rho_{\mathrm{m}}$. Estimating this expenditure, with the average values, we will gain $E_{\mathrm{op}} = 49{,}366$ J.

Let us estimate the dynamic expenditures of the energy on the channel's closing. We accept the mass of the barrier material, which moves at the opening and a collapse of the channel, as equal to the cone mass. In this case, the size of the barrier material mass that moves in a dynamic regime will be also underrated if the dynamic mass transfer of a cone material, without involving the nearby layers to a cone, is supposed in the collapsing process. A collapse velocity of the channel element can be calculated according to Eq. 4.4 [5]:

$$V_{\mathrm{com}} = \sqrt{\frac{2E_{\mathrm{s}}}{3M_{\mathrm{f}} - 2M_{d1}}}, \tag{4.4}$$

where E_{s} is the energy spent on the channel's collapse.

As the barrier can be divided, along its thickness, into two zones, that is, from 0 up to 0.2 m and from 0.2 up to 0.3 m, the data can be divided into two parts. The volume of an individual cone of the high-pressure zone is $V_{z(0-3)} = 0.347 \cdot 10^{-9}$ m^3 (the zone 0–0.3 m). A part of the cone in the zone 0.2–0.3 m (diameter 44.33 μm and length 0.1 m) will have the volume $V_{z(2-3)} = 0.154 \cdot 10^{-9}$ m^3 and

$$\Delta V = V_{z(0-3)} - V_{z(2-3)} = 0.192 \cdot 10^{-9}\ \mathrm{m}^3.$$

To neglect the influence of the static strength of the barrier material, we have made an assumption that in the cone volume, the high-pressure $\Sigma E_p = 1.42 \cdot 10^6$ J is produced. For the channel elements' opening, the energy $E_{op} = 49{,}336$ J is spent. Hence, the residual energy is:

$$\Delta E = \Sigma E_p - E_{op} = 1.37 \cdot 10^6 \text{ J}$$

Taking into account characteristic properties of the zone structures, this energy is distributed proportionally to the volumes of the high-pressure zones:

$$\Delta E = \Delta E_{0-2} + \Delta E_{0-3}$$

The estimated parameters of the collapsing channel elements are shown in Table 4.2 [1, 24].

The energy expenditure on the channel elements' collapsing in the barrier is:

$$E_{com} = 459{,}245 \text{ J} \quad \text{and} \quad E_{com} + E_{op} = 508{,}611 \text{ J}$$

Table 4.2 Parameters of a collapsing process of the channel elements

Zone depth	Mass of material under high pressure M_f	Mass of defective residual M_d	Collapsing velocity (v_{com})	Energy (E_{com})	Quantity of channel elements
m	kg	10^{-3} kg	m/s	J	Pcs
0–0.2	0.751	0.2028	824	254,955	480,812
0.2–0.3	0.213	0.0411	1385	204,290	170,737
00.3	0.964	0.2439	976	459,245	

Source: [1, 24]

4.5.3 Change of the Barrier Microstructure

In the barrier volume, a dislocation pattern characteristic for usual explosive loading appears. The changes of geometry and sizes of the grains and their twinning are observed.

The experiments with the explosive compression (in a cylindrical scheme) have shown that the observable structural changes in the case of explosive compression and for loading of the particles' clot give analogous patterns. The total energy of the explosive charge used for the preparation (sample) compression was

$E_{com \cdot ex} = 594 \cdot 10^4$ J. We have accepted that the energy spent on twinning, change of geometry, and sizes of the grains was only 5% of the total compression energy $E_{com \cdot ex}$. Thus, the energy spent on these microstructural changes is $E_{ex} = 29.7 \cdot 10^4$ J [22].

4.5.4 Other Factors Causing Energy Expenditures

Other factors causing the energy expenditures are formation of microjets of dense plasma at the back surface of the barrier [24, 25], pulse electromagnetic radiation, and formation of the streams of high-energy ions [24]. Because now there are no reliable quantitative data concerning these processes, thus, according to a principle of minimization, the energy expenditures on these processes were neglected.

4.5.5 Energy Balance

The introduced and spent energies will be compared:

$$E_p = E_a, \tag{4.5}$$

where E_p is the energy introduced during superdeep penetration (SDP) and E_a is the energy spent on the SDP process.

We consider the expenditures of energy on:

- Formation of the macrocrater ($E_k = 2548$ J)
- Barrier settlement ($E_d = 401{,}821$ J)
- Changes of a structure of the barrier material ($E_{ex} = 297{,}000$ J)
- Production of the high-pressure zones ($E_{\Sigma f} = 1.42 \cdot 10^6$ J)
- Opening of the channel elements ($E_{op} = 49{,}336$ J)
- Collapse of the channel elements ($E_{com} = 459{,}245$ J)

We also consider other expenditures of energy, for example, on radiation E_{ad} and jets formation.

The sum of the rated expenditures

$$E_a = E_k + E_d + E_{op} + E_{com} + E_{ex} + E_{\Sigma f} \approx 2.63 \cdot 10^6 \text{ J}.$$

However, we will apply a principle of minimization of energy expenditures. According to this principle, it is possible to neglect the expenditures of energy on formation of the macrocrater E_{k}, settlement of the barrier E_{d}, and the changes of a structure of the barrier material E_{ex}, since the energy generated in these processes could be repeatedly used in the processes of the channel formation $(E_{\mathrm{op}} + E_{\mathrm{com}})$.

Accordingly,

$$E_{\mathrm{op}} + E_{\mathrm{com}} > E_{\mathrm{k}} + E_{\mathrm{d}} + E_{\mathrm{ex}}.$$

However, as it has been shown earlier, the processes of the channel element formation can occur as a result of transformation of the energy of high-pressure fields:

$$E_{\Sigma\mathrm{f}} \geq E_{\mathrm{op}} + E_{\mathrm{com}}$$

We assume that E_{ad} results from the transformation of the energy $E_{\mathrm{op}} + E_{\mathrm{com}}$. Therefore, the minimal necessary energy expenditure for the process of SDP is assumed as

$$E_{\mathrm{a}} = E_{\Sigma\mathrm{f}} = 1.42 \cdot 10^{6}\ \mathrm{J}.$$

Thus, in dependence on the extent of the use of the minimization principle,

$$\frac{E_{\mathrm{p}}}{E_{\mathrm{a}}} = (5.6 - 10.5) \cdot 10^{-2}, \text{ that is, 5–10\%}.$$

Because for estimation the approach of understating of the expenditures of the energy and overestimation of the power consumption has been accepted, the possible mistakes in the given approach cannot be significant. It is obvious that even 5–10 times mistake in calculation of the energy expenditures does not change qualitatively the energy balance at the SDP. The probability of the following assumption is high:

$$E_{\mathrm{p}} + E_{\mathrm{unk}} = E_{\mathrm{a}}, \tag{4.6}$$

where E_{unk} is the additional energy emitted during the interaction process.

In this case, even the additional expenditures of the energy for radiation, heating, and emission of the barrier material in form of microjets can be compensated:

$$E_{\mathrm{p}} + E_{\mathrm{unk}} = kE_{\mathrm{a}}, \tag{4.7}$$

where k is the coefficient that does not include the energy losses during the SDP process.

4.6 SDP Method for "Green" Production Technology of Nanocomposites

Capabilities for obtaining a new material are determined by processing conditions. To observe a narrow group of composite materials—fiber materials—the manufacture process can be divide into three main stages. Selection and preparation of a large matrix material concerns the first stage. Selection and manufacture of filaments concerns the second stage for composite material reinforcement. The process of assembly of a composite material from the elements created during the first two stages of the process concerns the third stage. At each stage of the manufacture process of a composite material there are specific engineering and scientific problems. The main problem in composite material creation, as a rule, is essential contradictions between engineering decisions that are used for various stages of the process.

The effective solution for a problem of creating a fiber (reinforcing) material with a reinforcing skeleton is making of a frame material from a matrix material with a structure on the micro- and nanolevel. The material at the micro- and nanolevel has properties that essentially differ from properties of the same material at the macrolevel. The effective solution is a composite material with the macrozones fabricated at various structural levels (Fig. 4.8).

The composite material can be produced to reinforce a macrostructure device with nano- or microparticles.

SDP allows the introduction into the volume of a solid body of any inorganic substance (Fig. 4.9) and to change the structure of a solid body material (Fig. 4.10).

4.6.1 Features of Dynamic Reorganization in Steel at Superdeep Penetration

A qualitative difference of the production of composite materials through the use of SDP from traditional powder metallurgy is that the basic physical tool is the high-speed clot of discrete particles (powder). At SDP, a matrix is the massive high-strength material. At a loading in static conditions in a solid body, a uniform field of

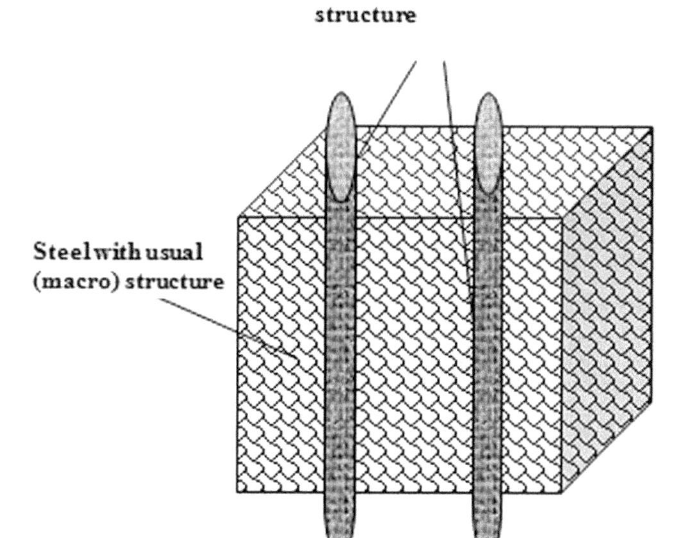

Figure 4.8 The view of a new composite material [1].

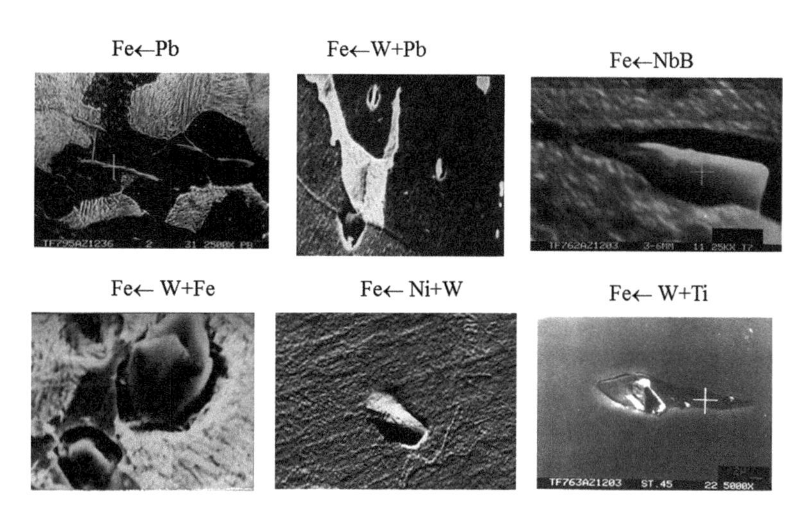

Figure 4.9 Microslices of the steel samples after SDP treatment by different kinds of strikers [1].

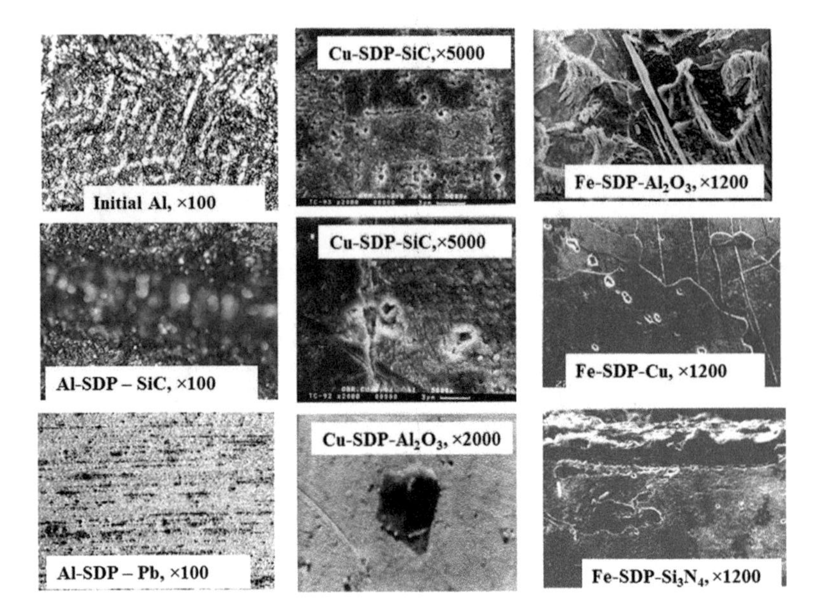

Figure 4.10 Microslices of the metal samples after SDP treatment by different kinds of strikers [1].

pressure is produced. Pressures in static conditions are created up to $\leq 10^5$ N/m^2. On the basis of this information it was assumed that at the distribution of shockwaves and deformation waves into a solid body, an approximately uniform field of pressure ($\leq 10^9$ N/m^2) will be produced.

Let us consider a real experiment on shock interaction (Fig. 4.11).

Registration of an area of high pressure will execute due to the changes of the structure and physical and chemical properties of a barrier material. For the approach to a real geometry blow the surface of the steel sample was deformed preliminary by a steel sphere with a diameter of 30 mm. Clots of steel powder particles were used as strikers. After the processing of steel preparation (concentration $C \leq 0.45\%$) a barrier was cut in a longitudinal plane and the cut surface was polished and etched with a nitric acid solution (Fig. 4.12 [1, 28]). The photo of areas of a high pressure (B_1) is shown in Fig. 4.12a. The background level of the pressure in the massive steel barrier (B_2) was $\leq$0.2–1.0 GPa. Thus, in a dark area

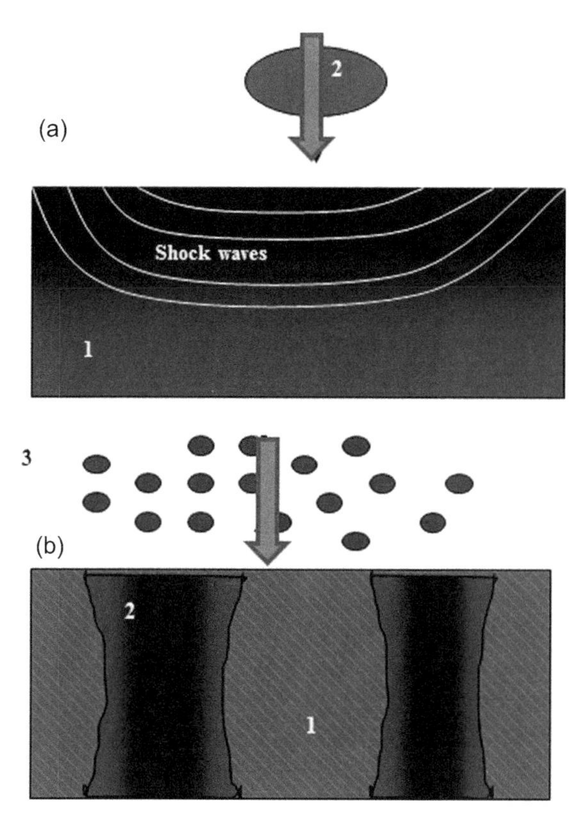

Figure 4.11 Formation of macrozones of high pressure: (a) usual impact (1, barrier; 2, striker); (b) SDP impact (1, zone of high pressure; 2, zone of low pressure; and 3, disperse striker) [1].

of a section of the barrier (B_1), the level of pressure of more than 8–12 GPa is obtained.

Between the macroareas of high and low pressure (B_1–B_2), the sharp border is visible (Fig. 4.12b). Therefore, in a metal solid body, at the action of clots of dust particles (in SDP mode), simultaneously there are the macroareas of both low and high pressure. The high-pressure area in a solid body volume is surrounded by an area of low pressure. In Ref. [13] it is foretold that at SDP in a solid body there is a steady wave—a soliton. Experimentally such a soliton corresponds to the pulsing area of high and ultrahigh pressures. The length of the area of a high pressure (soliton) corresponds to the thickness of a

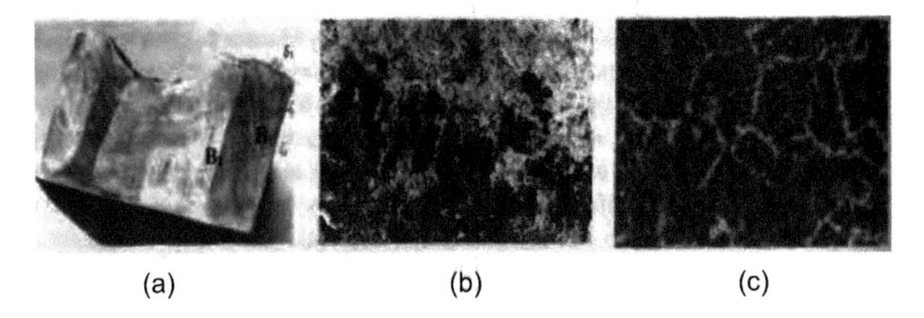

Figure 4.12 Distribution of different fields of pressure at impact: (a) ×0.5; (b) boundary between sections of high and low pressure, ×200; and (c) crushing of structure elements into the area of the high pressure of steel barrier at heat (1000°C, 1 h), ×200 [1, 28].

barrier. Tracks have been found in the areas with various levels of pressure It, apparently, proves that the narrow area of high pressure (soliton) arises as an independent object.

A wide area of the high pressure (B_1) was formed on the combination of a considerable quantity of high-pressure narrow oscillating solitons. Therefore, the area B_1 has a high level of defectiveness. For the proof, after SDP processing, a metal solid body was heated up at the temperature of 1000°C within 1 h. After that the changes in structures of areas of high and low pressures were compared. In a high-pressure zone, at the subsequent heating, the set of dot defects has arisen, grains were split up, and numerous centers of recrystallization appeared (Fig. 4.12b,c). Growth of grains (Fig. 4.12c) occurred only after an additional stage of allocation of new structural defects (recrystallization). Therefore for an increase in the sizes of grains, in a high-pressure zone at the subsequent heating, additional time is required. In the field of low pressure, the grains of a barrier material at heating increase in size faster because there is no stadium of additional structural defects. Therefore, stability on heating (red-hardness) of the steel processed in a field of high pressure is higher than the same steel processed in a field of low pressure.

Cavities in a cross section of the preparation are formed after chemical etching by a nitric acid solution. Formation of visible defects (Fig. 4.13 [1, 28]) is explained by removal, at etching, of the activated zones of structure (Fig. 4.13a). Specificity of these

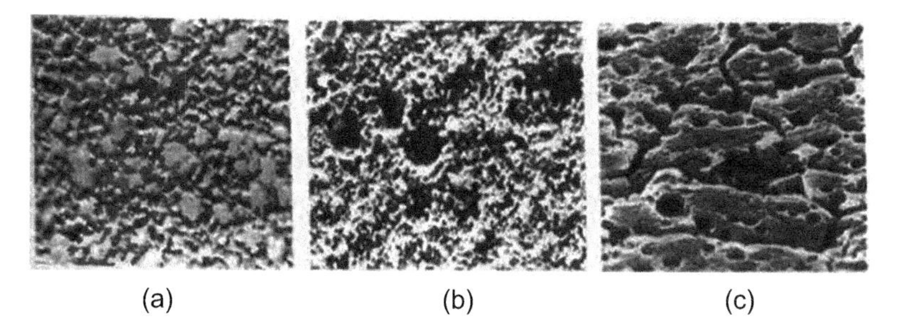

(a) (b) (c)

Figure 4.13 HSS steel at different stages of processing: (a) before processing, ×1000; (b) after SDP, ×1000; and (c) after SDP and heat treatment, ×5000 [1, 28].

defective zones is also high resistance of the activated material to thermal influence (Fig. 4.13b). The experiments have shown that defects in steel, created at SDP, do not disappear even after heating for 10 h at a temperature of 1000°C. After thermal processing in a HSS matrix, borders of grains' section are formed and the cross-section size of the activated zones decreases by 10–25 times (Fig. 4.13c).

The qualitative difference of some new elements of the structure, that is, the zones of the strikers' movement (channel zones), are local sections of overheated and quenched metal (Fig. 4.14 [1, 28]).

Such elements of a structure are not present in an initial material. These elements can be used for the identification of a process of SDP. Zones of local fusion can be formed round several closely located "channel" zones (Fig. 4.14a). A rather large zone is observed when the thermal energy from an individual channel

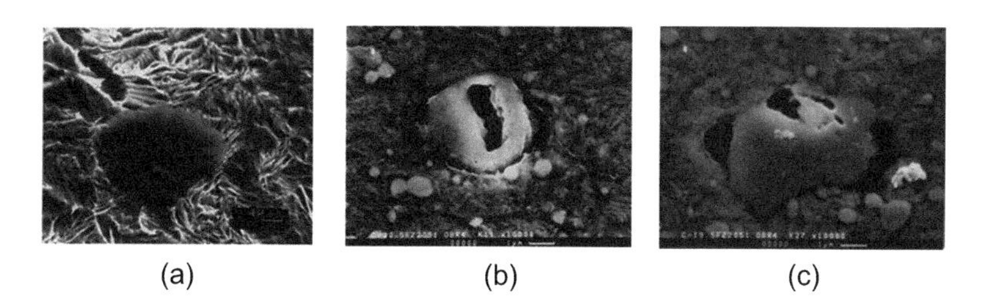

(a) (b) (c)

Figure 4.14 Zones of local fusion into HSS steel: (a) ×2200 and (b, c) ×10,000 [1, 28].

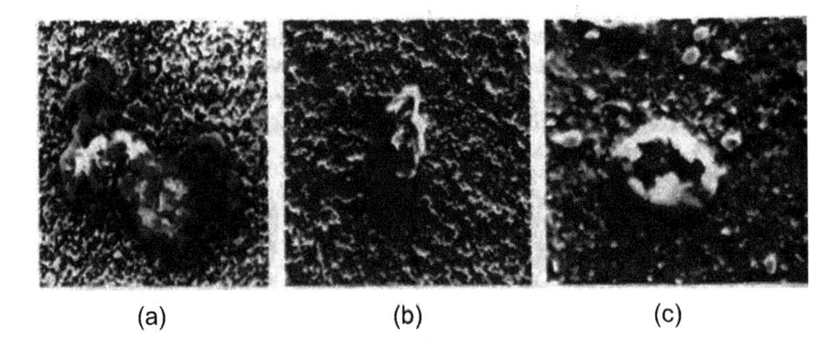

Figure 4.15 HSS steel fusion zones: (a) trail from driving of microparticles, ×2000; (b) trail from driving of microparticles, ×1000; (c) and zone after heating (1200°C, 1 h), ×4000 [1, 28].

element is insufficient for fusion. But at the big magnifications (≥10,000) individual channel elements with overheated elements (Figs. 4.14b,c) are observed. Such elements of a structure have a different reflective ability and reaction to etching than a matrix material.

In a longitudinal direction inside the HSS matrix, the channel sections in a plane do not exceed 0.5 mm. These results are explained due to the fact that in combined steel there are available high gradients of hardness and density in a volume. During the movement into a steel matrix, the strikers oscillate around the axis of their penetration (Fig. 4.15 [1, 28]).

Because of microparticle oscillation, the way of powder particles far exceeds the length of the strengthened preparation. The strengthened elements form a skeleton similar to a spring. By using special etching it can be seen that the diameter of a zone of interaction between moving particles and a matrix has a variable size (Fig. 4.15a). Chemical and physical properties of a material from the synthesized channel zone essentially differ from the matrix properties. The channel section, which after chemical etching on 4–5 μm projects from HSS matrixes, is shown in Fig. 4.15b. Intensive heating of channel zones causes occurrence of smaller structural elements (Fig. 4.15c). The complex consisting of the synthesized fiber elements and the plates from matrix steel, bound with them, creates zones of "influence" that, depending on the mode of SDP, constitute 3–20% of the volume of a composite material.

During the movement of a particle in a barrier, the whole complex of physical processes is realized: deformation of materials of a particle and a matrix, friction, pulsation of a field of high pressure (oscillating soliton), radiation, and heating [2]. After the striker penetration into a skeleton material, the processes of relaxation occur. As a result, the thermal energy from zone interactions (channel) is rejected to a matrix material. At heat rejection (heat-sink cooling), the synthesized material is quenched in a steel matrix. Depending on the total thermal energy and the intensity of heat removing, various structural conditions can be received in a channel material. The channel element created in the steel ($C \leq 0.45\%$) with a particle from Si_3N_4 is shown in Fig. 4.16 [1, 28]. Materials

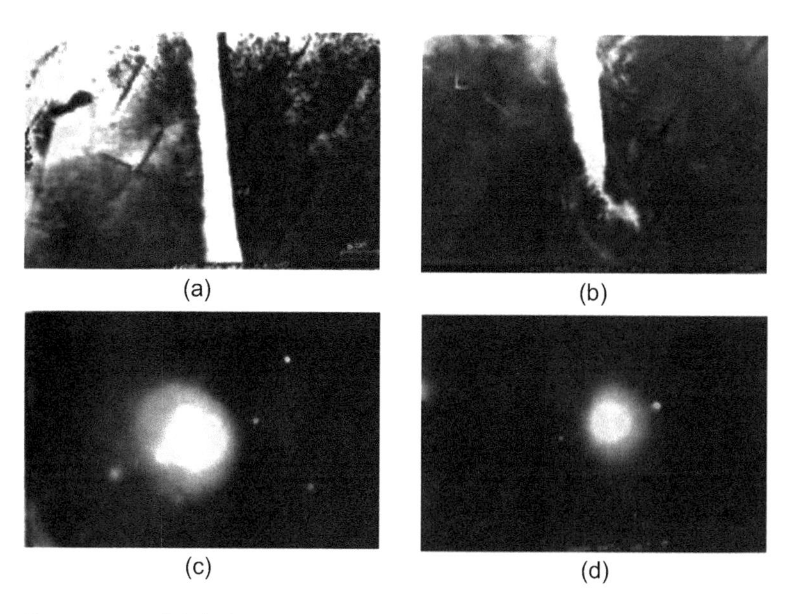

Figure 4.16 Solid-phase amorphization of a material inside a zone of penetration, ×20,000: (a) the top part of a channel element (length ~1.5 μm) with a zone of diffraction (+1); (b) a zone of braking and the braked striker with a zone of diffraction (+2); (c) diffraction picture (diffraction pattern) zones (+1); and (d) diffraction picture in the braked striker (+2) [1, 28].

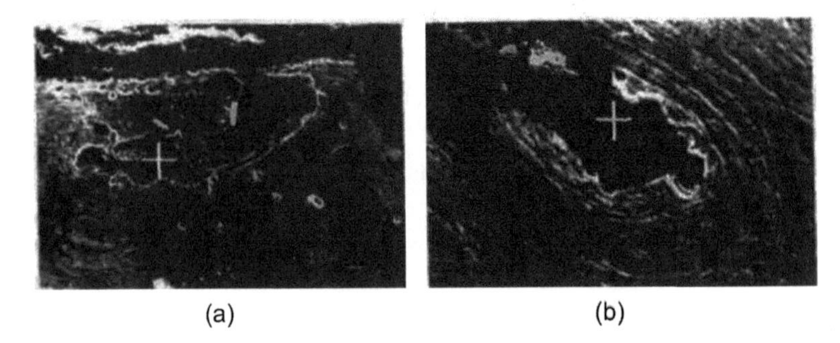

Figure 4.17 Zones of interaction of Si_3N_4 particles inside an iron barrier: (a) a new structural element to a depth of ~70 μm, ×300 and (b) a new structural element on a depth of ~4.3 mm, ×750 [1, 28].

of a channel zone and the braked striker are electron amorphous (Fig. 4.16c,d).

In a mode of SDP, Si_3N_4 particles intensively interact with an iron matrix and synthesize metastable compounds in an interaction zone (Fig. 4.17 [1, 28]).

The interaction of a striker and an iron matrix is accompanied by additional release of heat. At quenching of the overheated liquid material from an interaction zone in iron, the synthesized material became amorphous (Fig. 4.16).

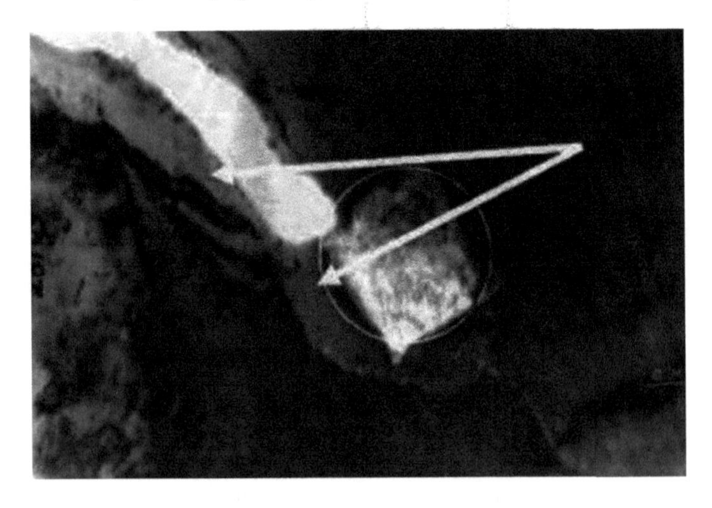

Figure 4.18 The zone of a Si striker of the cubic form, ×60,000 [1, 28].

To check this assumption, in SDP mode, we will enter silicon particles into an iron matrix. Initial particles of silicon had approximately a cubic form. Interaction of the particles of silicon with an iron barrier is shown in Fig. 4.18 [1, 28].

Around the etched long aperture and the braked striker, the remains of a material of a channel zone (white arrows) were situated. This material has more light shade and differs from a matrix material (Fig. 4.18). The structure of a material of a channel zone is nano- and microcrystalline. The striker has decreased in size approximately by 100 times, but it has kept the cubic form.

SDP proceeds in the time shorter than 1 s. This time is not enough to finish steel reorganization in a composite tool material. The increase in speed of a mass transition in a solid body volume leads to increase in the energy consumption of a process. Therefore, it is purposeful to finish the formation of a composite material at thermal processing.

4.6.2 Nanostructured Composites Based on a Metallic Matrix

Using of unusual physical features of the SDP method for the decision of a practical problem—creations of metal composite materials—is an actual problem for modern science. The basic features of this method, allowing the prediction of high competitiveness of technology, are high speed of process of change of structure of a massive metal material, ecological safety of processes of synthesis of reinforcing materials (synthesis in the closed system of a massive matrix), low power consumption of technology, and modification of a massive metal body in a solid state with special micro- and nanostructural elements. The decision of technological problems we will consider on examples of production of composite materials on the basis of an aluminum and iron matrix and method of strengthening of tool materials.

Generally, the use of tungsten carbide (WC) or cobalt (Co) alloys as materials for the cutting inserts of cutting tools employed in the mining industry is limited by physical properties of the tool per se as well as by relatively low resistance of the WC-Co-based cutting inserts to impacts and flexural loads. Moreover, in

Europe WC and Co alloys are regarded as carcinogenic materials unsuitable for production and use. In many cases, the use of tool materials strengthened by reinforcing coatings depends on operating conditions. For example, in the mining industry, coating-reinforced cutting tool inserts cannot be efficiently used because the need for frequent change of such inserts significantly decreases efficiency of the cutting process and mining and impairs operating conditions for workers. For the last 70 years, cutting tools have been equipped with cutting inserts made predominantly from WC-Co alloys, which have low resistance to dynamic loads and are ecologically hazardous.

Physical and mechanical properties of known tool materials limit the design possibilities for the development of new design tools and for saving energy consumed by cutting and mining processes [29].

The SDP method of strengthening of tool materials consists of impinging the surface of a body of a blank with a high-speed and high-energy pulsating jet of a specific working medium penetrating into and passing through the matrix of the treated material, thus restructuring and reinforcing the material with hard particles (Figs. 4.19 and 4.20).

Use of SDP for reception of new tool materials allows in 10^{-3} to 10^{-7} s the introduction into solid body volume (a steel of type HSS) of alloying elements to depths of tens of hundreds millimeters [24].

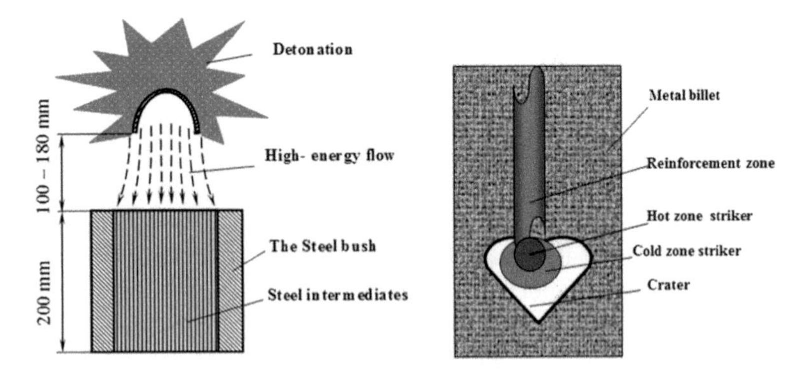

Figure 4.19 Schematic view illustrating penetration of the flow of a working medium in the volume of a preform of initial tool steel for forming the composite steel material [1].

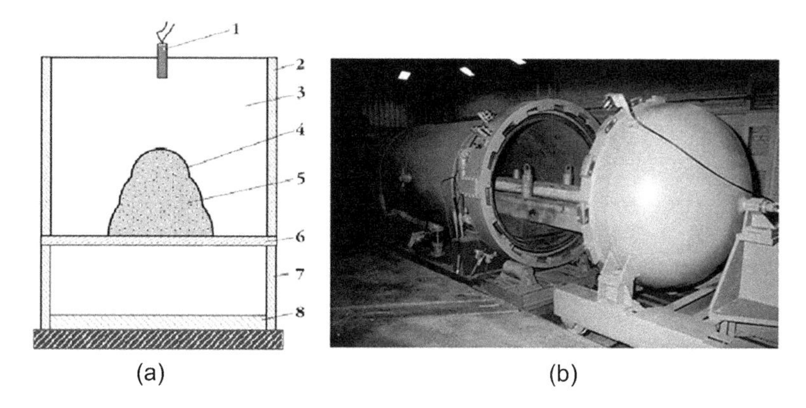

Figure 4.20 (a) The device for bursting alloying: 1, detonator; 2, metal becket for explosive material; 3, charge of explosive material; 4, metal becket for the cone of the explosive material; 5, powder composition; 6, plate—bottom of cartridge; 7, regulative abutment; and 8, barrier. (b) Installation for SDP treatment [1].

At formation into of steel preparation of the fibers having nano- and a microstructure, receive a composite tool material. The material of channel zones (fibers) is alloyed by an introduced substance. There is an anisotropy of mechanical properties, characteristic of a composite material. In the process of production it is possible to use cheap impulse accelerators on which are received streams of powder particles with speeds of ~1000 km/s.

The important mechanical property of tool steel is wear resistance. Usually the increase of level of wear resistance at production of tool steels is reached at the expense of substantial growth of concentration of alloying elements (from 5 to 40 mass percent). On dynamic alloying the concentration of introduced alloying elements does not exceed 0.01–0.1 mass percent. Therefore, it is possible to explain the increase of wear resistance of tool steel by tens of hundreds of percent only by the specific structure of a material.

Preparations from a tool composite material are easily processed. The increase in level of mechanical properties occurs after definitive thermal processing. It is assumed that the structural defects arising at pulse processing in metals and alloys are eliminated at diffusion processes. Heating of the metal preparations subjected to explosive hardening leads to a fast decrease in the level of hardness of the defective structure of surface layers. SDP does not lead to an increase of hardness of surface layers before thermal processing. If steel preparations with the raised level of hardness are exposed to a dynamic alloying then SDP leads to an appreciable reduction of hardness by depths to 10 millimeters. Research has shown that the new structural elements resulting from SDP are thermally very steady. For elimination of these defects annealing of a composite material at a high temperature ($\geq$1000°C) within some hours is necessary [30]. Quenching of the tool from steels type HSS lasts minutes and does not destroy the composite structure. In the process of tempering of the processed steel there is a possibility of additional hardening due to low-temperature synthesis of strengthening threads (whiskers). The maximum level of mechanical properties of the tool steel exposed to a dynamic alloying is reached at the complex approach, including development of introduced alloying composition, SDP modes, and optimization of modes of thermal processing.

Figure 4.21 demonstrated the effective thickness of a metal barrier allowing to stop dust flow.

Composite materials have anisotropy of properties in various directions. Anisotropy of physical-mechanical properties increases after thermal processing. In a direction of introduction of a stream of powder particles in preparation from HSS it was possible to raise wear resistance by 1.8 times and in a cross-section direction by 14% (Fig. 4.22 [1, 31]).

If the tool basically wears out along the length its service durability increases repeatedly. Use of this feature in the strengthened tool steel, characteristic of composite materials, allows one to solve successfully questions of creation of new constructions of the tool for processing of metals and cutting of rocks. Such approach has been realized at manufacturing of the rotating tool of a mining machine for salt extraction. Unexpected results have been received on use as criterion of change of mechanical properties such parameters as shock durability (viscosity). Reception in steel preparations of strengthened and activated zones should lead to wear resistance increase. Such elements of structure should reduce the level of shock strength (impact resistance) and bending

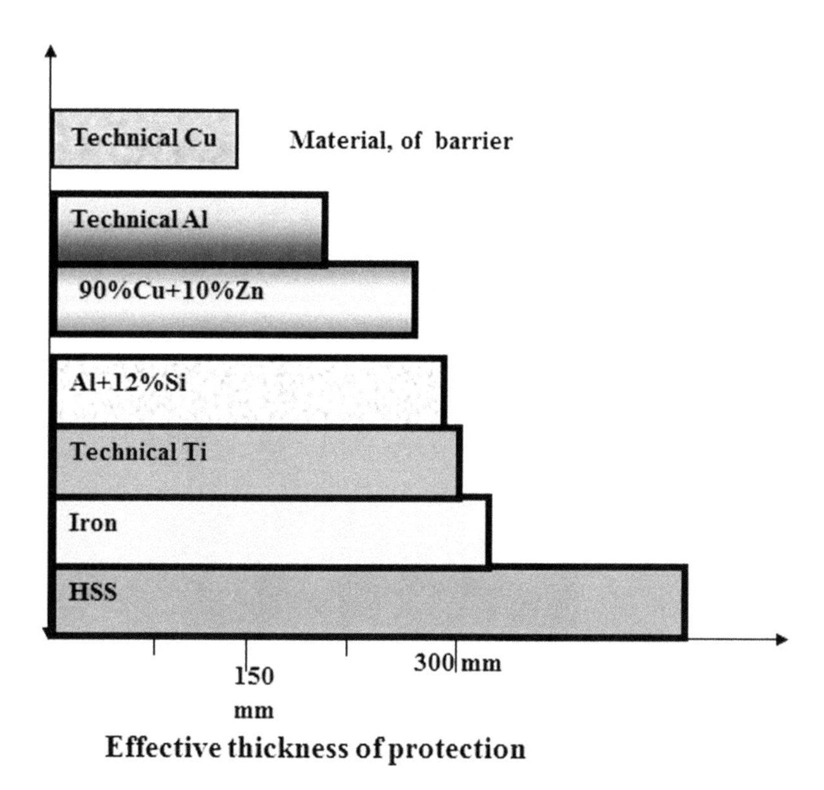

Figure 4.21 Penetration distance of dust strikers depending on the metal barrier [1].

durability. The skeleton material could be a source of formation of cracks. The increase in the specific density of these elements of structure could lower the level of physical-mechanical properties. However, SDP processing has raised shock durability (viscosity) of a composite material in comparison with initial steel by 20–40% and bending durability by 50%.

An interesting technological problem is the process of zone hardening of large-sized details and change of the thickness of a zone of high hardness on heat treatment [29]. On volume dynamic alloying of large-sized products the structure in big zones (to 200 mm) of the large-sized tool changes. Thus service durability of stamps has been increased by 20–60% (Fig. 4.23) [1, 30].

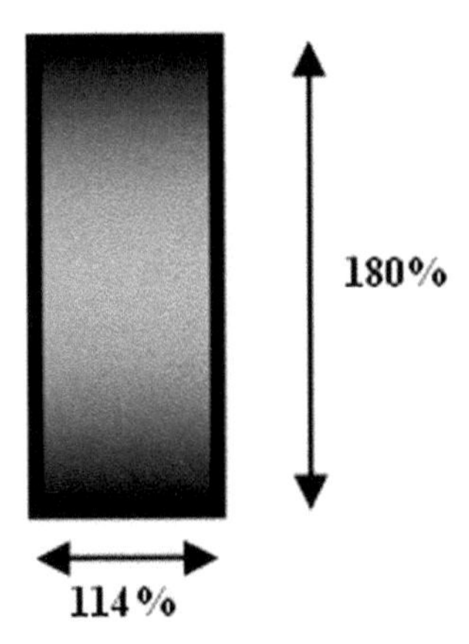

Figure 4.22 Change of wear resistance [1, 31].

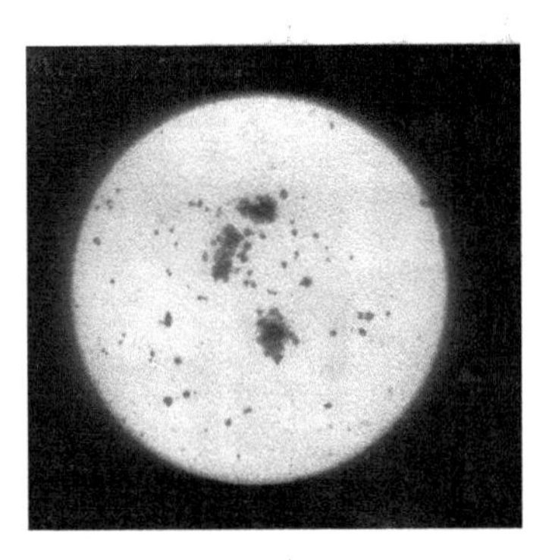

Figure 4.23 Radioautograph image of the steel 45 specimen at a of depth 4.3 mm dynamically alloyed by powdered particles W $(C + ^{14}C) + Ni$ [1, 30].

Let's take two examples of application of the SDP method for strengthening of the steel tools (Tables 4.3–4.6) [1, 29].

A metal-cutting tool from this material on processing with high-strength titanic alloys has shown resistance 1.8–3 times higher than a similar tool from an initial tool steel. The rotating tool has been made of composite steel for mining machines for potash salt extraction (Figs. 4.24 [1, 31] and 4.25). The tool was used in potash mines of Italy and Belarus. Service durability of the tool has appeared 1.5–5 times higher than that of a similar tool with cutting

Table 4.3 Compositions of working-medium mixtures used for treating steel blanks

Test no.	Composition
1	TiCN (1–100 μm) 60% + Ni (1–100 μm) 30% + Si_3N_4 (0–60 μm) 10%
2	TiCN (80–100 μm) 60% + Ni (10–30 μm) 30% + Ni (10–30 μ + Si_3N_4 (60–80 pm) 10%
3	TiCN (1–100 μm) 100%
4	TiCN (1–100 μm) 50% + Ni (1–100 μm) 30% + Si_3N_4 (0–60 μm) 10% + ethyl alcohol 10%

Source: [1, 29]

Table 4.4 Mechanical properties of a composite tool material after treatment according to above (Table 4.3) compositions of working-medium mixtures

	Strength with reference to untreated steel			
Test no.	Composition no.	Resistance to wear	Flexural strength	Impact strength
1	–	1	1	1
2	1	1.3	1.15	1.2
3	2	1.05	0.8	0.9
4	3	1.1	0.7	0.65
5	4	1.35	1.1	1

Source: [1, 29]

inserts from hard alloy on the basis of tungsten carbide (Table 4.7) [1, 30].

Aluminum is one of the most widespread and cheap metals. Without it, it is difficult to imagine modern life. Aluminum alloys play a huge role in the space industry. Many constructive elements of space devices are made of alloys of aluminum, including a system of aluminum-silicon. It is asserted that wings of planes are kept in air only by metastable zones and particles. If on heating instead of zones and particles there will be stable phases, the wings will lose their durability. It is important that the change of properties of aluminum alloys is realized at infusion of additional doping elements with concentrations of 0.001–0.1 mass percent.

Table 4.5 Compositions of working-medium mixtures used for treating steel blanks

Test no.	Composition
1	SiC (3–250 μm) 50% + Ni (1–100 μm) 40% + $A1_2 0_3$ (20–50 μm) 10%
2	SiC (3–250 μm) 100%
3	SiC (3–250 μm) 10% + Ni (1–100 μm) 20% + $A1_2 0_3$ (20–50 μm) 70%
4	SiC (3–250 μm) 50% + Ni (1–100 μm) 50%
5	SiC (3–250 μm) 10% + Ni (1–100 μm) 20% + TiB_2 (40–50 μm) 70%

Source: [1, 29]

Table 4.6 Mechanical properties of a composite tool material after treatment according to above (Table 4.3) compositions of working-medium mixtures

		Strength with reference to untreated steel		
Test no.	**Composition no.**	**Resistance to wear**	**Flexural strength**	**Impact strength**
1	–	1	1	1
2	1	1.55	1.1	1.25
3	2	1.05	0.7	0.6
4	3	1.1	1.3	1.22
5	4	1.07	1.0	0.7
6	5	1.4	0.5	0.2

Source: [1, 29]

Such magnitudes of concentrations can be created in conditions of dynamic processing. Using of effects of SDP allows one to provide additional doping in a volume of aluminum and its alloys and intensive dynamic loads simultaneously [32]. Studying of structural transformations in aluminum details at a stepping action of dust particles actual is. Because of clots of space dust moving with a high speed (above 5000 km/s) in orbits of the Earth, the probability of their impact with a spacecraft is high. The defectiveness of an alloy's structure determines the reliability of saving of physical-mechanical properties. Therefore, the dynamic changes in the structure of aluminum materials can essentially affect the survivability of aircraft.

Figure 4.24 Self-sharpening mining tool and metal blocks from a composite tool material [1, 31].

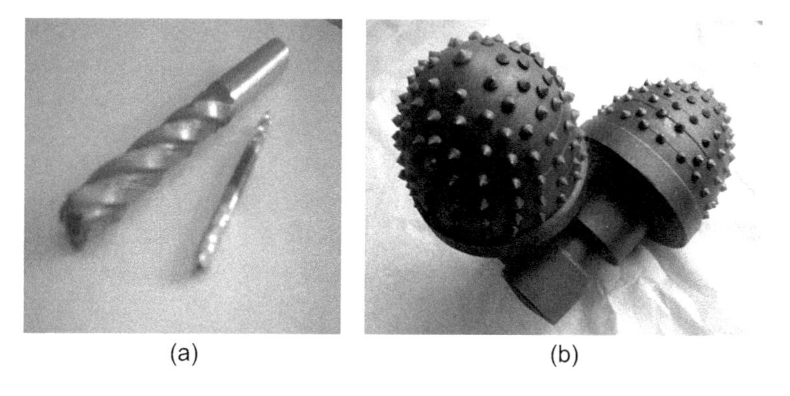

(a) (b)

Figure 4.25 Strengthen tools: (a) metal cutting tools and (b) the tools for cracking of coal.

Modification of aluminum and its alloys' micro- and nanoelements in the third stage of manufacture of massive material (detail) in traditional powder metallurgy faces a problem of growth of structural elements. On use of the SDP process of volumetric modification of massive material (detail) occurs for shares of second that mechanical properties providing at a level of an initial metal matrix. If as a criterion for use of a new composite material, not mechanical but physical or chemical properties are used, the necessity of sintering and high-heat treatment is eliminated. The

Table 4.7 Advantages of the new nanocomposite tool steel

	Tools	Wearing resistance	Power consumption	Fire safety	Cancerogenic factor
Analog	Tool steel HSS	1			
	Alloy: WC + Co	1	1		1
Nanocomposite steel tool		1.5–2	1.2	50	0

Source: [1, 30]

increase in structural elements by the manufacture of massive aluminum material within the limits of complex SDP is not realized.

The basic problem in the study of changes of an aluminum structure and its alloys after SDP is working-off techniques of preparation of samples from the activated material. Because of high plasticity of aluminum there is a puttying of the sample's surface. The definition of a mode of etching is executed in view of the specific activation of a material. The use of an optical microscope and the correct technique for preparation of samples allow the revealing of new elements of structure (Fig. 4.26 [1, 31]).

During research the basic problem was the detection of structural changes in aluminum and its alloys. An irregular sampling of powder composition for SDP can intensify the process of etching on a surface and in the cross section of a detail. Therefore false representation is created that the new composite material has up to 50% of the closed porosity (Fig. 4.26c). In a host material the closed porosity was less than 0.5% (Fig. 4.26a).

As in a condition of superdeep penetration (SDP) the powder particles with sizes less than 100 microns were used so the size of a channel zone in a cross section is less than the initial size of the striker [24]. The study of such objects can effectively be executed by means of transmission electron microscopy.

Attempts of study of SDP aluminum exemplars without the use of etching were not successful (Fig. 4.26b). On use as a powder of the substance decelerating etching (Pb) visible porosity after etching is not observed (Fig. 4.26d). Therefore, techniques for research of these materials have been modified. At SDP alternatives with use inhibitory material (Pb) and activate material (SiC) and electrochemical etching are observed. The use of optical microscopy

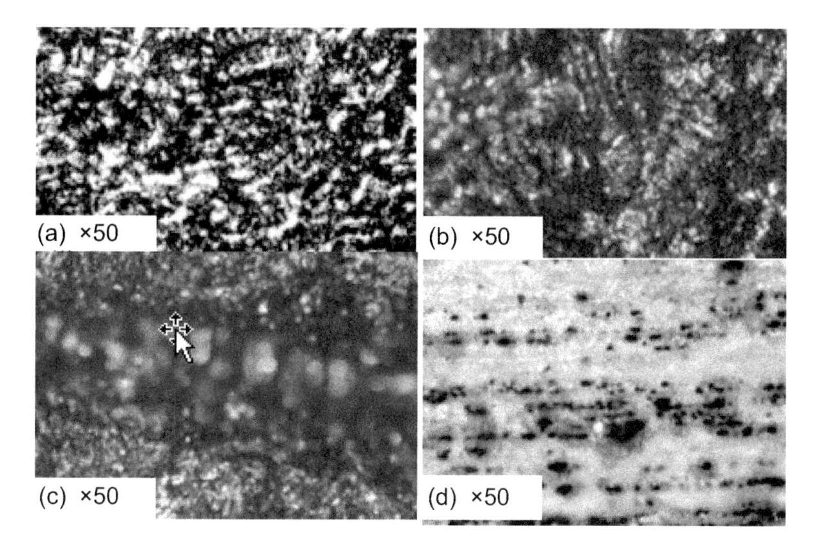

Figure 4.26 Structure of aluminum alloy Al + 12% Si: (a) after casting; (b) without etching after SDP; (c) after SDP-SiC and after electrochemical etching; and (d) after SDP-Pb and after electrochemical etching [1, 31].

has allowed even to determine noticeable structural changes in the treated materials (Figs. 4.26c,d). On the use of an inhibitor it became obvious that in the volume of aluminum matrix there are unusual zones of influence (Figs. 4.26c,d). These zones represent matrix aluminum or its alloy stitched by the assemblage of tracks (filaments) from a material that is synthesized on interaction of a matrix and powder particles. Zones of influence can constitute 2–50% of the volume of a host material. Created through SDP, the modifying micro- and nanoelements of the structure are cooled inside of a metal matrix. Therefore, an increase in the sizes of these structural elements on the shaping of the composite material is not observed.

On the broaching of an aluminum matrix the discrete particles initiate strong distortions in a zone of penetration. Results of initiation are presented in Fig. 4.27 [1, 31].

Electric and electrochemical research of processed specimens (details) has been executed. Aluminum and its alloy details were cut in macroplates. Detailed research has shown that there is an anisotropy of the examined properties in mutually perpendicular

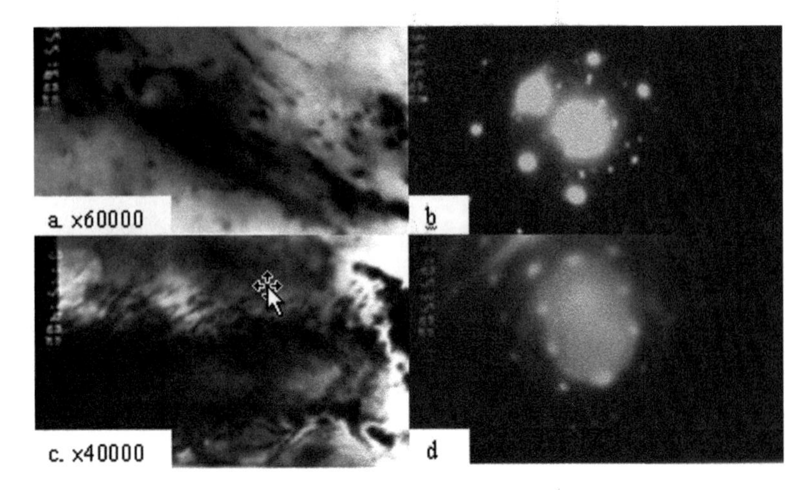

Figure 4.27 Thin structure of the aluminum processed at SDP: (a) zone of high concentration of tracks and congestions of dispositions; (b) electron diffraction pattern of the site (a); (c) structure of the central channel microzone; and (d) dim ring electron diffraction pattern in the form of a halo of dispersion from the site (c) [1, 31].

directions twice. In a volume of massive detail from technical the aluminum macrozones with a various level of a work function electron [32] are registered. Change of SDP regimes, compositions of a powder material, and a metal matrix allows the creation of new massive composite materials and in a wide range effectively to change their properties.

Production of new materials for electrical engineering and electronics is limited by a level of properties of initial materials. An attempt to bypass these restrictions is the creation of nanostructured materials. The greatest problems arise on the production of fiber-reinforced massive composites with high physical mechanical properties. Traditional technological processes do not give the effective decision because of their high cost. New opportunities of massive composite materials preparation arise on the use of physical effects that earlier were considered as improbable and exotic.

Reorganization of the structure of a massive solid body is possible due to the creation in it fields of pressure and gradients of pressure. The more gradient of pressure in a solid body, the more diverse material forming this body, and the higher the probability of local deformations and dispersion of the structure. Formation of fluctuations in a solid body at pulse processes, and also gradients of energy, pressure, temperatures, and deformations, is a rule in distinction to stationary processes [2].

In the experiments carried out on the accelerator of heavy ions in Darmstadt (Germany), the beam of ions of uranium was directed on the sample placed in the chamber. In usual conditions, the high-energy ion, passing through a substance, spends a part of energy during braking and causes destructions along the way. As a result the sample of a material after an irradiation appears with reinforced parallel and very narrow (10 Å) channels filled with amorphous substance. At the moderate dozes of irradiation these thin channels are located enough far from each other and don't influence on the common structure and properties of a material. However, at irradiation under high pressure another picture appears. At irradiation of graphite under pressure of 80,000 atm, channels have not been found. Dispersing of the structure of a material (phase transition) was observed [33].

Physical phenomena such as intensive electromagnetic radiation and pressure in the range of tens of thousands of atmospheres are combined is known. In the end of seventies was found an anomaly in behavior of a clot of powder (dust) particles during impact with a metal barrier. A clot of dust operating under SDP gets into the barrier to the depths of tens of hundreds of millimeters. At the usual impact the ratio of the depth of penetration to the caliber of a striker (the defining size) does not exceed 6–10. In the case of SDP the resistance of a barrier material to penetration of dust particles decreases by hundreds or thousands of times [2].

Research has shown that in the field of penetration of particles a high level of pressure sufficient for dynamic phase transitions is observed. Three-dimensional (type of soliton) zones of a material with pressure not less than 80,000–120,000 atm is revealed for an iron barrier [19]. The pressure in the basic part of a barrier material did not exceed 10,000 atm. Therefore, the structure of a composite material is formed at the SDP of clot powder particles into a massive metal barrier. Reinforcing fibers in the SDP process are products of interaction of penetrating particles and the matrix material. Special conditions of formation of these fiber structures allow one to obtain nanomaterials with physical-chemical properties that cannot be predicted on the basis of known data. The durability of such composite material can be above that of the initial metal and an alloy. Additional thermal processing may not be applied.

As a perspective material for the creation of an ultradisperse structure we shall consider aluminum and its alloys. Aluminum and its alloys have found wide application in the electrical engineering and electronics industry. However, opportunities for the qualitative improvement of its properties with alloying at molding are already exhausted by methods of traditional metallurgy. Therefore, it is represented actual to realize the processes, allowing to produce details from these materials with the formation of their structure on nano- and microlevels.

The purpose of the given research was to reveal the specificity of changes of the structure and properties of aluminum and its alloys in conditions of dynamic loading in the mode of a SDP.

The cycle of experimental research of behavior of various metals and alloys has been lead during interaction with high-energy clots of dust. As the material for these clots are used, for example, powders of silicon carbide. Under the accepted scheme of processing the mode of a SDP was realized. As a result of research it has been established that the aluminum barrier with a thickness of 0.1 m stops a stream of dust particles (fraction 10–100 microns). Aluminum with silicon and zinc as materials of protective barrier of alloys has been used for the achievement of the same purpose—a thickness of 0.16–0.18 m. It is known that static and dynamic durability of aluminum alloys is noticeably higher than the same characteristics of technical aluminum. It is obvious that the known dependence of the depth of penetration for a striker (a clot of discrete particles) from initial static and dynamic durability of a material of the barrier, the macro-objects certain at use, it is not carried out. Research of the materials, subjected to pulse influence by a clot of dust particles, has shown changes of aluminum and its alloys on submicro- and microlevels.

The channel zones arising at processing consist of a material with essentially distinct initial physical and chemical properties. Owing to it, during preparation of metallographic samples from a material of a protective shell, due to a difference in durability

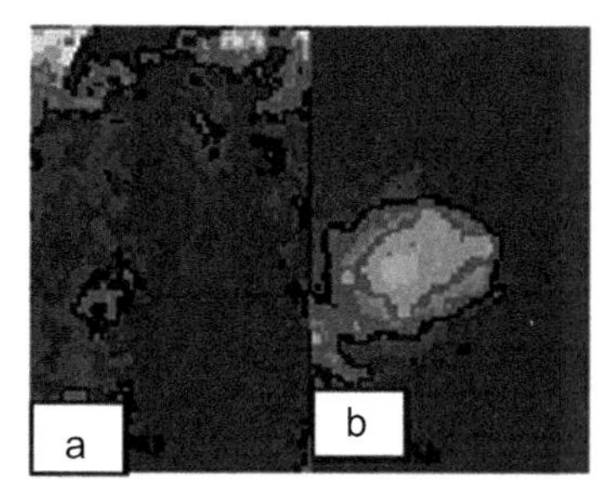

Figure 4.28 Structure of artificial fiber (canal): (a) ×40,000; and (b) electron picture [1, 34].

and in electrochemical potential (etching), these local sections are noticeable from the volume of the material of a barrier, for example, due to reflective ability.

A qualitative difference in the structure of aluminum and its alloys on processing by clots of dust in SDP mode has been revealed at greater increases. Study of the channel zone formed at the SDP has allowed one to find out in volume of massive preparation "amorphization," nano- and microstructural sections. In Fig. 4.28 [1, 34] the nanostructure of a channel formation in alloy Al + 12% Si is shown.

A typical mistake when studying channel zones is their perception as cavities. It is connected to the fact that the material of a channel zone, as a rule, possesses at preparation of metallographic samples the chemical activity differing from a matrix material. Therefore, at the same mode of processing of a surface of metallographic samples solutions of an acid or alkali on a place of a channel zone arises a cavity. Simplistically the channel zone can be considered the composite material consisting of several coaxial located zones. Therefore, if a cross section of the central zone is accepted as the unit, the cross section of a zone with a defective microstructure makes 4–10 units. A qualitative scheme of a channel zone is shown in Fig. 4.29 [1, 34]. Along an axis in this zone (section 1) can be observed amorphous structures or nanostructures. In a volume of this section it is possible to register also the result of the interaction of entered and matrix materials.

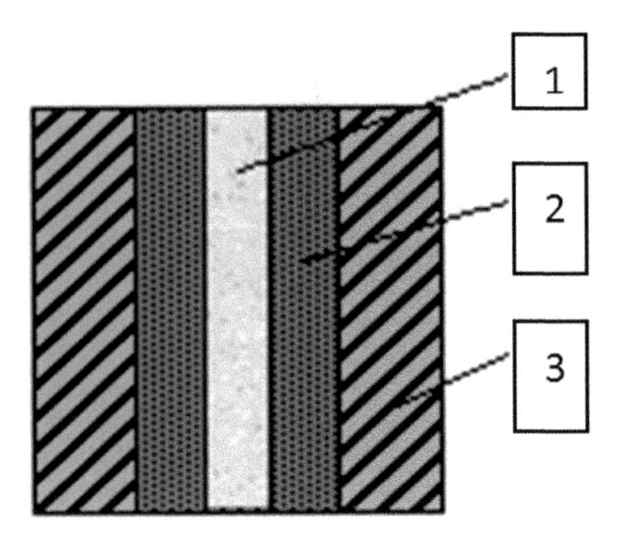

Figure 4.29 Scheme of canal zones: 1, amorphous structure or nanostructure; 2, microstructure; and 3, imperfect structure [1, 34].

Depending on modes SDP and the central section of a zone is possible not to find out used materials. It explains that at SDP due to cumulating of fields of pressure there is an emission of a defective material of the channel zone which are being a condition of dense plasma, in a kind of direct and return microjets. Depending on the level of pressure and time of reduction, the material of the central zone can leave with a different degree of efficiency. This process is influenced also by the chosen mode of SDP [35].

On the basis of calculation of experimental data [34] it is established that in local volumes the phase of a high pressure has the density $\rho = 1.89 \cdot 10^3$ kg/m^3 and the relation of the density of phases of low and high pressure makes $\rho_1/\rho_i = 0.70$. Such character of change of density of a phase of high pressure till now was observed only for thorium and uranium. Also it has been established that the phase of high pressure in an alloy from 12% Si has a density less than initial $\rho_1/\rho_i = 0.53$ or makes 53.2% from the initial density of an alloy. The relation of the density of initial phases in aluminum and an aluminum alloy (12% Si) makes $\rho_{A1}/\rho_{AK12} = 1.01$. The relation of the density of phases of high pressure makes $\rho_{A1}/\rho_{AK12} = 1.34$.

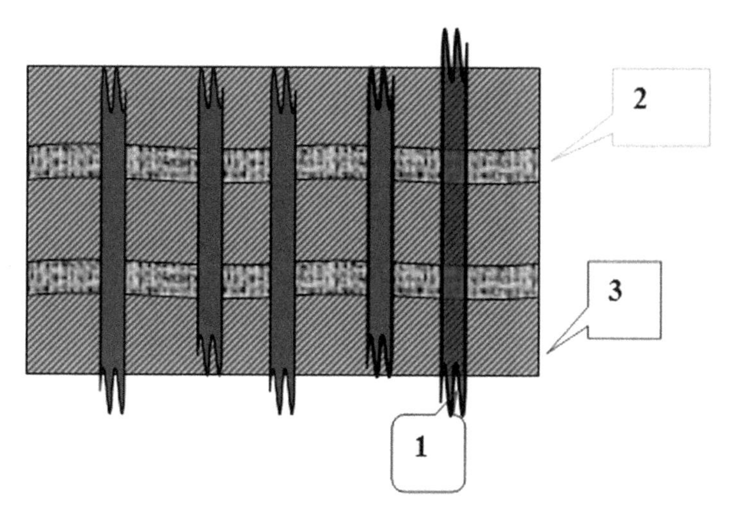

Figure 4.30 Scheme of the structure of a composite material: 1, zones of canals along to motion of astir dust; 2, zone across to moving of astir dust; and 3, matrix material [1, 34].

Thus matrix aluminum has appeared reinforced, so-called zones of influence (5–10 volume %). The chemical compound of these zones practically meets the structure of an initial material; however, the physical and chemical properties essentially differ from each other. In channel zones (the central site of a zone of influence) the occurrence of an ultradisperse structure alloyed by the entered material of a clot is observed. The analysis of this site in technical aluminum has shown that it contains up to 3 mass percent Si. During SDP in a material of a barrier the structure of a composite material is formed. The reinforcing skeleton of this material consists of fiber zones of the reconstructed structure. The scheme of such a material is shown in Fig. 4.30 [1, 34].

Occurrence of the zones reinforcing a material in a cross-section direction, apparently, is connected with the turn of a microstriker at breaking in a material of a barrier. The share of cross-section channel formations in aluminum and its alloys is 20–30% of the quantity of longitudinal zones.

Aluminum and its alloys are effectively used as elements of electric machines and electric schemes. In this area the competition

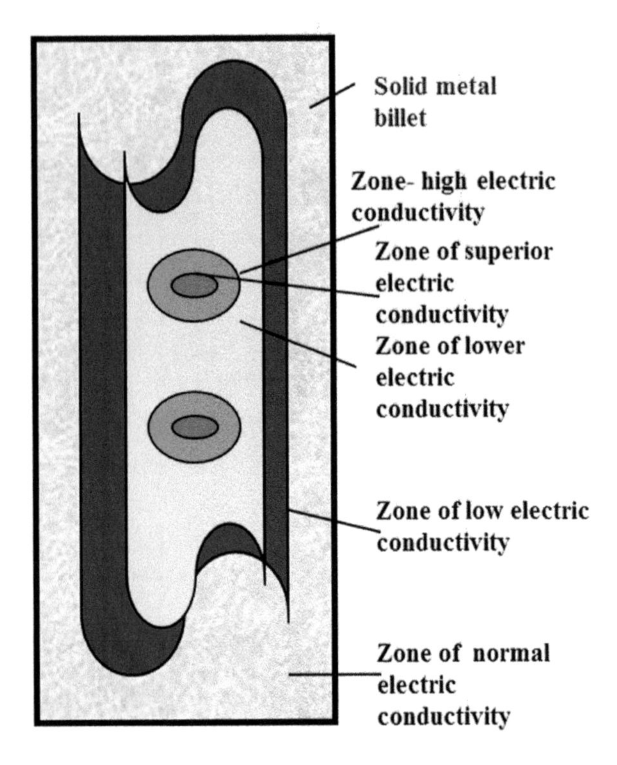

Figure 4.31 Distribution of electric conductivity zones at impact [1].

to aluminum is only from copper and silver. It is represented essentially important to receive an additional opportunity of management of its physical properties. Therefore, the composite material on the basis of the technical aluminum, received in SDP mode has been used by processing by a dust clot for change of such physical parameter, as electric resistance (Fig. 4.31).

As is known composite materials possess anisotropy of properties. The definition of electric resistance was made in mutually perpendicular directions: in the longitudinal section (along a direction of a dust stream) and in the cross-section section (Fig. 4.32 [1, 34]).

Cutting of samples for research was carried out by means of electro spark processing. From each sample in a cross-section and longitudinal direction 4–5 plates were cut out. Electric resistance

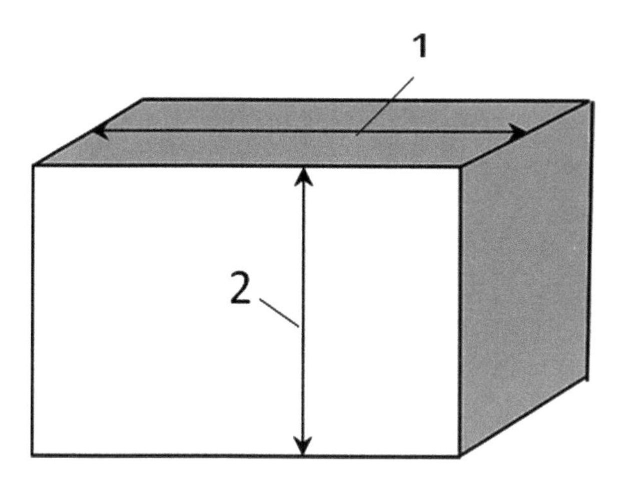

Figure 4.32 The scheme of measurement of electric resistance in a composite material: 1, along; and 2, across [1, 34].

was defined as the average value of the made measurements. Comparison of the electric resistance, received on the processed sample, was made with measurements on an initial material.

In direction 1, the value of the electric resistance of technical aluminum $\rho_{\text{Al-1}} = 5.27 \cdot 10^{-6}$ Om·cm; for a composite material $\rho_{\text{K-1}} = 4.41 \cdot 10^{-6}$ Om · cm ($\rho_{\text{K1}}/\rho_{\text{Al-1}} = 0{,}835$). Thus it is experimentally established that ρ_1 after processing has decreased by 16.4%.

In direction 2, the initial average $\rho_{\text{Al-2}} = 6.42 \cdot 10^{-6}$ Om · cm; for composite $\rho_{\text{K2}} = 9.08 \cdot 10^{-6}$ Om · cm ($\rho_{\text{K2}}/\rho_{\text{Al-2}} = 1.41$). Electric resistance after processing has increased by 41.2%.

Note that for initial technical aluminum $\rho_{\text{Al-2b}}/\rho_{\text{Al-1b}} = 6.42 \cdot 10^{-6}/5.27 \cdot 10^{-6} = 1.21$. Thus the difference of electric resistance in longitudinal and transversal directions in the initial sample is 21.74%. For a composite material after SDP processing anisotropy of electric resistance $\rho_{\text{K1}}/\rho_{\text{K2}} = 2.05$ times. Therefore, transversal conductance in the processed sample exceeds longitudinal conductance by 105%.

The obtained results have allowed the assumption that electric properties of a new composite essentially differ among themselves on zones. Because of this testing plates for scanning with Calvin's device have been prepared. Results of the scanning are presented in Fig. 4.33 [1, 34].

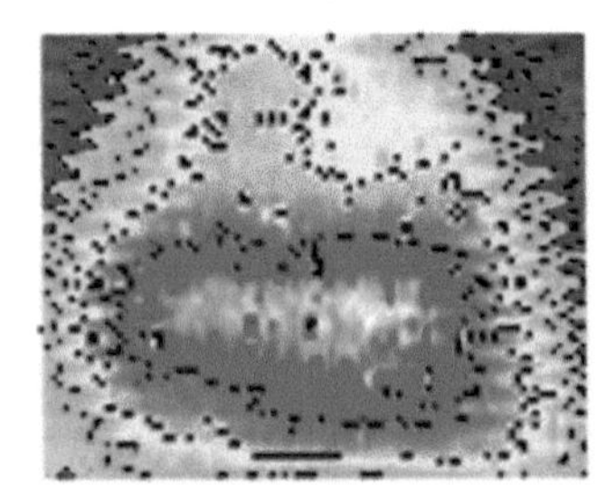

Figure 4.33 Change of electric resistance in a cross section of a composite [1, 34].

The central zone, which is designated by red color, corresponds with the lowest work of an output of electrons. Scanning is executed on a scale of 2:1. Attitude conductance of the zones designated by different colors in Fig. 4.33 makes 4–5 times.

4.6.3 Interaction of a Stream of Particles with Ceramics

As a subject of research used a plate of a monocrystal of silicon and a micro devices on their basis. Structural-specific changes on a surface of a plate of silicon monocrystal are examined. Conditions of interaction allow a stable effect of SDP on a metal barrier. Extensive experimental studies were carried out in this direction. Plates of silicon monocrystal have high hardness and high fragility. Nevertheless, in experiments using rigid regimes plate crushing was observed. The example of the destruction of a microdevice in a rigid regime in a direction perpendicular to the action of a stream of particles is shown in Fig. 4.34 [1, 35].

Damages to the plastic envelope of the microdevice have also not been detected on use of visual and x-ray methods. Damages have been detected after removal of the protective envelope. Cracks have arisen between knots of an inflection in a plane of the section of a plate of a silicon monocrystal. Knots of an inflection have been investigated (Fig. 4.35 [1, 35]), and zones of impact are detected.

In a silicon plate the zone of punching with a diameter in a cross section less than 1 pm and depth ~180 μm was generated. As a barrier was the fragile material then it is possible to consider that

Figure 4.34 A view of the damaged monocrystal of silicon, ×75 [1, 35].

the cross-section size striker does not exceed the size of a zone of punching.

In the work in Ref. [36] it is shown that in a metal barrier there are microjets with a speed in the longitudinal direction $v_t = 1482$ m/s and in a cross-section direction a jet has a compressed speed $v_{com} = 1048$ m/s. In such conditions ($P = 11.3 \cdot 10^{11}$ N/m^2) the material of a jet exists in a condition of dense plasma. On processing the container with a microdevice the plasma jets pass through gaps. Such gaps exist in a system between an internal surface of a cover and an ampoule and between an ampoule and a covet and in a microdevice between an internal surface of a cover and a surface of a plate from a monocrystal of silicon. The driving of a jet in gaps is realized in a interval of time from $0.67 \cdot 10^{-6}$ to $3.37 \cdot 10^{-6}$ s. During this period of time there is an unloading of a jet in a cross section and pressure in a point of contact varies in a range $1.1 \cdot 10^9 \div 11.3 \cdot 10^{11}$ N/m^2. If the hardness of a Si monocrystal is no more than 10^9 N/m^2 then at pressure decrease up to 10^9 N/m^2 the penetration of a jet into a chip stops. Executed calculations [21] show that the speed of driving of a jet in a silicon plate is

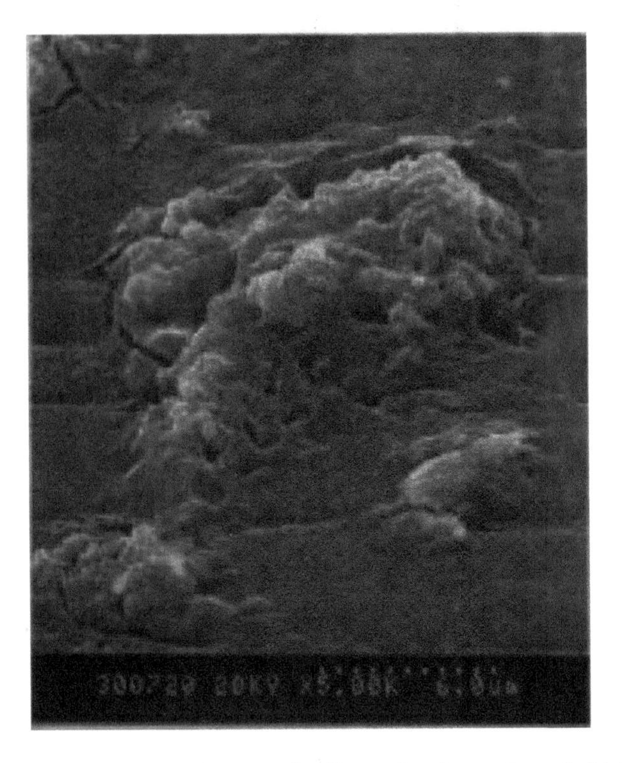

Figure 4.35 A zone of damage of a silicon plate in a soft mode [1, 35].

$U = 961$ m/s and the length of a jet is $L \geq 98$ μm. If to execute transition from rigid to soft regime of operation of microjets then punching of a plate stops. Affecting of plasma jets on silicon plates gets other kind. Operation in a soft regime is shown in Fig. 4.36 [1, 35].

In a soft regime the jet acts on a local area and heats it up. Such action evaporates from the surface of a plate a covering—the circuit of the microdevice. In this zone it is possible to see numerous small holes.

Heating of a local area depends on microjet parameters. The temperature of local heat can be above the temperature of fusion for silicon (Fig. 4.36). On intensive heating there are bubbles of melt (Fig. 4.36a). Significant overheating of a local area of a silicon plate initiates growth of new monocrystals (Fig. 4.36b).

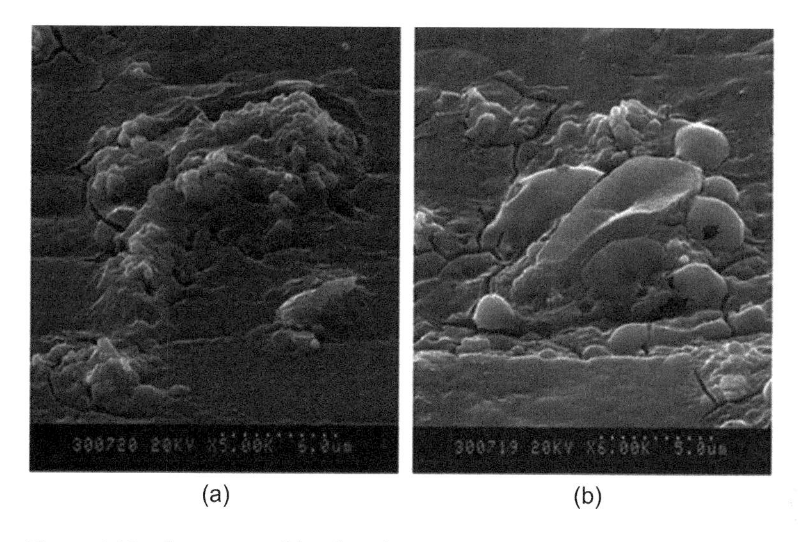

Figure 4.36 Segments of local melting on the surface of a silicon plate: (a) a zone of formation of melting; and (b) a zone of initiation of growth of new monocrystals of silicon [1, 35].

Bubbles of the fused silicon have traces from punching by microjets. Hence these microjets act on a barrier after formation of bubbles. Bringing in a SDP mode particle and its chemical compounds, the interaction of elements of a stream and a barrier (silicon plate) will be always realized at high temperature and high pressure. The spectroscopic analysis of a surface of a plate before and after processing shows interaction of carbon and silicon. The thick-walled container allows one to keep products of interaction in the set atmosphere and also to protect the working personnel from radiation.

Conditions of interaction ensure synthesis of chemical compounds. We observe heat removal into the silicon plate. Use of a barrier from carbon (diamond) on its surface enables successful synthesis of a wide gamma of metastable chemical compounds. The qualitative difference of the process of synthesis in SDP mode from other known processes show energy capacity at a stage of creation of a high-speed stream of discrete particles. An anomaly of this process is the fact that despite a rather low level of kinetic energy at synthesis high pressure and high heat are realized.

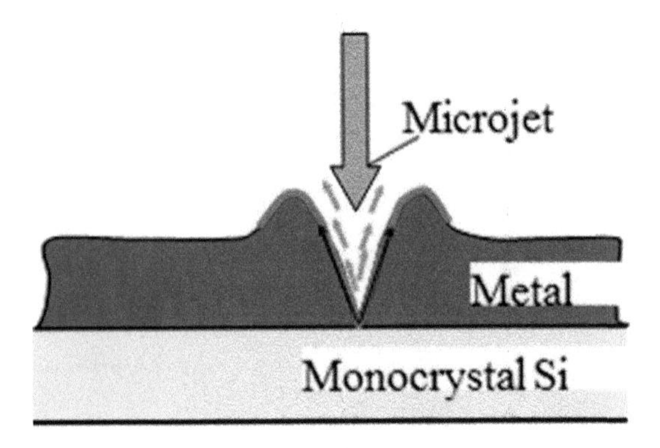

Figure 4.37 Damage of a microchip in SDP conditions [1].

Figure 4.37 illustrates damage by a microjet on the boundary "metal cover–base member" with subsequent explosive extrusion.

On using microchips as samples specific processes on metal surfaces are found (Fig. 4.38 [1, 35]). From the volume of a metal solid body under action of high pulse pressure the microtubes are squeezed out (lengths of 5–25 μm and diameters of 5–10 μm) (Fig. 4.38a). It is obvious that the process of extrusion happens due to energy from microexplosion of metal. The electroplating can be removed from a surface of a metal element only due to evaporation of a covering (Fig. 4.38b). The cross-section size of pinholes punching is not accompanied by the microexplosion is ~1 μm (Fig. 4.38c). Zones with racks, apparently, arise during a relaxation of residual stresses (Fig. 4.38d). The received experimental results have shown presence of significant additional expenditures of energy at synthesis. These energy expenditures were not considered earlier, during the calculation of the balance of energy of SDP.

4.6.4 Features of Interaction of a Stream of Discrete Particles with Plastic

Interaction of discrete streams of microparticles with plastic in SDP mode was investigated for the definition of properties of protective shells. So the study of plastic shells of microchips has shown that such shells in SDP mode do not lose their hermetic state. At small magnifications by visual observations it was not possible to find

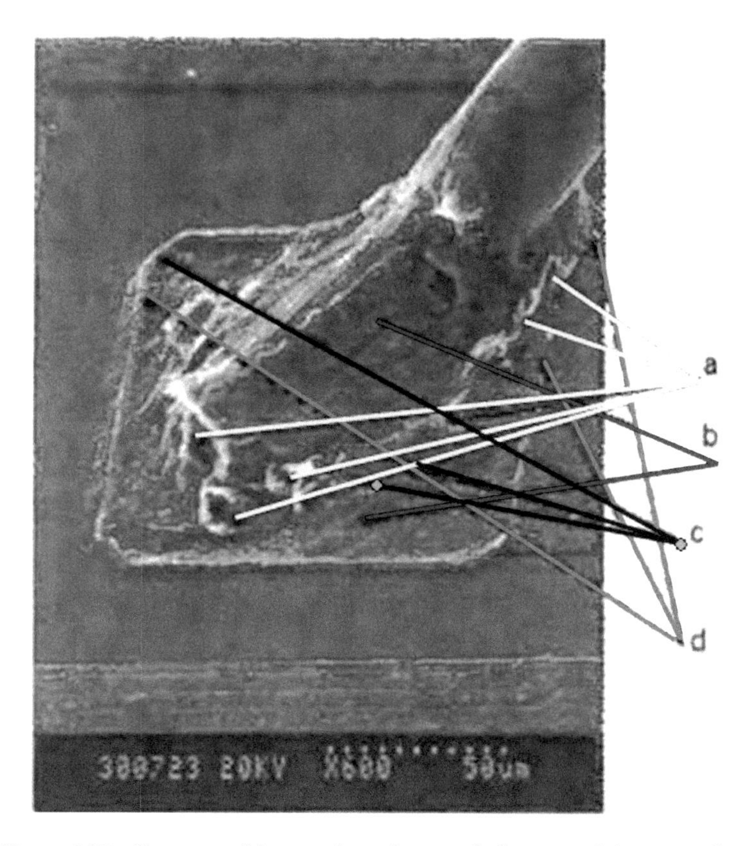

Figure 4.38 Features of interaction of a metal element of the microchip with a plasma jet at SDP: (a) zones of dynamic extrusion of metal micropipes (microexplosions); (b) a zone with the removed electroplating; (c) zones with microholes; and (d) zones of microcracks [1, 35].

out results of interaction. Therefore, the work in Ref. [37] traces the interaction between a stream of Ni particles and a plastic foil. The shape of traces that strikers have created in a foil is shown in Fig. 4.39 [1, 35].

Traces of inclusions contain material of particles and barrier in different percentages. Observable holes have been received only after etching by an acid solution. Selective etching of zones of puncture proves local activation of plastic in this zone. For a definition of SDP and efficiency of particles' penetration in composite barrier we had used the so-called foil method [37]. The

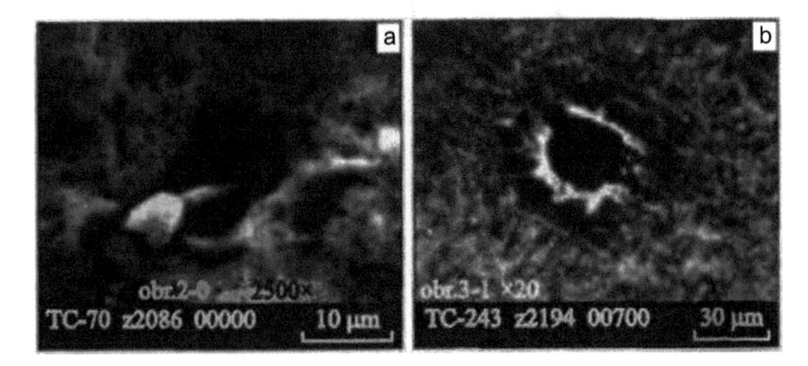

Figure 4.39 Tracks of damages in plastic films, received in a mode of superdeep penetrating, at processing in the protective container: (a) thickness of a barrier of 200 mm, 2 foil, and (b) thickness of a barrier of 50 mm, 31 foil [1, 35].

choice of this technique explains the simplicity of its application. This technique registers traces of penetration. There are traces that leave particles in nonmetallic materials. If such materials are used for a composite barrier as a fluoroplastic and a cardboard, then the analysis of changes of their structure will be difficult to execute. The channel that the particle creates "slams" and has a diameter from 1 μm up to 1 Å. The open cavity can be absent completely. Then only a track—the deformation area with inclusions of a material of the penetrating particle—is conserved. Etching of this deformation zone of a track till the size when the track is well visible using an optical or scanning electronic microscope (SCM) is necessary. Teflon has a high chemical resistance, the cardboard inversely easily etching. In carried out research were used three kinds of barrier.

The barrier in the first container was made from steel 45 and had a thickness of 50 mm. The second barrier was two-component "steel-fluoroplastic." The thickness of the layer of the steel located on the side of the collision with a stream of particles was 20 mm, and the following layer, fluoroplastic—teflon (CF_2–CF_2 – ...)n—was 25 mm. The three-component barrier consisted of the following layers in sequence: steel (thickness 20 mm), four sheets of a dense cardboard with a common thickness of ~6 mm, and a layer of fluoroplastic (25 mm). The definition of the efficiency of penetration of particles into barriers on the determined depth

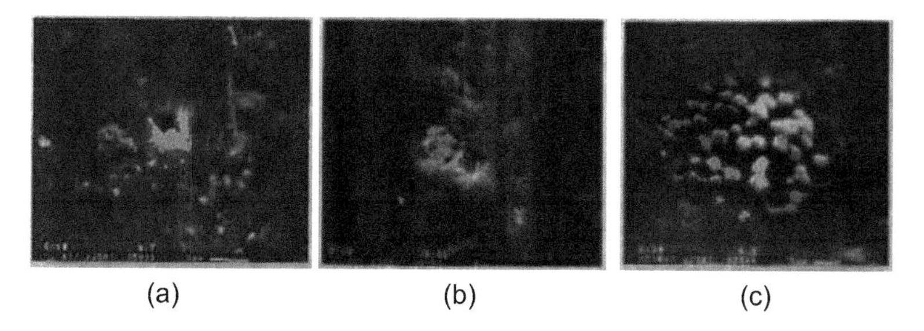

(a) (b) (c)

Figure 4.40 Variants of tracks from the interaction of the striker with the tenth foil: (a) through punching of foil, ×5000; (b) rest from the striker, ×3000; and (c) the beaten-out "plug" from the ninth foil, ×3000 [1, 35].

(equal to thickness of a barrier) was made by the amount of traces in a foil after penetration. Traces (inclusions) that qualitatively differed from the initial defects of foils were [37] calculated.

The foils were analyzed under an optical microscope. The second, fourth, and tenth foils from the back side of the barrier were in addition investigated under a scanning electronic microscope "Com-Scan" with the micro x-ray spectrum analyzer. The typical traces of interaction of strikers with a barrier, it is possible to divide into three kinds. In Fig. 4.40 [1, 35] a characteristic trace "through" punching of a foil is visible. One can see the "cork" from the beaten-out material of the previous foil after the punching of the striker (Fig. 4.40c). It confirms results of the point analysis of the inclusions that show the presence of aluminum in the cork—a material of foil.

For two- and three-component barriers on foils the new elements of structure with the spherical form are recorded. Under action of a scanning electronic microscope there is strain on the surface of a sphere. The surface of a sphere seems to be "spreading." The dynamics of change are displayed in Fig. 4.41 [1, 35], where inclusion through different time intervals from the beginning of the action of the scanning electronic microscope is displayed.

The analysis has shown that spheres consist of light elements. It is natural to assume that this material—fluoroplastic (elements of fluorine-carbonic chain (CF_2) of polymer). Fluoroplastic congeals from a liquid state, taking form with a minimal surface. At a temperature above 260°C fluoroplastic softens (melts at 327°C).

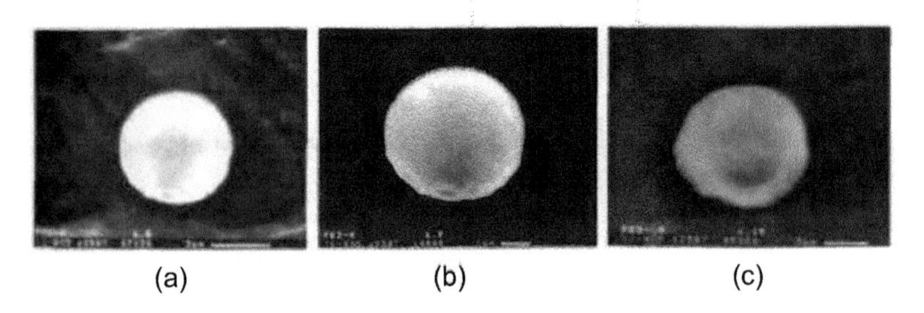

(a) (b) (c)

Figure 4.41 Inclusion on a foil behind the barrier consisting of three layers of "steel-cardboard-Teflon": (a) after its detection, ×7000; (b) in 10 s, ×10,000; and (c) in 20 s, ×5000 [1, 35].

Then this assumption well explains the occurrence of new spherical inclusions.

The composite barrier regulates the process of penetration in SDP mode [37]. During interaction of a discrete stream of particles with plastic in SDP mode intensive processes of mass transfer are observed. The intensive irradiation of electromagnetic fields and streams of high-energy ions in addition change properties of plastic.

4.6.5 Production of Polymer Nanocomposites

Polymer nanocomposites are the most effective advanced materials for different areas of application. Polymer-based reinforced nanocomposites that the polymers fill with small quantities (about 5–7 mass parts) of nanoparticles demonstrate great improvement of thermomechanical and barrier properties. The most part of published works is devoted to polyolefins filled with nanoparticles of laminated silicates, mainly montmorillonite (MMT) or bentonite.

The main methods used for the production of nanocomposites are the following: polymerization in situ, intercalation from a polymer solution, mixing in the melt, sol-gel technology, and others. The role of technology of nanocomposite components' mixing is very significant. This is due to the small size of nanofillers particles. Providing good compatibility of nonpolar polymers and rubbers with polar nanofillers is especially difficult.

The second problem is the hydrophilic surface of natural laminated silicates, which decreases the degree of the components' compatibility. Because of these reasons the mixing in the melt of nonpolar rubbers with polar nanofillers does not provide the high modifying effect.

The unusual physical phenomena at which a complex of physical effects is simultaneously implemented is known as SDP: an intensive electromagnetic radiation, an intensive strain, pressure of 8–20 GPa, flows of "galactic" ions, and so on [24, 39].

The set of experimental conditions was determined for which the penetration to relative depths of 100–10,000 calibers proceeds stably [3]. After reception of the evidence that the phenomenon of SDP exists and that there is a necessity to use physical effects that are observed in SDP conditions, there was a requirement to comprehend the fundamental result. Special attention, for more than 30 years, has been paid to the modeling of a mechanism of effective utilization of the kinetic energy of the SDP process [40].

For the first time we use SDP for modification of nonpolar isoprene elastomer by polar organomodified nanofillers: MMT and wollastonite (WST). The WST surface was modified by alkylbenzyldimethylammonium chloride and MMT—by quaternary ammonium salts $[(RH)_2(CH_3)N]^+Cl^-$, where R is a residue of hydrogenated fatty acids C_{16}-C_{18}.

The nanofillers WST and MMT were mixed with rubber by use of explosion with ammonite bulk charge with a density of 0.8–0.9 g/cm^3 and a velocity of detonation 3800–4200 m/s. Samples of rubber were located in a special container (Fig. 4.42) [1, 41] to prevent their destruction.

As a shooting substance we use a filler (MMT or WST).

Figure 4.42 The container with rubber samples inside [1, 41].

By the method of differential scanning calorimetry was established that thermostability of rubber based on isoprene elastomer with MMT is essentially higher in the case of SDP use as compared with mixing in melt (Table 4.8 [1, 41]).

Table 4.8 The thermostability of rubber based on isoprene elastomer with MMT (derivatograph "Paulic–Paulic-Erdei" with a heat velocity of 0.5–20°C/min)

Method of preparing	*T* °C oxidation temperature	Mass losses (%)
SDP	348	28
In melt	339	39

Source: [1, 41]

Simultaneously, the conditional tensile strength, tear, hardness elasticity, and adhesion to steel cord (Table 4.9 [1, 41]) increase in case of SDP utilization in comparison with the traditional method.

Table 4.9 The physical-mechanical properties of rubber based on isoprene elastomer (SRI)

	Composition and method of processing		
Properties	SRI	SRI + 5 mass fr. MMT (in melt)	SRI + 5 mass fr. MMT (SDP)
Tensile strength, MPa	15	13	22
Tear, kN/m	43	35	52
Hardness, arbitrary units	59	71	78
Elasticity, %	52	60	70
Adhesion to steel cord, N	9	8	11

Source: [1, 41]

By the method of x-ray analysis (at diffractometer D8 advance) it was estimated (Fig. 4.43 [1, 41]) that in rubber samples filled by the SDP method, MMT reflexes at diffractograms are absent. It indicates that the exfoliation of MMT in a rubber matrix takes place.

The modification of rubber mixtures by MMT leads to a great decrease of the ratio of intensivity of the first and second maximums. This is connected with an increase of dispersion of distances distribution between neighboring polymer chains and therefore with the penetration of them into the interlayer space of MMT.

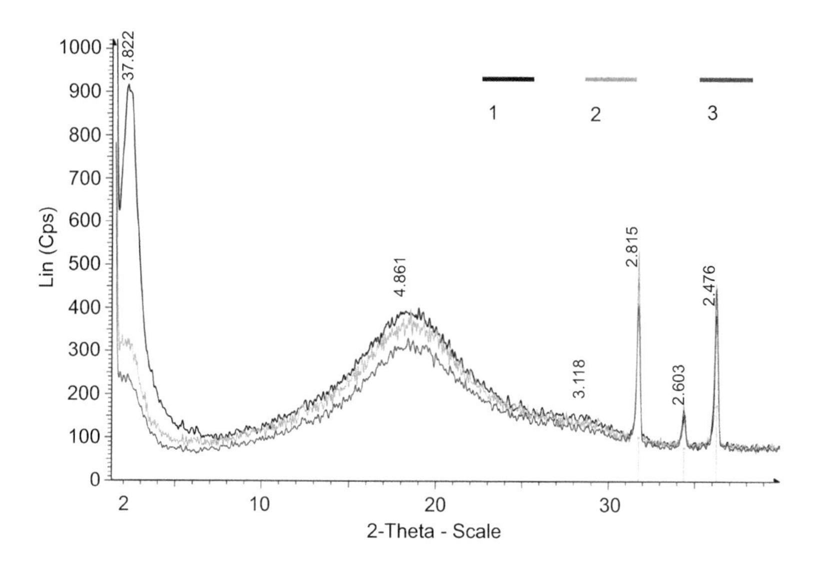

Figure 4.43 Diffractogram of rubber mixtures based on isoprene elastomer (1) and its composition with 1 (2) and 3 (3) mass fr. MMT (x-ray method at diffractometer D8 advance) [1, 41].

As a nanofiller in isoprene rubber mixtures WST was used also, which has the needle-like shape of its particles. The surface of this mineral was organomodified by alkylbenzyldimethylammonium chloride (Catamine AB).

The structure of rubbers with modified WST was investigated by electron microscopy at the Auriga device, Zeiss.

A comparison of the structure of rubbers manufactured by superdeep penetration (SDP) and by mixing in melt shows nanofiller particles irregularly distributed in the polymer matrix independently of the production method. At the same time a greater amount of filler particles with a smaller size is formed by the use of the SDP method as compared to the traditional way of component mixing. This naturally increases the surface of phase separation, which positively influences the complex of physical-mechanical and other properties of rubbers [42].

The maximum of strength and adhesion properties of rubbers is achieved at 3 mass fr. of the described filler content (Fig. 4.44

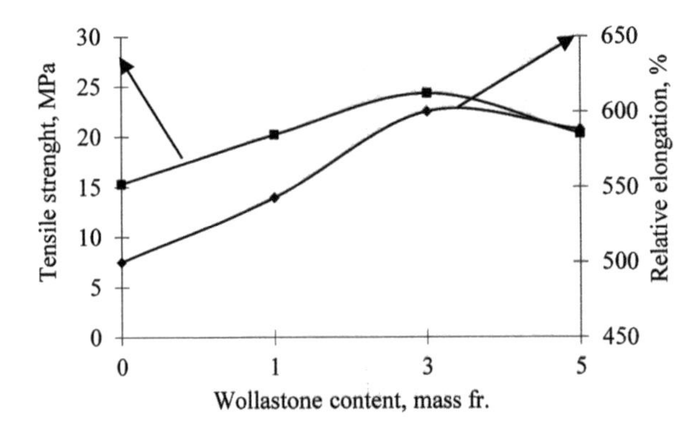

Figure 4.44 The concentration dependence of the tensile strength and relative elongation of rubber mixtures based on isoprene elastomer [1, 41].

[1, 41]). The rubber mixtures of this composition are characterized by a smaller size of filler particles (Fig. 4.45 [1, 41]).

Due to the greater surface of phase separation the SDP method provides the best properties of rubber with modified WST in comparison to the traditional way of nanocomposite production (Table 4.10 [1, 41]). So the tensile strength increases by 15% and adhesion to the steel cord by approximately the same degree. The tear increases more than other properties of rubber with WST.

Table 4.10 Physical-mechanical and adhesion properties of rubber based on isoprene elastomer

	Composition and method of processing		
Properties	**Unfilled rubber**	**Rubber + 3 mass fr. WST (SDP)**	**Rubber + 3 mass fr. WST (in melt)**
Tensile strength, MPa	15	28	24
Tear, kN/m	43	53	42
Hardness, arbitrary units	59	68	61
Elasticity, %	52	53	52
Adhesion to steel cord, N	9	15	13

Source: [1, 41]

It was also important to estimate the influence of modified WST on vulcanization characteristics of rubber mixtures because they determine the behavior in the processing of nanocompositions. The

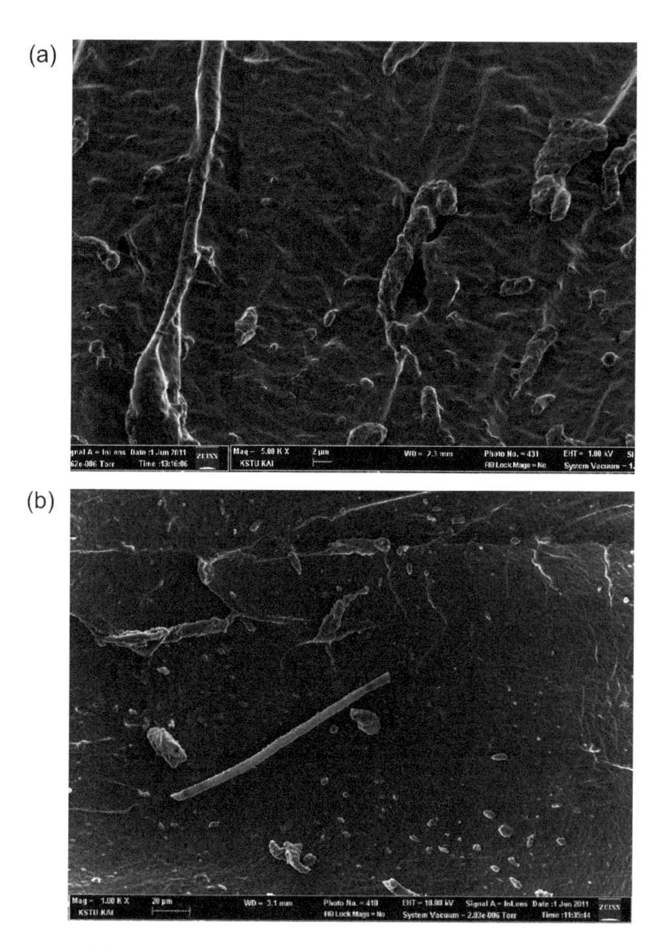

Figure 4.45 The scanning electronic microscope (the Auriga device of Zeiss) pictures of the structure of rubbers based on isoprene elastomer, modified with 3 mass fr. of wollastone by the SDP method (b) and mixing in melt (a) [1, 41].

data of Table 4.11 [1, 41] demonstrate that the time of vulcanization beginning ($T_{\text{beginning}}$) increases at 1 mass fr. of WST, and at its optimal content—3 mass fr.—it is at the level of an unfilled rubber mixture. The optimal vulcanization time (T_{optimal}) greatly increases at 1 mass fr. of WST and decreases a little at 3 mass fr.

So, the modified WST does not complicate a rubber mixture's processing.

Table 4.11 Rheometric characteristics of isoprene rubber mixtures

Composition	Min. torque	Max. torque	$T_{beginning}$, min	$T_{optimal}$, min
Unfilled rubber mixture	18	31	1.25	0.5
Rubber mixture + 1 mass fr. of modified WST	36	45	0.9	13.8
Rubber mixture + 3 mass fr. of modified WST	29	40	1.2	8.0

Source: [1, 41]

It is important to underline that polar-laminated silicates, such as MMT, practically don't increase the physical mechanical properties of nonpolar isoprene rubber at mixing in melt. At the same time the SDP method is more effective for the production of nanocomposites based on nonpolar elastomers and polar-laminated silicates.

In the case of WST a great amount of anisotropic particles of nanofiller is formed. They play the role of amplifying elements of the polymer structure. So we can say that by the SDP method a reinforcing effect can be obtained at small amounts of disperse filler.

4.6.6 Development of New Porous Materials

SDP processing takes place in the isolated volume as penetration of particles occurs even through a thick barrier. Creation of a particle flux and processing of specimens are carried out separately. But because of features of SDP interaction the additional dynamic loads inside of a specimen occur. Changes of interacting materials and the SDP regime change the process of synthesis of a carcass material. Reinforced skeleton material is created during synthesis inside of a solid body. The matrix material and a material of a skeleton have various physical and chemical properties. Due to this it is possible to delete selectively any element of a construction of a composite material. Let's consider a variant of such technological approach. For this purpose in a SDP regime we synthesize a skeleton whose material has increased chemical activity in comparison with a matrix material. As a matrix material we took glass.

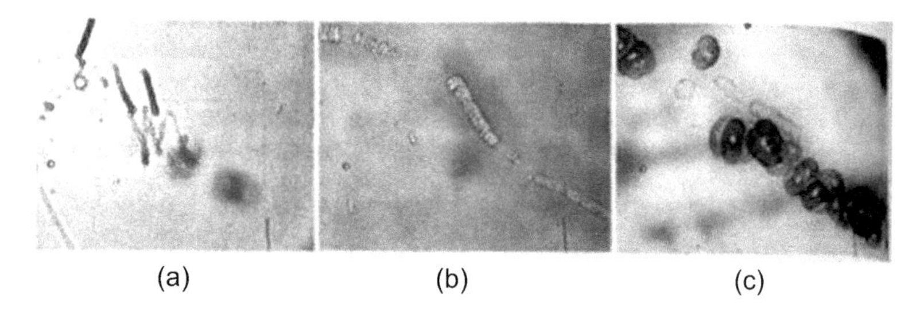

Figure 4.46 Structure processed in SDP mode of a glass sample after various modes of etching, ×1000: (a) 30 s, (b) 60 s, and (c) 120 s [1, 35].

The glass sample was etched at different dwell times in a solution of hydrofluoric acid. In Fig. 4.46 [1, 35] the change of the structure of the glass sample is shown at various conditions of etching.

The analysis of experimental results has shown that at a SDP regime in fragile nonmetallic materials there are specific defects [36]. Specific damages arise in fragile materials (glass, silicon, etc.) by the action of microjets inside of a protective envelop. Microjets have high penetrating ability. The interaction of a jet with glass has a high-energy character and leads to the occurrence of tracks and change of properties of glass. It is expressed in amplification of activity of defective zones in a glass specimen.

When processing glass materials it is expedient to execute for the purpose of shaping details in glass with a volumetric porous structure. The change of regimes of chemical and electrochemical etching in a range of real time allows one to create from a composite material a new porous material with the cross-section size of through pores from nanometers up to 10 μm.

4.6.7 Production of Polymer Tracking Membranes

Membrane technology is a rapidly growing field having large economic and ecological consequences and importance. In addition, you should know that track membranes are more effective [43].

A track membrane is a thin polymer film with through pores that are formed by penetrating a special substance into and through the material of a polymer plate and then removing the traces of penetrated particles from the matrix material, thus forming pores.

The track membrane may find use in various fields of industry as conventional membrane filters for purification of liquid substances from solid contaminants. In view of low manufacturing cost and only a slight deviation of the holes from the rated diameter (within the limits of 10 to 20%), the track membrane of the invention may be advantageously used as a dialysis filter.

A multitude of pores with straight openings in sheets of polymeric materials, formed by homogeneously bombarding the sheet with a source of heavy energetic charge particles to produce damage tracks, have been described in Ref. [44]. In subsequent stages radiation-damaged materials are removed by chemically etching, that is, by immersing the irradiated solid in an etchant. Different chemical reagents (etchants) and etching methods are known as a rule as etchants are used as the alkali solution. Without of destroyed materials as produced by high toxic solvents [44]. Its makes the industrial methods of track membrane production nonecological and less technological. Besides, these methods demand the usage of a nuclear reactor or accelerates, for example, the cyclotrons [43, 44].

At the same time, there are various methods of treatment of different materials and products, including polymers, with the use of explosive energy. For example the work in Ref. [45] discloses a treatment of synthetic polymeric materials by contacting endless sheet-like, ribbon-shaped, or filiform polymeric products with 0.1 to 2 mm size particles of sand, glass, corundum, or a metal by directing onto the surface a stream of gas carrying the aforesaid particles. This gives the textile structures a rough, woolly, soft feel and they are mat, while films become rough and mat and have a low transparency.

The advantages of track membranes, such as high pore density and uncial selection, combine with negative factors, for example, high absorption activity [43, 44].

According to this the great scientific and practical interest has the use for track membrane production the method [24, 39]. This method permits the realization of a complex of physical effects such as intensive electromagnetic radiation, intensive strain, pressure of 8–20 GPa, and flows of galactic ions [39].

So we can propose the possibility of SDP method use for making of open pores in a polymer matrix [47].

The method of SDP is carried out by using a matrix material of the membrane and special working substances that interact with the matrix in the form of a high-speed jet generated and energized by an explosion of explosive material.

The special working substance comprises a saturated or supersaturated aqueous solution of water-soluble organic salts or a saturated or supersaturated aqueous solution of water-soluble inorganic salts. The organic salts are selected from the group comprising tartrates, acetates, salicylates, and benzoates of alkali metals, for example, potassium tartrate, sodium acetate, and sodium salicylate. The inorganic salts are selected from the group comprising halides of the alkali metals and alkaline earth metals, for example, sodium chloride, sodium bromide, potassium fluoride, and calcium chloride.

The matrix material comprises an organic polymer material in the form of a solid plate.

As a polymer matrix we can use polyolefin (polyethylene, polypropylene, etc.), polyvinylchloride, fluorinated polyolefin (polytetrafluoroethylene, polyvinylidene fluoride, etc.), polyamide, polycarbonate, polyester, polysulfone, etc.

A device for the realization of the method comprises a shell in the form of a tube one end of which contains a cartridge with an explosive material and working substance in the form of a solution of solid water-soluble salt or salts. Inserted freely into the other end of the shell is a holder that contains a membrane matrix to be treated in the form of a plate. The open end of the holder is closed by a cover that is attached to the holder, for example, by screws, whereby the membrane matrix is secured in the holder. The shell with the cartridge that contains the explosive material and the working substance as well as the holder with the matrix of the material to be treated is placed into an explosion-proof chamber, and the explosive material is detonated to cause an explosion.

As a result, the working substance is expelled from the cartridge by an explosive wave in the form of a high-speed jet and penetrates deep into and through the polymer material of the plate. Under the effect of the explosion, the holder with the polymer plate and cover is ejected from the shell into the explosion-proof chamber. The cover is disconnected from the holder and the matrix is extracted and subjected to treatments with water that dissolves the water-soluble particles or washes them out from the membrane matrix, thus forming microscopic openings that pass through the polymer plate. Then the polymer plate is sliced into thin pieces that can be used, for example, as filter plates.

For tracking membrane production by the SDP method it is necessary to optimize the following parameters:

- The chamber size
- The type and charge construction
- The explosion power
- The velocity of detonation
- The thickness of charge
- The type and dispersion of the working substance
- The distance from the charge end till the polymer sample
- The solvent's composition
- The material and size of the screen

So we optimize the above-mentioned parameters of explosion chamber, which provide the demanded quality of tracking membranes. As a first step we choose the conditions of preservation of the sample during bombardment. The parameters of the explosion chamber used previously for the creation of open pores in the ceramic matrix cannot be used due to the difference of elasticity modules of the polymer material and ceramics. So we have produced the special protective steel screen with one central and several scattered holes. It provides the part of impact wave energy consumption for destruction of the steel screen.

For preservation of the polymer sample during bombardment a special steel container was produced with a hole at the bottom. During bombardment the particles of working substances penetrate into the polymer matrix. Ammonite of bulk density 0.8–0.9 g/sm^3 (Fig. 4.47 [1, 47]) was used as a charge.

The device contains a tubular plastic shell with both ends open. The height of the plastic tube is 200 mm. The device contains a cartridge with the detonatable explosive material and the working

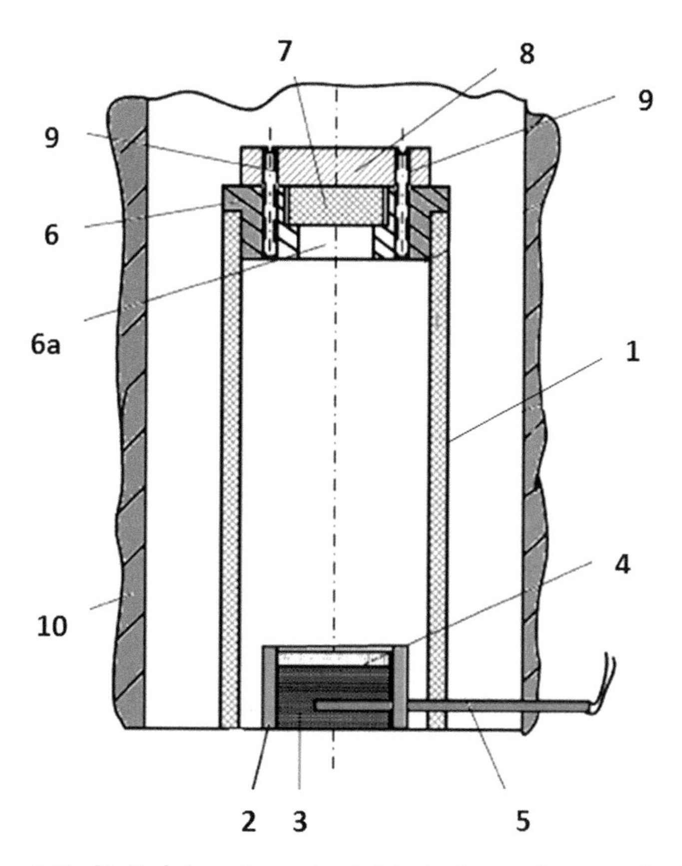

Figure 4.47 Vertical view of a constructed device for membrane production by the SDP method: 1, tubular shell; 2, cartridge; 3, detonatable explosive material; 4, supersaturated solution of a water-soluble solid salt; 5, detonator; 6, membrane holder; 7, cavity of the holder; 8, cover of the holder; 9, fasteners; and 10, explosion-proof chamber [1, 47].

substance in the form of a supersaturated solution of water-soluble solid salt. The cartridge is inserted into the lower open end of the tubular holder. The detonator is used for detonation of the explosive material. The membrane holder with an open-bottom cavity for receiving a membrane matrix is inserted into the upper open of the shell. The device is placed into an explosion-proof chamber.

The explosion wave that has a detonating nature should impart to the solid particles of the working substance a velocity in the range of 3800 to 4200 m/s.

Some particles deeply penetrate into the membrane matrix material, and some particles pierce the body of the membrane matrix from its exposed side. Under the effect of the explosive wave, the holder, together with the membrane matrix and the cover, is expelled from the shell into the explosion-proof chamber. The cover is then disconnected from the holder, and the treated membrane matrix is extracted from the holder 6. However, the membrane matrix will still contain residue of the water-soluble particles of the working substance. Removal of the residual trace particles of the solid substance from the membrane may be carried out in the running flow of water, leaving a plenty of small-diameter holes. As a polymer matrix firstly we use impact-strength polyethylene.

At the same time perspective materials for the production of polymer membranes by the SDP method are polyethyleneterephthalate, polycarbonate, etc.

The solid plate of matrix polymer materials may have a total thickness in the range of 10 to 20 mm. After removal of the residue of the working substance, the solid plate is sliced into the track membrane having a thickness of 5 to 50 μm by means of a microtome.

The microstucture produced by SDP method membranes is presented in Fig. 4.48 [1, 47].

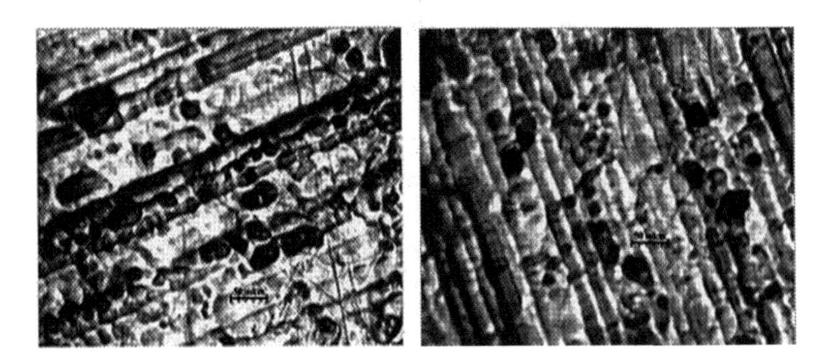

Figure 4.48 The electronic microscope pictures the structure of nanomembranes based on polyethylene [1, 47].

The picture shows a section of samples taken parallel to the diameter of the polymeric cylinder by means of a microtome. The black dots in the photographs represent the pores of the sliced samples, which have diameter sizes in the range of 80–100 nm.

The actual diameters and the range of the diameters of the holes depend mainly on the velocity of the particles, diameter of the shell, and distance from the cartridge with the explosive material and the particles to the membrane matrix material in the holder. The through holes produced in the track membrane are oriented in the direction of the jet of particles and occupy from 10 to 20 vol. percent of the membrane material volume.

4.7 Method for Treating Thin-Film Materials with a Flow of Solid Particles in the Electric Field[a]

The main drawback of the SDP method of polymer tracking membrane production [46, 47] is its very design, that is, all the risks associated with the use of explosives. The method cannot provide uniformity in the distribution of track holes and their diameters. Another significant disadvantage of the explosion method is a reflected wave upon detonation. At the same time the shockwave is reflected from the shell and moves to the center, carrying with it a significant part of the energy, while the pressure around the shell rapidly falls off (faster than instantaneous detonation). As a result, the acceleration of the shell is reduced more rapidly than with instantaneous detonation. This significantly reduces the efficiency of the explosive impact on the membrane matrix.

A new method for treating thin-film materials with a flow of solid particles in an electric field is developed. More specifically, the method for manufacturing track membranes is based on piercing a matrix of a thin-film material with a flow of hard particles generated by an electric field. The essence of the method consists of charging and accelerating particles of a powder that constitutes a working material for, for example, perforating a thin-film object

[a]With contributions from G. Makrinich and A. Fruchtman (Holon Institute of Technology, Holon, Israel).

intended for manufacturing, for example, track membranes for use as membranes for dialysis, filtering gases, etc. The particles are accelerated and acquire a kinetic energy under the effect of an electric field developed between two metallic electrodes such as a continuous charging electrode and a perforated electrode, for example, in the form of a net. The object being treated may comprise a replaceable thin-film sheet or a belt periodically shifted and fixed in a working position for exposure to the action of the moving particles.

Realization of the method is based on the use of an apparatus that consists of a closed chamber that contains two mutually spaced and electrically separated metallic electrodes. One of the electrodes is an acceleration electrode in the form of a net with a plurality of openings or cells for passing the particles to the exposed object, while another electrode, referred to hereinafter as a charging electrode, is continuous.

The method is carried out as follows.

First, a specific powder of a selected material, shape, and size is supplied to the interelectrode space by means of the powder supply unit or injector. Next, a cyclic voltammetry (CV) is supplied to the metallic electrodes. A part of the powder particles should already have a noncompensated charge, but neutral particles will acquire the noncompensated charge under the effect of the electric field. As a result, under the effect of the electric field (EF), which is generated between the electrodes in the interelectrode space, the charged particles begin to move with acceleration to the acceleration electrode and when they reach this electrode, the particles develop a significant kinetic energy that depends on the value of the charge, the particle mass, and the potential difference between the electrodes.

Realization of the proposed method is based on the use of an apparatus that consists of a closed chamber that contains two mutually spaced and electrically separated metallic electrodes. One of the electrodes has a plurality of openings, for example, cells. If this electrode is a net, another electrode is continuous. The apparatus is provided with a powder supply mechanism or injector for the supply of a specific powder into the interelectrode space formed between the electrodes, preferably closer to the continuous electrode.

4.7.1 Charging and Acceleration of Solid Micro- and Nanoparticles by the Action of an Electrical Field

First, a specific powder of a selected material, shape, and size is supplied to the interelectrode space by means of the powder supply mechanism or injector. Next, a voltage is supplied to the metallic electrodes. A part of the powder particles should already have a noncompensated charge, but neutral particles will acquire the noncompensated charge under the effect of the electric field. As a result, under the effect of the electric field the charged particles begin to move with acceleration to the net-like acceleration electrode. And when they reach this electrode, the particles develop a significant kinetic energy that depends on the value of the charge, the particle mass, and the potential difference between the electrodes.

More specifically, the particle charge can be evaluated by the following equation:

$$Q = \varepsilon_0 A E, \tag{4.8}$$

where Q is a dust particle charge, ε_0 is dielectric permeability of vacuum, A is the surface area of the particles, and E is the intensity of the external electric field.

A charge particle experiences an effect of the electric field, and when the electric force is greater than the weight of the particle, the latter may levitate. The value of this critical electric field E_c is evaluated from balance of forces in the following manner:

$$mg = \varepsilon_0 A E_c^2, \tag{4.9}$$

where m is a mass of a dust particle and g is acceleration of gravity. For a spherical particle a value of the critical electric field is equal to

$$E_c = \sqrt{\frac{\rho d g}{6\varepsilon_0}}, \tag{4.10}$$

where ρ is a density of the particle material and d is a particle's diameter.

Figure 4.49 shows theoretical and experimental values of a critical electric field for various solid materials. The values of these critical electric fields for various solid materials are shown in Table 4.12. Particles used in the experiments had a cubical or

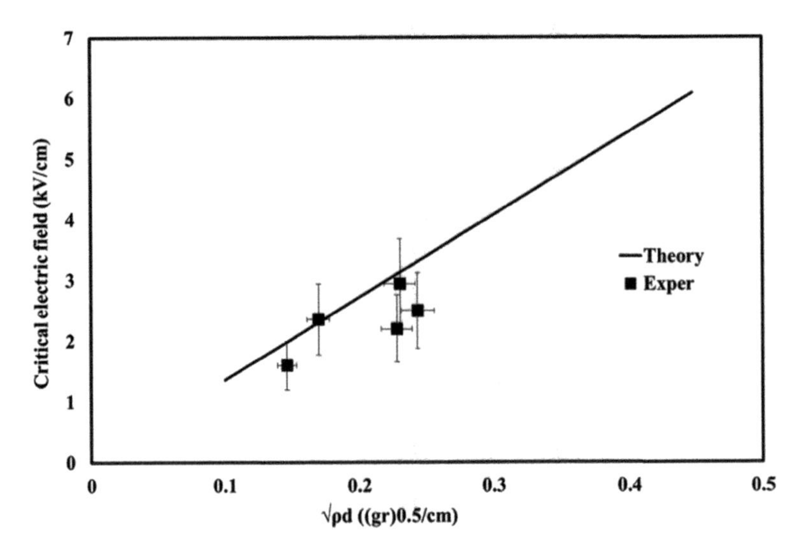

Figure 4.49 Theoretical and experimental values of a critical electric field for various solid materials.

spherical shapes, and the dimensions of cubical particles were recalculated to diameters of spherical particles of the same surface area. It can be seen from Fig. 4.49 that the measurement data and the theoretical data correspond each other.

When the intensity of the electric field is higher than its critical value for the given particle type, the particles are accelerated. If the friction forces are neglected, the pulse equation can be written as follows:

$$m\frac{d\upsilon}{dt} = \varepsilon_0 A E^2 - mg, \tag{4.11}$$

where υ is the velocity of the charged solid particle.

A charged particle is accelerated by the electric field in the direction opposite to the direction of the gravity force. If we assume that the electric force intensity $U = El$ is constant along the entire way of particle acceleration between two electrodes, the accumulated kinetic energy acquired by the particle can be expressed by the following equation:

$$K = \left(\varepsilon_0 A \frac{U^2}{l^2} - mg\right) l, \tag{4.12}$$

Table 4.12 Critical electric fields for various solid particles

Material of particle	Density (g/cm^3)	Dimension (μm)	Critical electric field (kV/cm)
Al_2O_3	3.2	100	1.6
Cu	8.9	100	2.2
Fe	7.8	100	2.3
NaCl	2.16	200	2.3
SiO_2	2.65	300	2.9

where l is the acceleration path and U is the applied field intensity between the electrodes.

When the particle is accelerated it collides with the material of the membrane and may be reflected from the surface of the membrane or may penetrate into the material, leaving a trail of the material in the form of pores—track. The geometry of the pores formed corresponds to the flight path of the particle within the material. The kinetic energy of the particle accelerated by the electric field of particulate matter is converted into energy of destruction of the material. It is assumed that the hardness of the material particles is much greater than the hardness of the membrane material and energy is not spent on the destruction of the particles themselves. For particles with a size comparable to the thickness of the bulk material (membrane material), the material may be destroyed when the brake force of inertia balances the destructive power.

The conducted experiments show the fact that in an electric field a solid particle can be charged and that its movement can be accelerated. This is an experimentally observed phenomenon that can be used as a basis for developing various devices and processes.

4.7.2 Production of Tracking Membranes by Accelerated Particles of Powders

When the accelerated particle collides with the material of the membrane, the particle may be reflected from the surface of the membrane or may penetrate into the material, leaving in the material a trail in the form of a pore known as a track. The geometry of the pores formed corresponds to the flight path of the particles within the material. The kinetic energy of a particle accelerated by the electric field is converted into energy of destruction of the material.

Let us assume that the hardness of the material particles is much greater than the hardness of the membrane material and that the energy is not spent on the destruction of the particle itself. For particles with a size comparable to the thickness of the bulk material (membrane material), the material may be destroyed when the brake force of inertia is in balance with the destructive power. Thus, the critical condition can be written as follows:

$$ma = S\sigma, \tag{4.13}$$

where a is the acceleration of a solid particle in the membrane material and S is a cross-sectional area of the destruction. In the condition expressed by Eq. 4.6, the friction force between the solid particle and the membrane material is neglected.

If it is assumed that on the other side of the membrane (i.e., on the side opposite to the bombarded side) the velocity of the particle is equal to 0), then the acceleration of the particle can be expressed as follows:

$$a = \frac{\upsilon_0^2}{2h}, \tag{4.14}$$

where υ_0 is the velocity of the solid particle after acceleration in the electric field and h is the thickness of the membrane.

The destruction area can be expressed as follows:

$$S = Lh, \tag{4.15}$$

where L is the total length of destruction. For a spherical solid particle the total length of destruction can be expressed by the following formula:

$$L = n\frac{d}{2}, \tag{4.16}$$

where n is a destruction number. The value of the destruction number n is dimensionless and can be obtained experimentally. In other words, when a solid particle that moves with a high velocity passes through a thin film and makes a hole in it, such a hole is normally surrounded by a number of thin cracks that extend radially outward from the periphery of the formed hole. The applicants decided to evaluate the destruction number n as a ratio of the total length of such cracks to the radius of the hole formed in the thin-film material. Normally the destruction number $n$is in the range of 3 to 6 and the final value is calculated as an average value, for example, from 10 to 20 measurements.

The value of the destruction number n is dimensionless and can be obtained experimentally. In other words, when a solid particle that moves with a high velocity passes through a thin film and makes a hole in it, such a hole is normally surrounded by a number of thin cracks that extend radially outward (Eq. 4.12). The upper limit of the membrane thickness can be determined from the following equation:

$$h = \sqrt{\frac{\left(\varepsilon_0 A \left(U/l\right)^2 - mg\right) l}{\sigma L}} \tag{4.17}$$

For spherical particles, the formula for the membrane thickness can be converted into the following expression:

$$h_s = \sqrt{\frac{\pi l d \left(6\varepsilon_0 \left(U/l\right)^2 - \rho d g\right)}{3\sigma n}} \tag{4.18}$$

Calculation by means of the upper limit of the membrane thickness by using Eq. 4.18 for a spherical particle of aluminum oxide (having diameter $d = 10^{-4}$ and density $\rho = 3200$ kg/m^3) accelerated in the electric field generated by the potential difference $U = 6000$ V at the interelectrode distance $l = 0.02$ m, for a strength limit of the membrane material $\sigma = 10^6$ Pa and a destruction number $n = 3$, gave the upper limit of the membrane thickness at which the membrane could be pierced with the formation of through openings $h_s \approx 10^{-6}$ m.

Realization of the proposed method is based on the use of an apparatus (Fig. 4.50).

The apparatus consists of a closed chamber (1) that contains two mutually spaced and electrically separated metallic electrodes (2) and (3). The acceleration electrode (3) is a net and has a plurality of openings, another charging electrode (2) is continuous. The apparatus 100 is provided with an adjustable high-voltage power supply unit (4) for the supply of CV to the electrodes, a powder

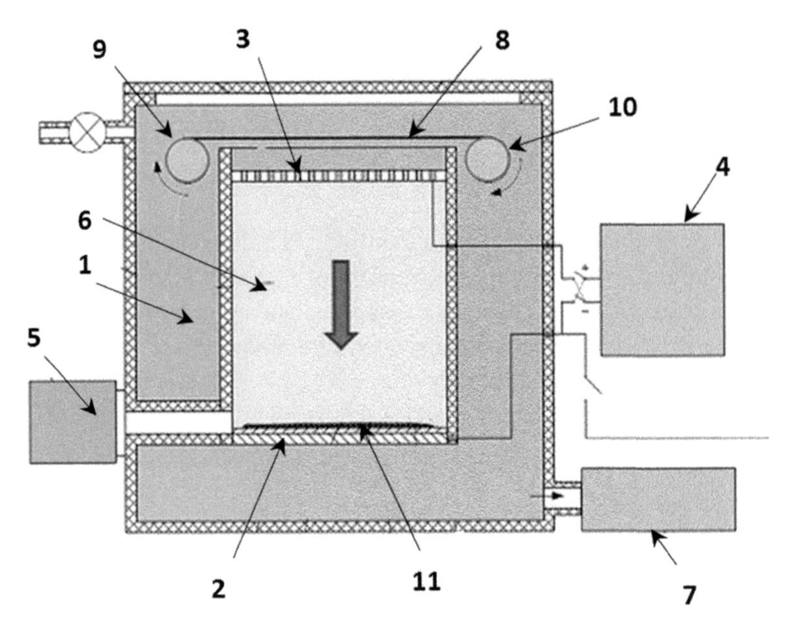

Figure 4.50 Apparatus for the production of tracking membranes by accelerated particles of powders: 1, chamber ($L = 100$ mm, $W = 30$ mm, $H = 100$ mm); 2 and 3, electrodes; 4, voltage power; 5, powder injector; 6, interelectrode space; 7, vacuum pump; 8, film: 9 and 10, bobbins; and 11, specific powder.

supply unit or injector (5) and the supply of a specific powder into the interelectrode space (6). The apparatus may be equipped with a vacuum pump (7) for evacuation of air from the interior and with a device for injection of inert gas into the space (6). A thin-film material (8), which in Fig. 4.49 is shown as a continuous belt movable from the supply bobbin (9) to the receiving bobbin (10), is located over the acceleration electrode (2).

The process of production of tracking membranes is described below.

Solid particles, which in the interelectrode space (6) are loaded onto the charging electrode (2), acquire a charge that in its sign corresponds to the sign of the charging electrode. Under the effect of the electric field, the particles start moving toward the accelerating electrode (3), which bears the charge reverse to the charging electrode. As a result, solid particles of the powder are accelerated by the electric field generated between the charging and accelerating electrodes. The powder particles pass through the open cells of the net and impact the thin-film material (8).

The exposure time $\tau(s)$ is determined by the intensity $J_{\mathrm{m}}\left(\frac{\mathrm{kg}}{\mathrm{m}^2\mathrm{sec}}\right)$ of the particle stream at the output from the interelectrode space and by the desired density of holes σ $\left(\mathrm{m}^{-2}\right)$ in the membrane to be produced. The exposure time $\tau(s)$ is calculated by means of the following formula:

$$\tau = \frac{\sigma \rho k_{\mathrm{f}} D^3}{J_{\mathrm{m}}}, \tag{4.19}$$

where ρ is a density of particle material, k_{f} is a coefficient that depends on the shape of particles, and D is an average side of the particles.

The practice examples of obtaining a tracking membrane via a stream of powder particles accelerated in an electric field are given in Table 4.13.

Samples were rectangular-shaped high-density polyethylene films. A working substance for treating the sample comprised:

- Aluminum oxide powder
- Iron powder
- Copper power
- Silicon oxide powder
- Sodium chloride powder
- Sucrose powder

4.8 Conclusions

The process of shock interaction of a clot of high-velocity separate strikers with the metal barriers was analyzed and the expenditure of energy for superdeep penetration (SDP) was estimated. On this basis, the following conclusions were drawn and a new concept of the physics of SDP phenomenon is offered.

The new concept of the physics of SDP phenomenon is based on the successive realization of a set of well-known physical effects, stage by stage leading to the creation of a closed energetic system and to the realization of the cavitation process (collapse

Table 4.13 Examples of obtaining a tracking membrane

No.	δ[a] μm	S[b] mm	L[c] mm	Kind of particles	Δ[d] kV	V[e] m/s	t[f] min	$D_{average}$[g] μm
1				Al_2O_3		3.8		50 ± 15
2		100 ± 10		Fe		2.9		30 ± 10
			80	Cu	15	2.7	1.0	
4	15	12 ± 2		SiO_2		42		10 ± 1
5		35 ± 10				5.6		25 ± 5
6			20	NaCl	25	14	0.5	22 ± 5
7	24	35 ± 5			30	16		20 ± 5
8	15		80	$C_{12}H_{22}O_{11}$	15	6.3	1.0	

[a]Thickness of polyethylene film
[b]Particle size
[c]Distance between the charging electrode and acceleration electrode
[d]Air pressure
[e]Velocity of the accelerated particles
[f]Exposure time of the powder to the electric field
[g]Average diameter of the openings formed in the sample film

of microcavities in dense plasma) with the additional energy emission.

On the basis of complex research the opportunity for the effective manufacture of composite materials by clots of discrete powder particles is shown. Iron, aluminum, and alloys are used as the matrix materials.

The regime of SDP allows one to obtain new composite materials from a massive solid body by modified nano- and microelements. The presence of nano- and microelements in a skeleton of a composite material is the reason for significant changes in physical and chemical properties. The material on these levels of structure has physical and chemical properties that considerably differ from properties of this material on mezo- and macrolevels. The directed regulation of physical and chemical properties of massive composite materials is reached.

Manufacturing of a tool with a composite material on the basis of a high-alloy tool steel allows one to increase the level of properties of matrix steel by tens of hundreds of percent and to produce ecological and competitive metal-cutting and mining tools with hard alloy. Using the SDP process makes it possible to obtain reinforcing and hardening depths in iron exceeding 200 mm

Manufacturing of a composite material on the basis of technical aluminum provided anisotropy of electrochemical and electric properties up to 2 times and regulation of electron work function in the given zones of a product.

The SDP method opens great perspectives for the creation of advanced nanocomposites based on nonpolar elastomers and polar nanofillers. This method is more effective than mixing of components of rubber mixtures in melt. The optimization of explosion conditions while using the SDP method will provide for further improvement of nanocomposite properties. This new method of mixing nanofillers with polymer matrixes allows one to produce smart functional materials based on polymers of different chemical composition and polarity.

The SDP method opens great perspectives for the production of tracking membranes based on polymer matrix. This method is cheaper, simpler, and more ecologically friendly as compared to nuclear industrial methods. Its use excludes the application of highly toxic solvents. Simultaneously it permits the creation of polymer membranes with micro- and nanosize open pores and besides ensures the absence of oxidation products, which can migrate into filtrate.

A new method for manufacturing track membranes by piercing a matrix of a thin-film material with a flow of hard particles generated by an electric field is proposed. The essence of the method consists of charging and accelerating particles of a powder that constitutes a working material for treating, for example, perforating a thin-film object intended for manufacturing, for example, track membranes for use as membranes for dialysis, filtering gases, etc.

References

1. Figovsky O., Beilin D., Usherenko S., Kudryavtsev P. (2016). Environmental friendly method of production of nanocomposites and nanomembranes, *Journal Scientific Israel - Technological Advantages*, **18**(2).
2. Usherenko S. M. (2006). The phenomenon of superdeep penetration, *Journal Scientific Israel - Technological Advantages*, **8**(1,2), 83–94.

3. Usherenko S. M. (2002). Modern notions of the effect of superdeep penetration, *Journal of Engineering Physics and Thermophysics*, **75**(3), 753–770.
4. Roman O. V., Andilevko S. K., et al. (2002). Effect of superdeep penetration. State of the art and prospects, *Journal of Engineering Physics and Thermophysics*, **75**(4), 997–1012.
5. Usherenko S. M., Korshunov L. G., et al. (2002). Multifactor experiments under superdeep- penetration conditions, *Journal of Engineering Physics and Thermophysics*, **75**(6), 1249–1253.
6. Cherny G. G. (1987). The gear of anomalous resistance at motion of bodies in solid media, *Reports of Academy of Sciences of USSR, The Theory of Elasticity*, **292**(6), 1324–1328 (in Russian).
7. Grigorian S. S. (1987). About the nature of the "superdeep" penetration of firm microparticles in solid materials, *Reports of the Academy of Sciences of USSR, Mechanics*, **292**(6), 1319–1322 (in Russian).
8. Rakhimov A. E. (1994). Qualitative model of the superdeep penetration, *Messages of the Moscow University. Mathematics, Mechanics*, **Series 1**(5), 72–74 (in Russian).
9. Korshunov L. G., Chernenko N. L. (2005). About the mechanism and conditions of realization of effects of superdeep penetration of the accelerated particles in metals, *Physics of Metals and Metallurgical Science*, **101**(6), 660–667 (in Russian).
10. Aleksentseva A., Krivchenko A. (1998). Analysis of the conditions for ultradeep penetration of powder particles into a metallic matrix, *Technical Physics*, **43**(7), 859–860 (in Russian).
11. Makarov P. V. (2006). Model of the superdeep penetration of firm microparticles into metals, *Physical Mezomechanics*, **9**(3), 61–70 (in Russian).
12. Cherepanov G. P. (1996). An analysis of two models of superdeep penetration, *Engineering Fracture Mechanics*, **53**(3), 399–423.
13. Kiselev S. P., Kiselev V. P. (2002). Superdeep penetration of particles into a metal target, *International Journal of Impact Engineering*, **27**(2), 135–152.
14. Kolmogorov V. L., Zalazinskij A. G., et al. (2003). About the superdeep penetration of a particle into elasto-plastic medium, *Collection of Materials. VII Zababahin Scientific Readings*, Snezhinsk, 1–17 (in Russian).

15. Naimark O., Collombet F. A., et al. (1997). Super-deep penetration phenomena as resonance excitation of self-keeping spall failure in impacted materials, *Le Journal de Physique* IV, **7**(3), 773–778.
16. Kheifets A. E., Zeldovich V. I., et al. (2003). The shock-wave model of the effect of superdeep penetration of powder particles into metallic materials, *E-MRS Fall Meeting* (www.science24/com/conferences).
17. Sivkov A. (2001). A possible mechanism of "superdeep" particle penetration into a solid target, *Technical Physics Letters*, **27**(8), 692–694.
18. Usherenko S. (1983). Conditions of superdeep penetration and creation of process of hardening of tool steels by a high-speed flow of powder materials, *The Thesis on Competition of a Scientific Degree of Cand. Tech. Sci. Minsk. The Belarus Polytechnic Institute*, 308 (in Russian).
19. Zeldovich V. I., et al. (2001). Structural changes in iron-nickel alloys caused by action of high-speed flow of powder particles. Effects of impact loading, *Physics of Metals and Metallurgical Science*, **91**(6), 72–79 (in Russian).
20. Homskaja V., et al. (2006). Structural transformations and effects of localization of deformation into copper under action of a high-speed flow of powder microparticles, *News of the Russian Academy of Science. A Physical Series*, **70**(7), 1054–1056 (in Russian).
21. Altshuler L. V., Andrilevko S. K., et al. (1989). About model of the "superdeep" penetration, *Pis'ma ZETF*, **15**(5), 55–57 (in Russian).
22. Usherenko S. M., Koval O. I., et al. (2004). Estimation of the energy expended for superdeep penetration, *Journal of Engineering Physics and Thermophysics*, **77**(3), 641–646.
23. Rusov V. D., Usherenko S. M., et al. (2001). The simulation of dissipative structures and concentration waves of point defects in the open nonlinear physical system "metal + loading + irradiation", *Questions of a Nuclear Science and Technique*, Ukraine, National Centre of Science, Kharkov Physicotechnical Institute, **4**, 3–8 (in Russian).
24. Owsik J., Jach K., Usherenko S., Usherenko Y., Figovsky O., Sobolev V. (2008). The physics of superdeep penetration phenomen, *Journal of Technical Physics*, **49**(1), 3–29.
25. Usherenko S., Granberg A., Sobolev V. (2010). High-energy physical effects at formation of composite materials, *Journal Scientific Israel - Technological Advantages*, **12**(1).

26. Babeshko V. A. (2001). The phenomenon of localization of wave processes and resonances, *The Educational Journal of Soros*, **7**(11), 134–137 (in Russian).
27. Sobolev V. V., Usherenko S. M. (2006). Shock-wave initiation of nuclear transmutation of chemical elements, *Journal de Physique IV France*, 134, 977–982.
28. Owsik J., Usherenko S. (2008). SDP technology for "green" technology of metallic reinforced nanocomposites, *Proceeding of the NATO Advanced Research Workshop on Environmental and Biological Risks of Hybrid Organic-Silicon Nanodevices*, St. Petersburg, Russia, 18–20 June 2008, pp. 31–54.
29. Usherenko S. (2011). Method of strengthening tool material by penetration of reinforced particles, US Patent 7897204 B2.
30. Figovsky O., Usherenko S., et al. (2007). The creation of metal composite materials, *The Physics and Technics of High-Energy Processing of Materials*, Art-Press: Dnepropetrovsk, 218–235 (in Russian).
31. Figovsky O., Usherenko S. (2009). Nanocomposites prepared by SDP method: the physics of superdeep penetration phenomenon, *Journal Scientific Israel - Technological Advantages*, **11**(3), 59–73.
32. Usherenko S., Nozdrin V., et al. (1994). Motion and deceleration of explosively accelerated solid particles in a metallic target, *International Journal of Heat and Mass Transfer*, **37**(15), 2367–2375.
33. Glasmacher U. A., Lang M., et al. (2006). Phase transitions in solids stimulated by simultaneous exposure to high pressure and relativistic heavy ions, *Physical Review Letters*, **96**, 195701.
34. Usherenko S., Figovsky O., et al. (2007). Production of the new metalcomposite materials by method of superdeep penetration, *Journal Scientific Israel - Technological Advantages*, **9**(1), 28–32.
35. Usherenko S., Figovsky O. (2008). Novel environment friendly method of preparing nanoreinforced composites based on metallic, ceramic and polymer matrixes - superdeep penetration, silicon versus carbon, *Proceeding of the NATO Advanced Research Workshop on Environmental and Biological Risks of Hybrid Organic-Silicon Nanodevices*, St. Petersburg, Russia, 18–20 June 2008. pp. 31–54
36. Usherenko S. M., et al. (2005). Flows of "galactic" ions is arising at collision of the dust clots with protective envelopments, *2nd Belorussian Space Congress, Materials of a Congress, Minsk, OIPI NAN* Belarus, 33–38 (in Russian).

37. Roman O. V., Dybov O. A., Romanov G. S., Usherenko S. M. (2005). Damage of integrated by high-velocity microparticles penetrating thick-wall obstacles, *Journal of Technical Letters*, **31**, 46–47.
38. Dubov O. A. (2000). Technique of foils and its application at recording processes of the penetration, dynamic reorganization of structure of materials. *The Collective Monograph*, Minsk, 5–21 (in Russian).
39. Figovsky O., Usherenko S. (2009). Composite materials prepared by SDP-method: the physics of superdeep penetration, *Nanotechnics*, **3**(19), 27–37 (in Russian).
40. Usherenko S., Figovsky O. (2008). Superdeep penetration as the new: physical tool for creation of composite materials, *Advanced Materials Research*, **47–50**, 395–402.
41. Figovsky O., Gotlib E., et.al. (2012). Super deep penetration - new method of nanoreinforced composites producing based on polymer matrixes, *Journal Scientific Israel - Technological Advantages*, **14**(1), 74–78.
42. Gotlib E., Figovsky O., et al. (2011). Influence of mixing method of modified wollastonite with rubbers based on SRI-3 on their structure, *Newsletter of the Kazan State Technological University*, **15**, 141–145.
43. Apel P. (2001). Track membrane production, *Radiation Measurements*, **37**(16), 559–566.
44. Shirkova V. V., Tretyakova S. P. (1997). The ways of track membranes manufacturing *Radiation Measurements*, **28**(1–6), 791–798.
45. Koerber H. (1990). Method for manufactory of mat and rough, laminar, ribbon-shaped or fibrous polymeric products with a stream of particles, US Patent 4960430 A.
46. Figovsky O., Gotlib E., et al. (2015). Method of manufacturing a track membrane, US Patent 8980148 B2.
47. Figovsky O., Gotlib E., et al. (2012). The production of polymer tracking membranes by super deep penetration method, *Journal Scientific Israel - Technological Advantages*, **14**(3), 120–124.

Chapter 5

Nanotechnology in the Production of Bioactive Paints, Coatings, and Food Storage Materials

5.1 Introduction

In the last decades in the paint industry considerable attention has been paid to bioactive coatings. Such coatings are formed from various varnish-paint materials, impregnants, and polymer compounds with additions of different biocides as active components. As a rule, the biocidal additions to polymer compositions were made to obtain quite definite functional properties—for wood protection from mold or wood engraver, for the protection of concrete plaster from fungi, for the conservation of water compositions, etc.

The questions of long preservation of biological objects represent enough challenge. The manufacture of many products of food and medicines after removing microbic activity appears insufficient to preserve them. Therefore, stabilizing and preserving materials are the goal of many kinds of production.

Green Nanotechnology
Oleg Figovsky and Dmitry Beilin

ISBN 978-981-4774-10-9 (Hardcover), 978-1-315-22928-7 (eBook)
www.panstanford.com

In spite of their wide use and popularity, such bioactive compositions have some serious disadvantages, which stimulate new research in this field.

- First and foremost, the biocides applied are highly specific, that is, they are active against only one definite group of microorganisms; hence, their protective action is far from being as complete as is desirable for practical purposes. Besides, they are, in principle, directed against microbes dangerous for paint compositions themselves or painted materials, and are powerless against pathogens dangerous for man. Meanwhile, today the level of bacterial or viral contamination of environment is so high that the creation of paints and coatings active against a wide spectrum of pathogens becomes more and more urgent. At the same time, it is clear that the known disinfectants (chloramines, chloride of lime, etc.) cannot be applied for this purpose since their chemical properties do not allow to introduce them into paints and coatings.
- Secondly, there is a great problem with pollution of the environment. From this point of view, the bioactive coatings used so far are not satisfactory since, apart from the pollution with solvents, they are not inert and often have toxic effects on various media in contact with them.

The problems mentioned above may be overcome through the use of the new biocide of an entirely different origin, namely of silver nanoparticles. Nanoparticles are very small particles of different substances, including metals. As found in numerous studies during the last two decades, particles with dimensions on the nanometer scale (10^{-9}–10^{-8} m) possess peculiar properties, different from those of atoms and ions on the one hand and of bulk substances on the other. There is strong evidence that nanoparticles, especially those made from metals, may find a wide variety of applications in science, technology, and medicine. In particular, there are serious reasons for the statement that silver nanoparticles may serve as a very effective biocide [1–3].

It is known that introduction in the structure of products of some derivatives of a lactic acid of 1% allows the stabilization essentially of biochemical changes in a product, increasing its storage time [4–7]. It is known also that nano-suspended materials, in particular nano Ag, possess antimicrobic properties not studied in detail [8, 9]. In boiled meat products the basic protective barrier is the environment in which we place a product. It is interesting to estimate the effect of the action of nanosilver and also a complex of nano Ag and sodium lactate upon their use in technologies of meat production.

5.2 Production and Antimicrobial Characteristics of Nanoparticles of Silver

Research of metal nanoparticles plays an important role in the development of modern nanotechnologies. It is caused by a wide spectrum of their practical application, in which specific properties of the nanoparticles and materials modified by them are used. For the real progress in their studies and practical use it is important to obtain highly stable nanoparticles with reproducible characteristics in large enough quantities and cheap enough for the technical purposes. The main problem here is a high reactivity of metal nanoparticles, resulting in their rapid aggregation or chemical transformation (e.g., oxidation) in the native system.

The original method of biochemical synthesis of silver nanoparticles in reverse micelles[a] is proposed in Ref. [9]. The method is based on the application of natural biologically active materials from the group of flavonoids as reductants.

It is experimentally proved that natural flavonoids (quercetin, rutin, and morin) are capable of effectively reducing metal ions in a water core of the reverse micelles with the formation of silver nanoparticles [10].

---

[a]Reverse micelle is a micelle in which the hydrophilic groups form the core and the hydrophobic groups the outer shell.

The basis of the method is the reduction of metal ions to atoms in reverse micelles in the ternary system:

$$Me(H_2O)/AOT/HC,$$

where AOT is the surface-active substance that forms a micellar shell and HC is saturated liquid hydrocarbon (C_6–C_8). The reduction is realized by the addition of biological reductizers—some natural plant pigments from the group of flavonoids; therefore, the method is called the biological activated reduction, or BAR-synthesis.

BAR-synthesis allows one to obtain metallic nanoparticles stable in air or in a micellar solution for a long time (for a year or more). Optical and adsorption properties of nanoparticles in the native solution are studied by measuring optical absorption spectra in the UV-VIS range. Particle size distributions are determined by the photon correlation spectroscopy (PCS) technique [11].

It is believed that the following sequence of reactions occurs at the interaction of quercetin (Qr) with silver ions in a micellar solution [10]:

$$Ag^+ + Qr \rightarrow [Ag^+ \ldots Qr] \quad (5.1)$$

$$[Ag^+ \ldots Qr] \rightarrow Ag^0 + Qr^0 \quad (5.2)$$

$$Ag^0 + Ag^+ \rightarrow Ag_2 + Ag^0 + \ldots \rightarrow Ag_k^+ \quad (5.3)$$

Initially the complex of quercetin and cation of silver appears (Eq. 5.1), then this complex disintegrates with the formation a silver atom and oxidized quercetin (Eq. 5.2), and finally, silver atoms and ions associate with the formation of the silver nanoparticles (Eq. 5.3).

Silver nanoparticles are usually obtained by the addition of silver salt (water solution) to the micellar solution of flavonoid. The nanoparticles have an intensive absorption band (optical path length 1 mm, $\lambda_{max} = 420$–430 nm); the formation is practically completed within 2–4 days (Fig. 5.1).

To expand the range of applications of silver nanoparticles obtained by biochemical synthesis the novel method for preparation of water dispersion of nanoparticles from their micellar solutions was developed. Two procedures for the preparation of such dispersions—centrifugation of the two-phase system micellar solution/water and mixing operation of the micellar solution with

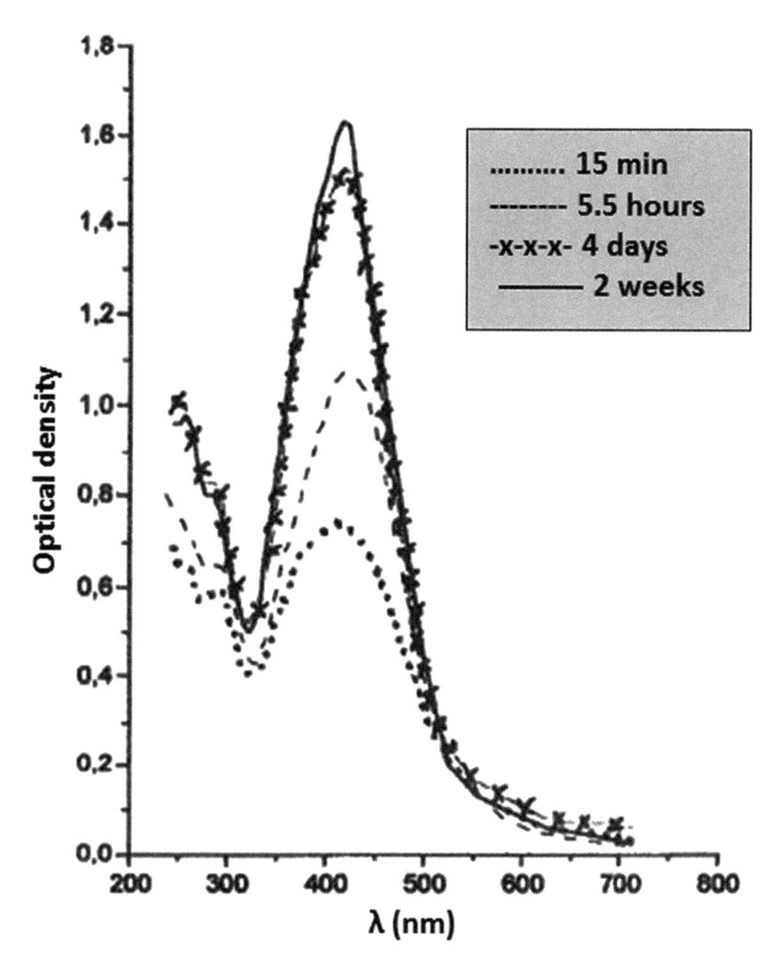

Figure 5.1 Synthesis of Ag nanoparticles. Optical absorption spectra of a micellar solution of the flavonoid at different times after the introduction of a silver salt solution [1, 12].

water and subsequent sedimentation—are presented in Refs. [10, 14]. The resulting water dispersions are stable for a long time (4–6 months) on storage in air at room temperature.

As found from the PCS measurements, silver nanoparticles in a micellar solution have sizes not exceeding 100 nm. The main contribution gives very small particles, 2–4 nm in size; one example of particle size distribution is given in Fig. 5.2 [1, 12]. From micellar

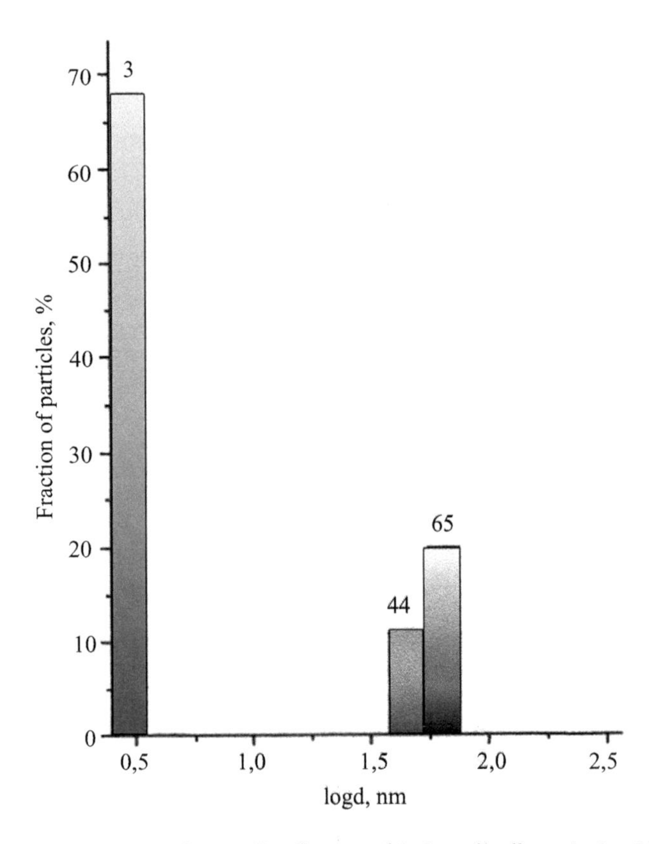

Figure 5.2 Nanoparticle size distribution of Ag in a micellar solution [1, 12].

solutions, water dispersions of silver nanoparticles may be obtained; as seen from the comparison of the absorption bands (Fig. 5.3 [1, 12]), the nanoparticle concentration in water dispersion is close to that in a micellar solution.

Studies of silver nanoparticles' interactions with various bacteria or viruses showed that nanoparticles demonstrate a high antimicrobial activity. As an example, we present here some data on the inactivation of the virus species (phage MS-2) in water after the addition of silver nanoparticles or silver ions (Table 5.1 [1, 12]). Note that for each silver concentration the inactivation extent for nanoparticles is distinctly higher than for ions, testifying to the obvious advantages of nanoparticles as a biocidal agent. The higher

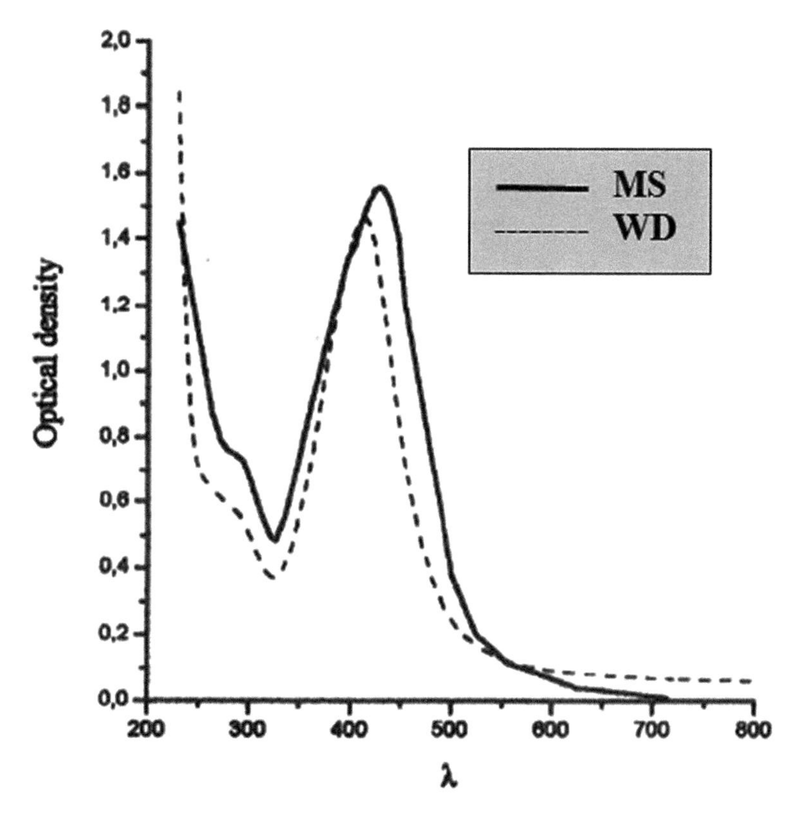

Figure 5.3 Optical absorption spectra of silver nanoparticles in a micellar solution (MS) and water dispersion (WD) [1, 12].

inactivation ability of nanoparticles compared to ions points also to the qualitative difference in these two silver states and the resulting difference in the mechanism of silver action on viral particles.

5.3 Experimental Investigation of New Bioactive Nanomodified Paints and Coatings

Micellar solutions of silver nanoparticles were introduced as small additives to the various compositions of paintwork materials (PMs) based on organic solvents.

Table 5.1 Comparison of the inactivation dynamics of coliphage MS-2: the effect of various concentrations of silver introduced into the infected water in the form of nanoparticles (NPs) in water dispersion and silver salt ($AgNO_3$) in a water solution

Time of the probe	Control	$1 \cdot 10^{-4}$ g-ion/l				$3 \cdot 10^{-5}$ g-ion/l			
		NP		$AgNO_3$		NP		$AgNO_3$	
	cells/ml	cells/ml	% inactive	cells/ml	% inactive	cells/ml	% inactive	cells/ml	% inactive
initial	185000	25000	–	100000	–	100000	–	200000	–
1 hours	182000	23000	90.8	70000	70	80000	20	120000	40
2 hours	179000	700	99.72	8000	92	11400	88.6	35000	85
3 hours	172000	620	99.75	6200	93.8	8800	91.2	32000	984
4 hours	175000	500	99.8	5000	95	8100	91.9	18700	90.7

Source: [1, 12]

Table 5.1 (*Continued*)

Time of the probe	Control	$6 \cdot 10^{-4}$ g-ion/l			
		NP		$AgNO_3$	
	cells/ml	cells/ml	% inactive	cells/ml	% inactive
initial	185000	50000	–	59000	–
1 hours	182000	7000	86	26200	55.6
2 hours	179000	2000	96	14800	77.9
3 hours	172000	900	98.2	5900	90
4 hours	175000	900	98.2	5000	91.5

Table 5.2 Dynamics of bactericidal action of the micellar solution of silver nanoparticles introduced into the water dispersion paint in the amount of 1%

Strain of bacteria	Control PM/ PM	Log number of the live bacteria			
		Hours			
		0	1	2	4
E. coli	Control PM	6.2 ± 0.1	6.2 ± 0.1	6.2 ± 0.1	6.2 ± 0.2
ATTC 25922	PM	4.8 ± 0.2	0	0	0
Salmonella typhimurium	Control PM	6.1 ± 0.1	6.1 ± 0.1	6.0 ± 0.2	6.0 ± 0.2
TMLR 66	PM	5.1 ± 0.2	0	0	0
Salmonella typhi	Control PM	6.0 ± 0.1	6.1 ± 0.2	6.0 ± 0.1	6.0 ± 0.1
Ty 2	PM	4.6 ± 0.1	0	0	0
Shigella	Control PM	6.2 ± 0.2	6.2 ± 0.2	6.2 ± 0.2	6.1 ± 0.1
flexneri 516	PM	4.5 ± 0.1	0	0	0
Staphylococcus	Control PM	6.0 ± 0.1	6.0 ± 0.2	6.0 ± 0.1	6.0 ± 0.1
aureus Wood-46	PM	5.4 ± 0.1	0	0	0
Enterococcus	Control PM	6.1 ± 0.1	6.1 ± 0.1	6.1 ± 0.1	6.1 ± 0.1
faecalis CO 110	PM	4.9 ± 0.1	0	0	0
Listeria monocyntogenes	Control PM	6.2 ± 0.2	6.2 ± 0.2	6.2 ± 0.1	6.1 ± 0.1
EGD	PM	4.6 ± 0.1	0	0	0
Pseudomonas	Control PM	6.1 ± 0.1	6.1 ± 0.1	6.0 ± 0.1	6.0 ± 0.1
aeruginosa 508	PM	5.6 ± 0.2	3.4 ± 0.1	1.2 ± 0.1	0

Source: [1, 10]

They testify to the fact that PMs with silver nanoparticles have a pronounced bactericidal action against a number of pathogenic bacteria whereas control PM samples without nanoparticles don't reveal such effect. Test results are demonstrated in Table 5.2 [1, 10].

Research of antimicrobial activity of a water dispersion of silver nanoparticles obtained by a centrifugation method was conducted on the model test microorganisms *Escherichia coli* bacteria and phage MS-2. The most effective inactivation of bacteria was observed under the influence of silver nanoparticles; 100% inactivation occurs within 2 h of exposure and under $AgNO_3$ only after 4 h [14].

A strongly pronounced biocidal effect is determined in an experiment with coliphage simulating contamination of the most common viruses (hepatitis A and influenza A and B) [15].

Further studies of the biocidal effect of silver nanoparticles in coatings were connected with research on organosoluble (alkyd and oil) and water-dispersed (polyacryl and polyvinyl acetate) paints. The nanoparticle-modified paints were compared with similar paints containing chemical biocides of the new generation—polyhexamethylene guanidine (PHMG) and with control specimens without biocidal additions. In development of the receptures (compositions) and technology of bioactive compositions, the main problems lay in:

- The incompatibility of the paint components with bioadditives
- The necessity to preserve high physical-mechanical and exploitation properties of the coatings

These problems were solved by working out the receptures of special concentrates with introduction of hydrophilic and hydrophobic (oleophilic) surface-active substances.

The biological activity of varnish-paint materials modified by silver nanoparticles or PHMG was estimated on the following microorganisms:

- *E. coli* 1257 as a conventional model of bacterial contamination of the environment
- Coliphage (RNA-phage MS-2) as a model of viral infection, including influenza A and B and hepatitis A
- Mold fungi (*Penicillium chrysogenum*) as a typical representative of microflora of the dwellings and a model of fungicidal contamination
- Spores as a model of spores and other microflora

In several series of preliminary experiments, the concentrations of biocidal additions were specified and it became possible to foretell the resource of their activity in the coatings. On the basis of these results, the industrial tests of bioactive alkyd enamels and water-dispersed paints were conducted in the infection hospital, tuberculosis section; in one of the prisons; and in several schools. As an example, we present the results obtained in the prison (Table 5.3 [1, 12]). From the coatings on various parts of a chamber,

Table 5.3 Surface concentration of microorganisms (cells/100 cm^2)

Object number	Sampling point	Indicators									
				Staphyloccus		Fungi					
		GMN*	*E. coli*	Overall number	Golden	Overall number	Mold	Yeast-like	Spores	Viruses	Coliphages
Control 1	Door	200	0	500	0	5000	2000	3000	200	–	100
	Toilet	320	40	2400	35	31000	15000	16000	4000	+	300
	Wall	100	0	1300	0	1400	400	1000	0	•	100
	Ceiling	200	0	200	0	1400	400	1000	0	•	200
Control 2	Door	500	0	3000	0	6500	5000	1500	10	–	150
	Toilet	2200	30	6600	20	12000	6000	6000	100	+	400
	Wall	100	0	100	0	4000	1000	3000	20	–	100
	Ceiling	200	0	500	0	5000	2000	3000	10	–	150
AgNP	Door	10	0	100	0	3000	1000	2000	20	–	20
	Toilet	200	15	2300	10	6500	2500	4000	100	–	50
	Wall	10	0	100	0	1200	200	1000	0	–	20
	Ceiling	20	0	300	0	4000	3000	1000	0	–	30
PHMG 1	Door	40	0	100	0	600	400	200	0	–	30
	Toilet	60	20	400	20	1600	600	1000	20	–	60
	Wall	30	0	100	0	500	300	200	0	–	20
	Ceiling	40	0	200	0	700	400	300	0	–	40
PHMG 2	Door	0	0	100	0	700	300	400	0	–	10
	Toilet	20	10	450	1	900	400	500	10	–	20
	Wall	0	0	100	0	300	200	100	0	–	10
	Ceiling	0	0	100	0	450	300	150	0	–	10
PHMG 3	Door	10	0	150	0	300	200	100	0	–	10
	Toilet	50	10	200	0	700	400	300	1	–	30
	Wall	0	0	100	0	200	100	100	0	–	10
	Ceiling	0	0	100	0	300	100	200	0	–	10

Source: [1, 12]
*General microbial number

the microflora was washed off periodically and the specimens obtained were analyzed for the presence of active microbial cells of various types. As seen from the table, bioactive coatings exert quite an obvious effect on the majority of microorganisms on various elements of the painted chambers.

An important point is that micellar solutions of silver nanoparticles were introduced into the paints in very small concentrations—the Ag contents in the compositions were in the range $(1 \div 4) \cdot 10^{-4}\%$. At these concentrations, on the basis of the data obtained in experimental and natural tests, one may guarantee that the coating will remain active for no less than 1 year. At the same time the PHMG concentration was in the range of 0.5–1.5%. Hence it follows that for the comparable biocidal activity the PHMG concentration must be at least 1000 times higher, that is, the PHMG expenditure for the paint unit volume is significantly higher than that of nanoparticles. Further experiments are supposed to be made on coatings with a higher concentration of nanoparticles; its optimization is conditioned, to a large extent, by the potentialities of the BAR-synthesis—in particular, by the possibilities of increasing the nanoparticles' concentration in micellar solutions and using solvents more compatible with paints components.

The experiments with bioactive coatings containing silver nanoparticles and PHMG showed that for the contents of these additions in paints studied so far, their antimicrobial activity is selective, that is, while nanoparticles are more active against bacteria and viruses, PHMG works better against fungi and spores. Hence their combination in a given composition may give the optimal effect against a wide variety of microorganisms. At the moment we are continuing to elaborate more effective biocidal coatings based on a combination of silver nanoparticles with special bioactive organic alkali-soluble silicates [16].

5.4 New Water-Dispersion Paint Composition with Biocide Properties

Biocide paint materials with biocide, virulecide, and sporocide properties are applied on the surfaces of walls, ceilings, concrete, bricks, wood, gypsum, etc., for reduction in the level of microbe infection of apartments in medical establishments, schools, kindergartens, nurseries, offices, etc.

As active biocide additives sulfur and nitrogen containing heterocycles, active chlor, and quaternary salts are used.

A biocide paint on the base of a binder, pigments, tillers, phosphate or acetate PHMG, water and compound of nanoparticles of silver were considered above. This paint has good bactericide and fungicide properties but not high sporocide properties.

The new paint composition with a reduced level of microbe infection of the coating was developed. Coatings on the base of the paint have good physical-mechanical and protection properties and high inactivation properties against microorganisms on the coating.

Table 5.4 The optimal formulations of the new biocide paint

Components	Paint no. 1	Paint no. 2	Paint no. 3 control	Paint no. 4	Paint no. 5
		Binder			
Acrylic dispersion	23.0	–	23.0	23.0	23.0
Polyvinyl acetate dispersion	–	18	–	–	–
		Pigment			
TiO_2	20	12	20	20	20
		Fillers			
Chalk	6.0	14.5	6.0	6.0	6.0
Powder talk (French white)	3.0	5.5	3.0	3.0	3.0
Ground mica	1.0	–	1.0	1.0	1.0
		Biocide additives			
Phosphate PHMG	1.0	1.0	1.0	1.0	1.0
Compound of stable silver nanoparticles	1.1	1.1	1.1	1.1	1.1
40% water solution of DBU-E $(SiO_2)_n$	0.25	2.5	1.25	0.1	3.0
Special additives (surfactants)	3.95	3.92	3.95	3.95	3.95
Water	40.7	41.48	39.7	40.85	39.75
Summary	**100**	**100**	**100**	**100**	**100**

Table 5.5 The biocide properties

Paint	Effect							
	Bactericide		Virulicide		Fungicide		Sporocide	
	Concentration E. Coli on surface 43000		Virus Polyfag MS-2 KOE/ml 1340		Fungus Penicilium Chrysodenim 20000		Spores Bacellus Cerous KOE/ml 300	
	After 10 min	Inactivation %	After 30 min	Inactivation %	After 3 days	Inactivation %	After 7 days	Inactivation %
Paint no. 1	0	100	0	100	0	100	45	80
Paint no. 2	0	100	0	100	0	100	35	88
Paint no. 3	0	100	0	100	0	100	40	87
Paint no. 4	0	100	0	100	0	100	56	70
Paint no. 5	0	100	0	100	0	100	45	89
Paint Ru*	0	100	0	100	0	100	56	70

*RU Patent 2195473 PHMG 0.8%; Nanosilver 1.1%

The paint consists of a binder (water polyacrylic of polyvinyl acetate dispersion), pigments (dioxide titan, oxide zinc, etc.), fillers (chalk, barite, talk, etc.), and biocide additives:

- Phosphate or acetate PHMG
- Compound of nanostructure particles of silver (300–500 nm)
- Water-solution DBU-E (1,8 diazabiclo [5.4.0]) and SiO_2

and surfactants

- Dioctylsulfosuccinate Na
- Quercetin (3,5,7,3′,4′-pentahydroxyphlavone)

Paint is prepared by a conventional way: preparing of a pigment paste in a high-speed dissolver with dispergation in a ball mill, mixing with the rest of the binder, and introduction of functional additives. The paint may be sprayed, brushed, or rolled.

The optimal formulations and the biocide properties are given in Tables 5.4 and 5.5.

5.5 Nanosilver as a Potential Protective Material for Foodstuff

The nanoparticles of Ag were used as an additive by means of the reaction of an equivalent amount of $AgNO_3$ with tannin E181 and $NaHCO_3$ on heating to 71°C within 2 h. The average size of Ag nanoparticles was about 30 nanometers according to the light scattering ccording to Rayleigh [16]. In this work, sodium D-lactate was also used.

The total amino acid content of proteins was determined in the hydrolysate obtained according to the standard method by treating with a 6M HCl solution at 120°C for 24 h under an Ar gas flow followed by thrice-repeated stripping to dryness of volatiles and the dissolution of the sample in the buffer with a pH = 2.2. Free amino acids were determined by the extraction with 85% ethyl alcohol solution in water. Then, the extract was stripped to dryness and the residue was resuspended in the buffer for sample dissolution with pH 2.2 [18].

The antimicrobial study was carried out with the Ag nanoparticles by the standard agar dilution technique [18]. The study of bactericidal properties of the silver nanoparticles was performed by applications on Sabouraud solid medium (RJA Sabouraud) composition (pancreatic hydrolyzed fish meal 10.0, pancreatic casein hydrolyzate 10.0, baking yeast extract or yeast extract imported 2.0, sodium dihydrogen 2.0, glucose 40.0, and agar 10.0).

As test cultures most frequently used were *Aspergillus niger*, *E. oli*, and *Mucor heterosporum*. The zone of growth inhibition of microorganisms was observed at a temperature of 30°C for 7 days. As bactericidal materials were used 0.05% aqueous dispersion of nanosilver, as well as particulate material containing 0.05% nano Ag and 1% sodium lactate. It was previously found that in a concentration of 0.05–0.1% of nanosilver there is almost complete inhibition of growth of these strains.

For the reception of an aggregate stable of metal nano Ag particles we used a chemical way of dispersion of this metal by the restoration of ionic silver from its nitrate salt up to a molecular condition in the water medium under the influence of an organic reducer, which was food tannin. The reaction of the reception of nanosilver was carried out in two ways.

The first way was to realize this reaction on the surface of the cover membrane of a sausage. The mix contained or did not contain the additive of 1% sodium D-lactate. The second way was the variant of the reception of nano Ag in a modeling solution in similar conditions and this was used for an estimation of the size of the formed particles by means of turbodimetriya, measuring easing of intensity of light on its passage through a liquid disperse system.

On passage of light through the colloidal solution containing small particles, the absorption practically is absent; intensity attenuation of an impinging light is due to its scattering in all directions by the colloidal solution. For the containing particle systems with sizes much less than the length of a light wave, the size of full light scattering submits to the Rayleigh equation. This principle has been used for an estimation of the sizes of particles of silver. Calculation on the Rayleigh equation for systems diluted up to

1:2000–1:10,000 was shown as it is done in the work in Ref. [17]. In a modeling system the average diameter of formed Ag nanoparticles was about 30 nanometers. As similar process was carried out on the surface of a cover membrane of foodstuff in comparable conditions, it is possible to assume that the size of particles of silver in both cases is comparable.

For an estimation of the antibacterial activity of nanoforms of silver in foodstuff, the following idea was used. Influence of microflora on a matrix of a product leads to the development of hydrolytic disintegration of its components, in particular the most important component of food—proteins. The intensive disintegration of proteins increases the share of free amino acids. The process can be supervised by an amino acid analyzer.

Previous studies have shown that the introduction into a sausage of a 0.5% preservative—sodium lactate—allowed the shelf life of meat products to double at +4°C [6].

The obtained material of nanosilver in the equimolar mixture of sodium D-lactate was applied to the envelope of sausages by dipping.

The results of the determination of free amino acids are shown in Table 5.6 [1, 20].

Table 5.6 shows a slow decay of the protein in the product due to hydrolysis. The full amino acid composition of the total protein was practically unchanged during the study period of storage and averaged (g/100 g protein): TAU – 0.2, ASP – 8.2, THR – 3.2, SER – 3.6, GLU – 20.9, PRO – 3.9, GLY – 8.0, ALA – 5.2, CYS – 0.9, VAL – 3.9, MET – 1.5, ILEY – 3.3, LEY – 6.2, TYR – 2.7, GABA – 0.05, PHE – 3.7, HIS – 4.2, OPN – 0.1, LYS – 7.5, and ARG – 7.7 (only 95 g/100g of protein or 12.5 g/100 g body weight of sausages). The greatest interest is the integral index of the contents of free amino acids. The nature of the curves shows that in the case of lactate, and especially materials with nanosilver particles, processes of unwanted chemical reactions in foods have been notably slow.

The hydrolytic nature of the changes is shown in Fig. 5.4 [1, 20]. The highest stabilizing effect was observed in the case of silver nanoparticles with sodium lactate.

Independent tests by atomic absorption spectroscopy showed that the silver content in the 3 mm layer of the sausage product, adjacent to the shell of the product, is less than the sensitivity of metal determination (<0.0001 mg/kg).

Table 5.6 The contents of free amino acids in meat products at various periods of storage (mg per 100 g of a product, +4°C)

Name	Sausages (control)			Sausages with lactate and nano-Ag		
	0	15 days	25 days	0	15 days	25 days
TAU	19.5	17.1	33.8	32.6	42.9	62.2
ASP	3.1	6.2	13.7	7.5	4.7	9.0
THR	5.0	4.8	19.6	7.7	5.9	8.3
SER	1.1	8.9	13.6	11.4	47.4	16.2
GLU	110.9	93.1	125.9	118.6	133.6	209.2
PRO	7.2	6.4	10.1	11.1	2.2	11.9
GLY	14.0	12.2	17.8	14.3	11.9	16.1
ALA	17.7	23.7	29.7	27.8	34.9	41.5
CYS	0.1	0.3	1.5	0.1	0.1	0.4
VAL	10.8	11.1	14.4	13.8	13.1	16.3
MET	5.1	3.1	2.6	4.5	3.3	4.1
ILEY	7.7	6.6	10.3	8.1	8.0	9.2
LEY	16.9	15.1	21.3	18.7	23.2	28.5
TYR	7.8	10.2	17.4	11.7	10.7	13.4
GABA	0.1	0.4	5.6	0	0.7	1.7
PHE	5.5	13.0	15.3	13.8	17.3	21.3
HIS	27.7	34.6	49.2	38.8	41.4	67.1
OPN	4.4	0.7	4.2	2.7	0.1	7.9
LYS	10.5	6.0	15.4	8.9	9.4	16.8
ARG	11.8	17.1	23.1	23.0	14.9	25.9
Σ	**286.8**	**290.6**	**444.5**	**375.0**	**425.6**	**587.0**

Source: [1, 20]

The literature describes the mechanism of the impact of nanoparticles on living cells of microorganisms. In particular is considered that Ag nanoparticles several tens of nanometers in size can be attached to the surfaces of cellular membranes and significantly disrupt the inherent features, such as permeability and respiration. In addition, metal nanoparticles can penetrate into the bacteria and cause damage at the biochemical level, probably due to interaction with sulfur- and phosphorus-containing residues of proteins or DNA. Ag nanoparticles form silver ions, which enhance the bactericidal effect, causing inhibition of transmembrane transport of sodium and calcium [8, 21].

Table 5.7 [1, 20] shows some results define zones suppress the growth of microorganisms in the presence of bactericidal materials

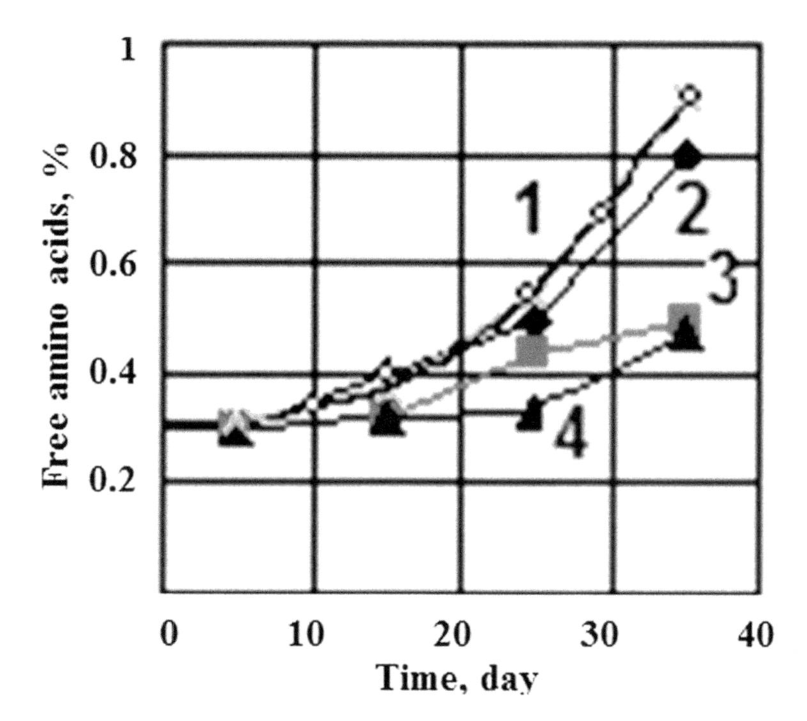

Figure 5.4 Dependence of the content of free amino acids in the food product on the storage time at $+4^{\circ}$C in the presence of lactate (1), nano Ag (2), and an equimolar mixture of sodium lactate and nano Ag (3) in comparison with the control—no additives (4) [1, 20].

obtained. Nanosilver exhibited a clearly pronounced antibacterial effect, but its composition with lactate had a somewhat larger effect of the action, apparently as a result of synergies. It is worth noting the synergy of metal nanoparticles observed with a variety of other substances, such as described in the literature, mixed with dihydroquercetin [8].

This part of the chapter presents the most typical food strains of microorganisms. More detailed studies of a wide variety of cultures make it possible to identify a real antibacterial activity of nanodispersed forms of metals of variable valence.

Of special interest is bactericidal polymeric films used as food packaging material of food products. Such films with Ag nanoparticles were prepared by introducing a water solution of the

Table 5.7 The change in the diameter of zones of suppression of microorganisms under the influence of antisepsis materials

	Zone of growth inhibition of a culture, mm			
Microorganism	**Sodium D-lactate**	**Mixture of sodium lactate and nano-Ag**	**Nano-Ag**	**Control**
A. niger	20	23	22	18
E. coli	21	27	23	19
M. heterosporum	23	29	25	21

Source: [1, 20]

nanoparticles in the water solution of the biodegradable polymer (based on chitinte).

Films' test results on antimicrobial activity are shown in Table 5.8 [1, 10].

Table 5.8 Test results of the bactericidal polymeric films

		Lg number of live bacteria			
		Time (hours)			
Tested strain, dose	**Content of Ag**	**1**	**3**	**6**	**24**
S. typhimurium	Initial bacterial culture	6.1	6.1	6.1	6.0
TMLR66	0% Ag	6.1	5.7	5.2	2.8
10"KOE	0.03% Ag	5.0	4.0	2.7	0
	0.06% Ag	0	0	0	0
S. aureus	Initial bacterial culture	6.0	6.0	6.0	6.0
Wood 46	0% Ag	6.0	5.8	4.9	2.8
10"KOE	0.03% Ag	5.0	4.1	3.0	0
	0.06% Ag	0	0	0	0
S. typhimurium	Initial bacterial culture	4.0	4.0	4.0	4.0
TMLR66	0% Ag	4.0	3.7	3.4	1.5
104KOE	0.03% Ag	2.8	1.8	0	0
	0.06% Ag	0	0	0	0
S. aureus	Initial bacterial culture	4.0	4.0	4.0	4.0
Wood 46	0% Ag	4.0	3.8	3.3	1.6
104KOE	0.03% Ag	2.9	1.9	0	0
	0.06% Ag	0	0	0	0

Source: [1, 10]

5.6 Conclusion

In contrast to the other biocidal additions widely used in the paint industry, new bioactive additions described above are absolutely nontoxic for man and environment, as confirmed by the corresponding certificates of sanitary-hygienic services. Therefore, coatings made from the new paints have applications as preventive means in public meal enterprises; children, sports, and medical institutions; and in all places crowded with people.

Apart from silver nanoparticles, interesting perspectives in the production of varnish-paint materials may be anticipated also for the nanoparticles of other metals. For instance, zinc nanoparticles used in protector anticorrosion grounds may allow one to obtain a coating with a thickness of 5–10 microns, which is very important in painting of rolled metal sheets (for coil coating). Also, cobalt, zinc, and calcium nanoparticles may be of interest for the intensification of auto-oxidation processes in alkyd and oil binder films.

It is established that a new preservative material based on silver nanoparticles and its complex with salts of lactic acid is a very effective preservative that can be used both to protect food products from the development of undesirable processes of chemical decomposition of biologically active components, mainly proteins, and prevent microbial spoilage of food products based on raw meat. This will prolong the shelf life of food products.

On the basis of these experiments it can be concluded that the inactivation dynamics of the investigated microorganisms are more intense on exposure to silver nanoparticles than silver nitrate in similar concentrations. This clearly demonstrates the higher bactericidal and virucidal activity of silver nanoparticles compared to silver ions [14].

References

1. Figovsky O., Beilin D. (2016). Nanotechnology in production of bioactive paints, coatings and food storage materials, *Journal Scientific Israel - Technological Advantages*, **18**(2).
2. Figovsky O. (2002). Interface phenomena in polymer coating, *Encyclopedia of Surface and Colloid Science*, Marcell Dekker, Inc.: New York, pp. 2653–2660.

3. Egorova E., Revise A., Rostovschikova T., Kiseleva O. (2001). Bactericidal and catalytic properties of stable metallic nanoparticles, *Vestnik of Moscow State University*, **42**(2), 332–338 (in Russian).
4. Neklyudov A. D., Ivankin A. N. (2006). *Processing of the Organic Wastes*, MFSU Publishers: Moscow, 380 p.
5. Yatsuta A. L., Nikitchenko D. V., Kostenko Yu. G. (2006). Ecological systems (*Ecological Systems and Devices*, Russia), 4, 40–43.
6. Ents G. Y., Neklyudov A. D., Oliferenko G. L. (1999). Storage and processing of agricultural raw material (*Storage and Processing of Agriculture Raw Material*, Russia), 9, 46–48.
7. Guerman A. B., Nekludov A. D. (1999). *Congress proceed. 4th International Conference on Ecomaterials*, 10–12 Nov., Gifu, Japan, pp. 177–180.
8. Serov A. V., Shipulin V. I., Shevchenko I. M. (2010). *Meat Industry (of Russia)*, 2, 29–32.
9. Sarkar S., Jana A. D., Samanta S. K., Mostafa G. (2007). *Polyhedron*, **26**(15), 4419–4426.
10. Egorova E. (2011). Nanoparticles of metals in solutions: biochemical syntesis m properties and application, Doctorat Thesis, Moscow, http://www.dissercat.com/content/nanochastitsy-metallov-v-rastvorakh-biokhimicheskii-sintez-svoistva-i-primenenie
11. Egorova E., Revina A. (2002). Optical properties and sizes of silver nanoparticles in micellar solutions, *Colloid Journal*, **64**(3), 334–345 (in Russian).
12. Kydryavtzev B., Figovsky O., Egorova E., Buzslov F., Beilin D. (2003). The use of nanotechnology in production of bioactive paints and coatings, *Journal Scientific Israel - Technological Advantages*, **5**(1–2), 209–215.
13. Figovsky O., Shapovalov L., Kydryavtzev B. (2004). Bioactive coatings with silver nanoparticles, *Proceedings of XXVII FATIPEC Congress*, 19–21 April, Aix-en-Provence, France, vol. 3, pp. 817–823
14. Egorova E. M., Revina A. A., Rumiaintsev B. V. (2003). Production and antimicrobial properties of water dispersion of silver nanoparticles, *Collection of Research Papers of Russian (International) Conference "Physical-Chemistry of Ultradisperse (nano)- Systems"*, Moscow, pp. 148–152.
15. Egorova E. M., Revina A. A., Rostovschiko T. N., Kiseleva O. I. (2001). Bactericide and catalitical propertyies of stable metal nanoparticles in reverse micelles, *Vestnik of Moscow Universitea*, **42**(5), 332–338.

16. Romm F., Karchevsky V., Axenov O., Potashnikov R., Figovsky O. (2002). Coatings based on quatemare ammonium silicates: properties and application, *Journal Scientific Israel - Technological Advantages*, **4**(4), 75–81.
17. Lisitsin A. B., Ivankin A. N., Yushina Yu. K., Gorbunova N. A. (2010). *Proc. The 56th International Congress of Meat Science and Technology*, 15–20 Aug., Jeju, Korea, p. 72.
18. Lisitsyn A. B., Ivankin A. N., Nekludov A. D. (2002). Methods of practical biotechnology. Analysis of components and microscopic amounts of admixtures in meat and other food products, VNIIMP Publishers: Moscow, p. 402.
19. National Committee for Clinical Laboratory Standards (NCCLS) (1999). Performance standards for anti-microbial susceptibility testing, Wayne, PA.
20. Ivankin A., Figovsky O., Evdokimov Y., Gorbunova N. (2011). Nano-silver as a potential protective material for foodstuff on the basis of animal raw material, *Journal Scientific Israel - Technological Advantages*, **13**(1).
21. Duran N., Marcato P. D., De Souza G. I. H., Alves O. L., Esposito E. (2007). Antibacterial effect of silver nanoparticles produced by fungal process on textile fabrics and their effluent treatment, *Journal of Biomedical Nanotechnology*, **3**, 203–208.

Chapter 6

Green Nanostructured Biodegradable Materials

6.1 Introduction

Conventional disposable packaging items, such as containers, trays, plates, and bowls, are commonly made from polystyrene or other synthetic hydrophobic plastics and also from paper or paperboard coated commonly with polyethylene (PE). These materials are durable, moisture resistant, and grease resistant.

The disposable items are produced by the industry in great quantities and are relatively inexpensive. After usage, these biostable goods are discarded in the environment and therefore create serious ecological problems [1–6]. Packaging waste forms a significant part of municipal solid waste and has caused increasing environmental concerns, resulting in a strengthening of various regulations aimed at reducing the amounts generated. Plastic waste imposes negative externalities, such as greenhouse gas emissions and ecological damage. It is usually nonbiodegradable and therefore can remain as waste in the environment for a very long time; it may pose risks to human health and the environment; in some cases, it can be difficult to reuse and/or recycle. The expected lifetime

Green Nanotechnology
Oleg Figovsky and Dmitry Beilin

ISBN 978-981-4774-10-9 (Hardcover), 978-1-315-22928-7 (eBook)
www.panstanford.com

of polystyrene materials, for example, is several hundred years. Recycling of plastic items is an expensive process. Moreover, in some countries recycled plastics are prohibited as a material for packaging for food.

Among other materials, a wide range of oil-based polymers is currently used in packaging applications. These are virtually all nonbiodegradable, and some are difficult to recycle or reuse due to being complex composites having varying levels of contamination.

The annual world consumption of nonbiodegradable plastic packaging is about 200 million tons. More than 150 million tons is the annual world consumption of paper and board packaging. The consumption of coated paper and board is about 45 million tons per year. The global paper packaging market was $213.8 billion in 2014 and is estimated to reach $306.73 billion by 2020, growing at a compound annual growth rate (CAGR) of 6.2%. Of the paper packaging market, paperboard is estimated to be the fastest-growing market, with a forecasted growth rate of 7.5% during the period 2014–2020 [5, 6].

Figures 6.1 and 6.2 [1] illustrate the stable growth of production and consumption of plastic packaging materials in Europe.

In 2008, the total generation of postconsumer plastic waste in EU-27, Norway, and Switzerland was 24.9 million tons. Packaging is by far the largest contributor to plastic waste, at 63%.

Due to the biostability of synthetic materials the tendency to use biodegradable disposable packaging materials has steadily increased in the last decade. Recently, significant progress has been made in the development of biodegradable plastics, largely from renewable natural resources, to produce biodegradable materials with a similar functionality to that of oil-based polymers. The expansion in these bio-based materials has several potential benefits for greenhouse gas balances and other environmental impacts over whole life cycles and in the use of renewable, rather than finite, resources. It is intended that the use of biodegradable materials will contribute to sustainability and reduction in the environmental impact associated with disposal of oil-based polymers.

The main approach has been the manufacture of inexpensive goods from biodegradable and compostable natural materials, such as starch, cellulose, and proteins [10–16]. However, these

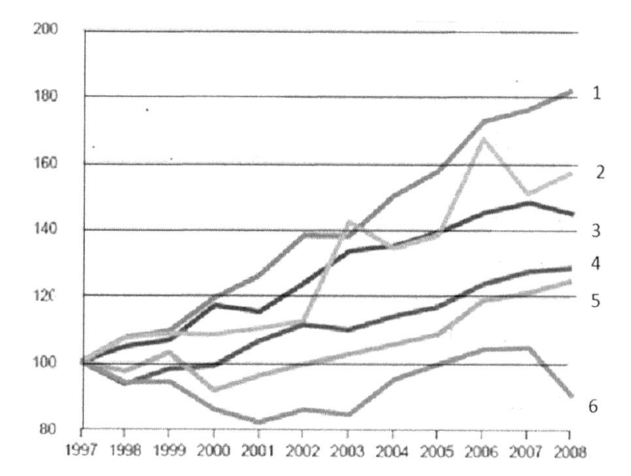

Figure 6.1 Tendency of production of plastic packages in Europe (index = 100 in 1997): 1, Germany; 2, Ireland; 3, Belgium; 4, Sweden; 5, Great Britain; and 6, Denmark [1]. Reprinted from Ref. [1] with permission from *Engineering Journal of Don.*

natural biodegradable materials have several shortcomings, the most important being susceptibility to water, grease, and various other liquids. The starch binder is water soluble and penetrable to grease.

This chapter discusses the potential impacts of biodegradable packaging materials and their waste management, particularly via composting. It presents the key issues that inform judgments of the benefits these materials have in relation to conventional petrochemical-based counterparts. Specific examples are given from new research on biodegradability in simulated "home" composting systems. It is the view of the authors that biodegradable packaging materials are most suitable for single-use disposable applications where the postconsumer waste can be locally composted.

Cellulose materials, papers, and cardboards are widely used in the manufacturing of packaging. These materials are ecologically safe because they can undergo biodegradation and repulping. However, cellulose packaging is hydrophilic and porous and, therefore, swells in water. When exposed to water or significant amounts of water vapor, this packaging material loses form stability and becomes susceptible to breakage. Therefore, it fails to protect

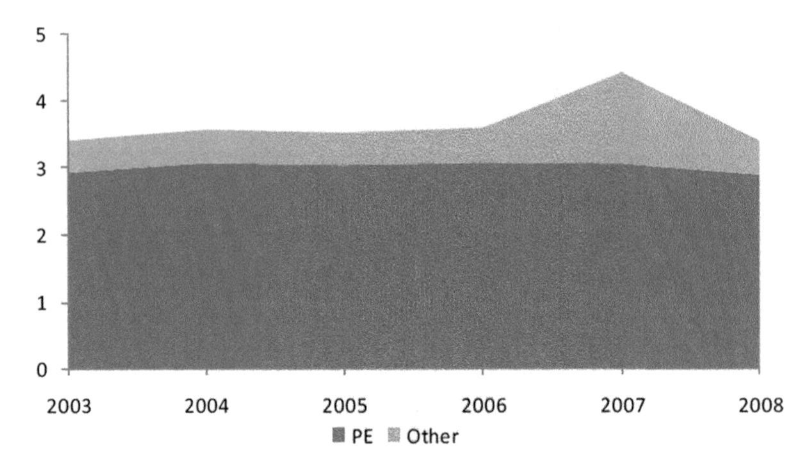

Figure 6.2 Volume of plastic and bags produced in EU, 2003–2008 (in Mt) [1]. Reprinted from Ref. [1] with permission from *Engineering Journal of Don*.

packaged products from the negative effects of water, greases, vapors, gases, and other environmental factors. To give them barrier properties, cellulose substrates are laminated or coated with polyolefins (PE, polypropylene [PP], etc.) and other synthetic hydrophobic polymers. Despite the fact that cellulose-polyolefin composites possess good barrier properties, their prospective application is problematic for ecological reasons. Since it is nonbiodegradable and nonrepulpable, unlike papers or cardboards, and nonrecyclable, unlike common plastics, cellulose-polyolefin composite waste accumulates in dumps in immense amounts and pollutes the environment. This creates a sharp ecological problem.

To overcome the shortcomings of the natural hydrophilic materials, it is necessary to develop advanced, environmentally friendly, hydrophobic polymer materials that are capable of biodegradation and being recycled.

The most attempts to manufacture biodegradable packaging were focused on developing biodegradable plastic films. Creation of biodegradable packaging products is based on improving low barrier properties of cellulose-based materials by coating them with special polymer composition. Moreover, the coated cellulose material is recyclable and biodegradable, like ordinary paper or cardboard. These products can be included also in pulp composition.

In this chapter, the technology of novel protective nanocoatings on the surface of natural biodegradable packaging materials is described.

One of the most developed fields of nanotechnology is nanoplastics containing polymer binders and inorganic nanofillers, such as clay, silica, chalk, titanium dioxide, and ceramics. However, inorganic nanofillers have the following disadvantages:

- A high density increases the weight of composites and articles
- High abrasibility decreases the lifetime of equipment
- High hardness hinders polishing of coatings
- Low bonding ability with organic polymers gets trouble gain in strength
- Biostability hinders the production of biodegradable plastics
- The settling ability in liquid systems causes phase separation

In contrast with nanoinorganic fillers, nano-organic fillers can contain various functional groups (FGs) allowing them bonding with an organic polymer that leads to an increase in the strength of the plastics. Moreover, organic nanofillers, such as nanocellulose, will have a low density, low hardness, and abrasibility, as well as increased settling stability of dispersions, etc. Besides, nano-organic fillers made from cellulose are biodegradable. Therefore, our aim was to develop a preparation method of nanocellulose and study main application fields of the new nanoproducts.

This chapter discusses the potential impacts of biodegradable packaging materials and their waste management, particularly via composting. It presents the key issues that inform judgments of the benefits these materials have in relation to conventional petrochemical-based counterparts. Specific examples are given from new research on biodegradability in simulated home composting systems. It is the view of the authors that biodegradable packaging materials are most suitable for single-use disposable applications where the postconsumer waste can be locally composted.

The skin has a difficult multilayer structure comprising an exterior layer (stratum corneum), a middle layer (epidermis), and an inner layer (dermis). For prophylactic skin care, cosmetic remedies can be used that act mainly on the skin exterior layer. In the case of skin injury or disease, a specific biocide (drug) is used that should penetrate inside the skin through pores having an average size of 50 μm. To prevent side effects, improve effectiveness, and impart the slow-release effect, as well as to extend application areas, a chemical attachment of the biocide to an appropriate carrier should be performed.

Restricted materials are suitable as carriers of biologically active substances (BASs) in biology, medicine, and cosmetics. Such a carrier should meet the following requirements:

- It should be generally recognized as safe (GRAS).
- It should contain fine particles having a gentle sensation.
- It should be insoluble in water, oils, and various organic solvents.
- It should be able to modify in order to form specific reactive groups capable of binding the BAS.
- It should be stable to the attached BAS.
- It should not inhibit the attached BAS.
- It should not interact with other ingredients of the drug composition.

Microcrystalline cellulose (MCC) belongs to the GRAS substances [17]. However, coarse particles (50–200 μm) and inertness hinder grafting of a BAS to MCC and transdermal delivery of the particles through skin pores. To use cellulose particles as biocarriers, the size of the particles needs to be reduced to nanoscale and the material must turn reactive.

A method for the preparation of the reactive nanocellulose biocarrier and discussion about some of its biomedical and cosmetic applications are presented.

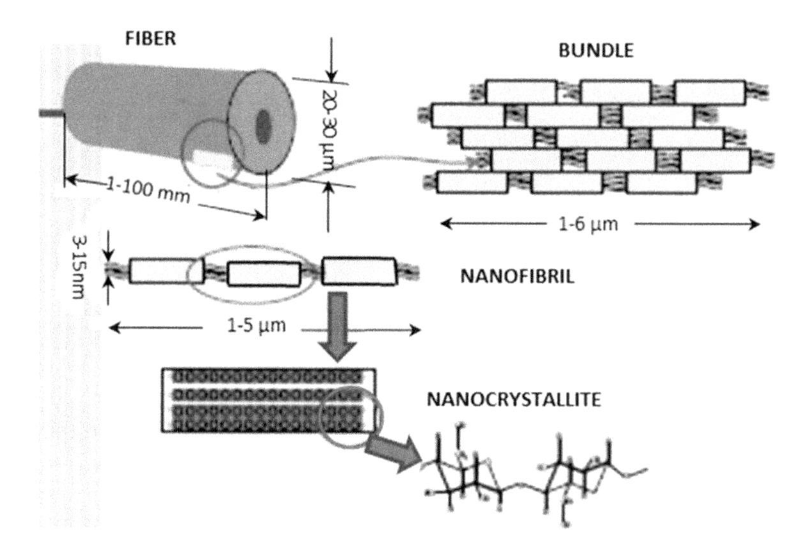

Figure 6.3 Structure of cellulose.

6.2 Nanocellulose and Its Applications

6.2.1 Materials and Methods

As is known cellulose is the main building material of flora that forms the cellular walls of trees and other higher plants. The structure of a molecule of cellulose (Fig. 6.3) is a high-chain polysaccharide consisting of n-glycosidic residues, linked by ether bridges (1,4-β-glycosidic linkages).

Bleached wood cellulose (95% alfa cellulose and degree of polymerization [DP] = 1180) was used as a raw material for the preparation of the nanocellulose (NanoCell).

NanoCell products were manufactured by the process showed in Table 6.1.

The initial cellulose raw material is cut into pieces, and they are put in a glass reactor. A water solution containing the acidic catalyst modifier is poured into the reactor. The reactor is hermetically closed with a cover, and the reaction system is kept at an increased temperature for a short time and then cooled. The cellulose slum is filtered and the wet cake is diluted up to 1–5% and transferred into

Table 6.1 Main stages for the manufacture of nanocellulose

Stages	Process	Product
1	Chemical depolymerization of the cellulose, washing and filtration	Wet cake
2	Treatment of the MCC water dispersion in a high-pressure homogenizer	NanoCell dispersions
3	Centrifugal concentrating and ultrasonic disintegration	NanoCell paste
4	Freeze drying and vortex superfine milling	NanoCell powder

Source: [1, 18]. Reprinted from Ref. [1] with permission from *Engineering Journal of Don*.

a Gaulin homogenizer and an inhibitor and a dispersing agent are added to the slurry. The dispersion is mechanically homogenized at 500–1000 bar, and the NanoCell dispersion is collected. This dispersion can be concentrated by centrifugation up to a 25–30% solid paste of NanoCell and on to a dry NanoCell powder. The NanoCell paste can be freeze-dried and disintegrated by a vortex superfine mill.

The process of NanoCell production is illustrated in Fig. 6.4.

Testing of NanoCell materials and compositions was carried out by means of the following methods:

- XRD (Rigaku-Ultima Plus diffractometer)
- Scanning electron microscopy (SEM) (Hitachi S-430)
- Viscometry (Ostwald's capillary viscometer)
- Laser-light scattering (LLS) (Malvern's Mastersizer-2000 tester)
- Mechanical tests (LLOYD LR 50K instrument)

6.2.2 Using of the Viscometry Method for NanoCell Investigations

This method is based on theoretical conceptions and experimental data about supermolecular structure of cellulose and its change during chemical treatments [E3-E5]. It is known that cellulose microfibrils contain amorphous regions and crystallites linked with intramolecular valence bonds. The amorphous regions are weak

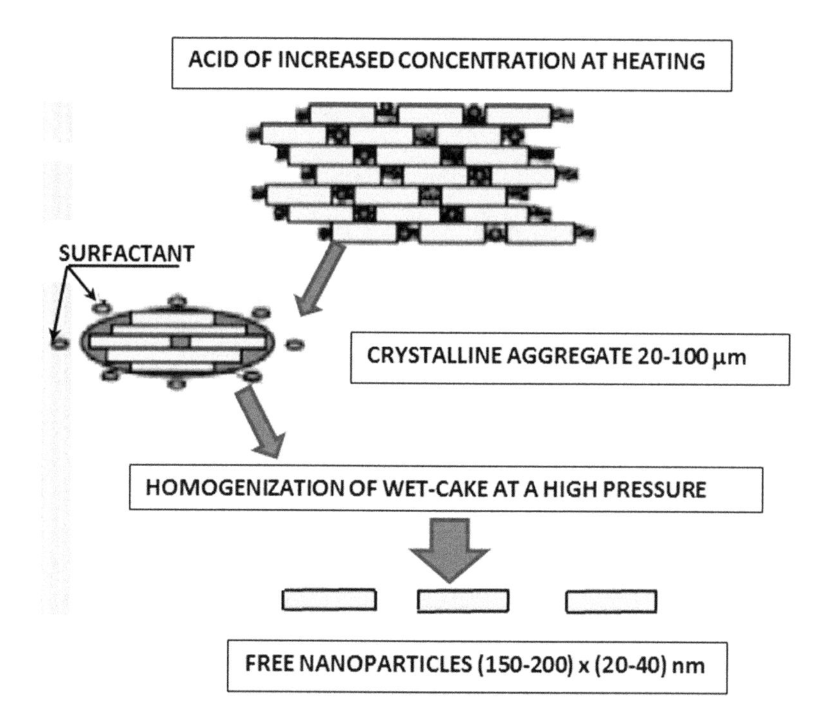

Figure 6.4 Process of NanoCell production.

places of microfibrils accessible to chemical reagents, while crystallites are strong and inaccessible structural elements. Therefore, chemical treatment of cellulose with destructive reagents leads to breaking of chains in weak amorphous regions only and to decrease in the DP of cellulose (chemical cutting). After finishing of the depolymerization process in amorphous regions, the about-constant DP value—"level-off DP" (LODP)—is reached. This value conforms to the average DP of cellulose crystallites. The viscometry method is usually used for testing DP and LODP. If LODP was determined experimentally, the average length (L) of elementary cellulose crystallites can be calculated:

$$L = l \cdot \text{LODP}, \quad (6.1)$$

where $l = 0.517$ nm is the length of the glucopyranose link of cellulose.

The dependence of DP on the time of chemical cutting of cellulose chains in amorphous regions was investigated. As can be seen from

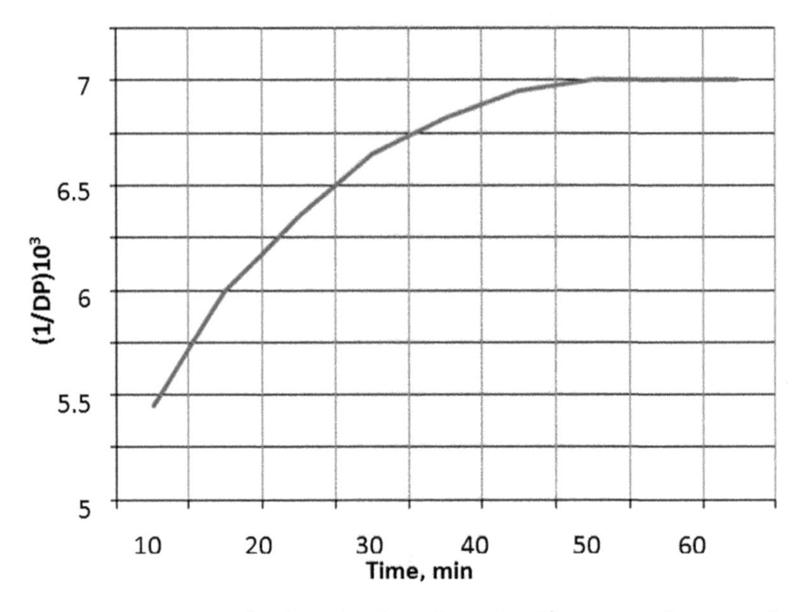

Figure 6.5 Degree of polymerization of wood cellulose as a function of chemical cutting time [1, 18]. Reprinted from Ref. [1] with permission from *Engineering Journal of Don*.

Fig. 6.5 [1, 18], DP of the studied wood cellulose decreases until it reaches an approximately constant value corresponding to the LODP. The determined LODP value of the studied NanoCell was 141. Then the average length of elementary crystallites of NanoCell is $L = 0.517$ nm $\cdot$ 141 $\approx$ 73 nm.

6.2.3 Investigation of the NanoCell by the XRD Method

The dry NanoCell powder was selected for x-ray analysis. The tablets were pressed from dry powder in order to obtaining x-ray diffractograms. The following characteristics of the sample can be calculated from XRD investigations [19, 21]:

- Degree of crystallinity (DC):

$$\mathrm{DC} = (S_c/S_0) \cdot 100\%, \tag{6.2}$$

where S_c is the total surface of crystalline peaks and S_0 the total surface of the diffractogram.

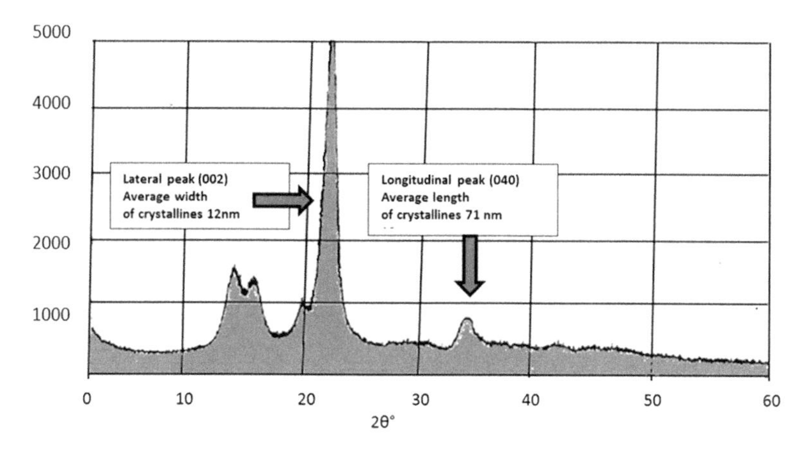

Figure 6.6 XRD of NanoCell with Cl-crystalline modification [1, 18]. Reprinted from Ref. [1] with permission from *Engineering Journal of Don.*

- Average size of an elementary crystal (length = L_0 and width = H_0) was calculated in accordance with the equation

$$L_0(\text{or } H_0) = [(B_0 \cos\theta)/\lambda)^2 - (\delta/d)^2]^{-1}, \tag{6.3}$$

where B_0 is the corrected width of the peak, θ is the diffraction angle, d is the interplanar distance, δ is the lattice distortion of the second type, and λ is the length of the CuK_a x-ray wave (0.15418 nm).

XRD of the NanoCell sample (Fig. 6.6 [1, 18]) was carried out using the Rigaku-Ultima Plus diffractometer.

- DC = 81%.
  Lateral peak (002) has the following parameters:
  $\theta = 11.25°$, $d = 0.395$ nm, $B_0 = 0.02045$ radian, and $\delta = 0.040$.
  The calculated width of the elementary crystallites of NanoCell:
  $H_0 = [(0.02045 \cdot 0.9898/0.15418)^2 - (0.04/0.395)^2]^{-1} = 12$ nm.
- Longitudinal peak (040) has the following parameters:
  $\theta = 17.35°$, $d = 0.2585$ nm, $B_0 = 0.02$ radian, and $\delta = 0.0318$.

The calculated length of the elementary crystallites of NanoCell:
$L_0 = [(0.02 \cdot 0.9545/0.15418)^2 - (0.0318/0.2585)^2]^{-1}$
$= 71$ nm.

As it follows from XRD results, the NanoCell has CI-crystalline modification. The elementary nanocrystallites are anisometric, having the length 71 nm and width 12 nm.

So, both the viscometry and XRD methods evidence the nanoscale structure of elementary crystallites of NanoCell samples.

NanoCell powder usually contains aggregates of the elementary crystallites. To find out the size of crystalline aggregates methods of LLS and SEM were used.

6.2.4 Method of Laser-Light Scattering

This method underlies Malvern's Mastersizer-2000 apparatus for measuring particle size distribution (PSD) and average particle size ($D_{0.5}$) of various dispersions and powders. As it follows from testing results (Fig. 6.7 [1, 18]), a NanoCell dispersion contains small nanocrystalline aggregates with an average size $D_{0.5}$ of about 200 nm.

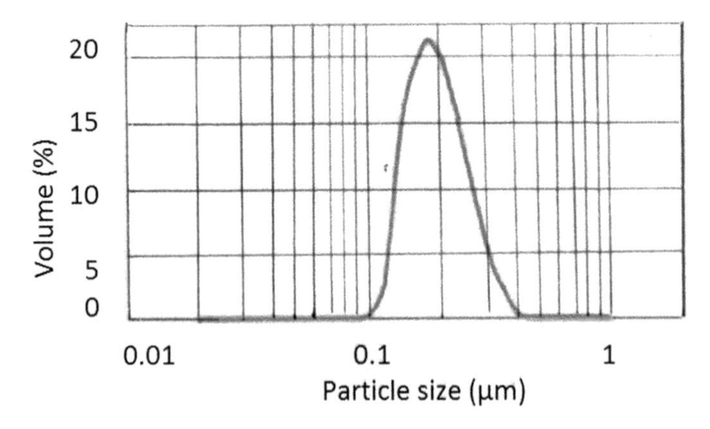

Figure 6.7 PSD curve of the NanoCell dispersion [1, 18]. Reprinted from Ref. [1] with permission from *Engineering Journal of Don*.

Figure 6.8 Scanning electron micrograph of the NanoCell sample [1, 18]. Reprinted from Ref. [1] with permission from *Engineering Journal of Don.*

6.2.5 Using of the SEM Method for NanoCell Investigations

A dry powder of NanoCell was prepared for SEM investigation by chemical cutting of the initial cellulose, blending, homogenization, freeze-drying, and mechanical disintegration. To improve contrast the NanoCell powder was preliminary coated with gold. As seen from Fig. 6.8 [1, 18] this sample contains ellipsoidal nanoparticles with an average diameter of 100–300 nm. This is correlated with LLS results.

6.2.6 Potential Application, Preparation, and Investigation of Composite Materials Based on NanoCell

Nanocellulose can be widely used in various fields of consumer industry, pharmaceuticals, and construction owing to its unique physical and chemical properties, including correspondence to requirements of environmental safety. Some potential fields of application of NanoCell is shown in Fig. 6.9.

The following NanoCell-containing materials were prepared:

- Polyvinyl chloride (PVC)-plastic filled with NanoCell powder
- Biodegradable polycaprolactone (PCL)-polymer filled with NanoCell powder

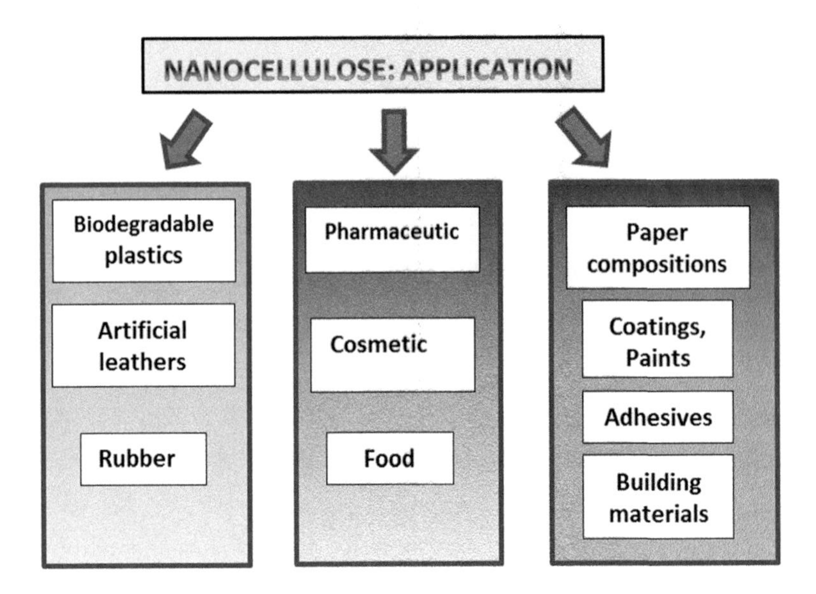

Figure 6.9 Some potential fields of application of NanoCell.

- Paper pulp water-based composition filled with NanoCell paste
- Glue composition containing polyvinyl acetate latex, dioctyl phthalate (DOP), and NanoCell paste
- Water-based paint composition containing UCAR™ Latex, TiO_2, ground calcium carbonate (GCC), NanoCell paste, and assistant agents (disperser, surfactant, thickener, etc.)

As it can see from mechanical testing (Figs. 6.10 and 6.11 [1, 18]). Introducing of the NanoCell leads to a considerable increase in the strength of PVC, PCL, and paper materials. Moreover, filling of PCL with NanoCell contributes to an increase of the biodegradation ability of the composite polymer material (Fig. 6.12 [1, 18]).

Introducing of NanoCell in the glue composition permits improving tackiness and gluing strength (Table 6.2 [1, 18]).

Paint coatings were applied on to opacity charts by "Sheen Instruments Ltd." using the 100 microns bar applicator. The opacity value (contrast ratio) of the dry paint coating was measured by Sheen Opac Reflectometer. The NanoCell showed excellent extender properties for the paint containing mineral pigments.

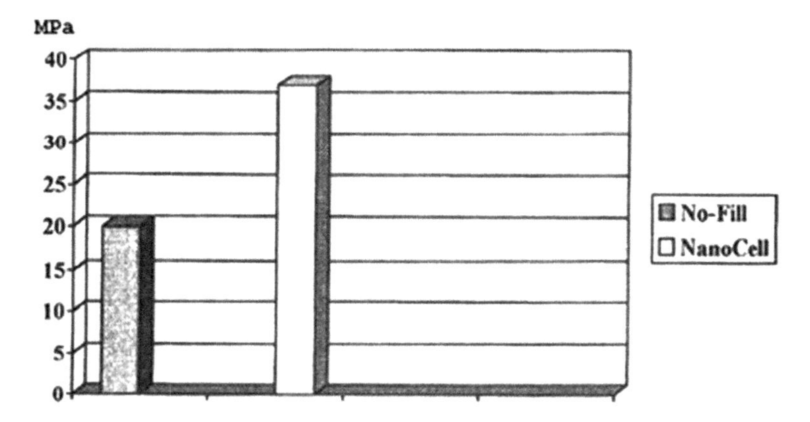

Figure 6.10 Tensile strength (Mpa) of initial PVC and PVC filled with 15% of NanoCell powder [1, 18]. Reprinted from Ref. [1] with permission from *Engineering Journal of Don*.

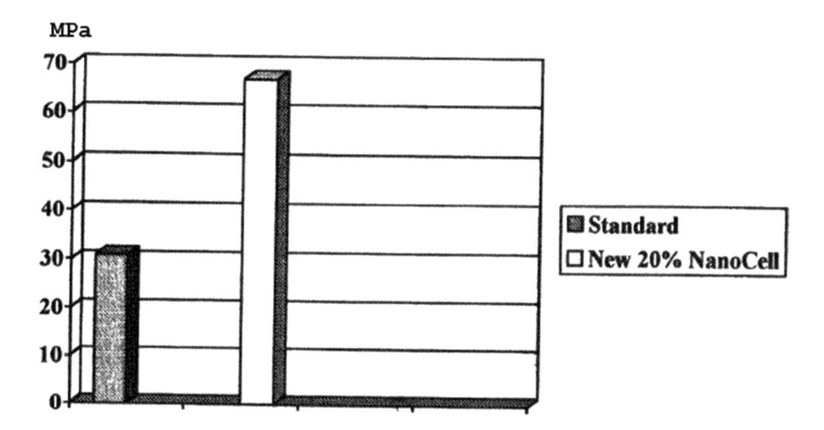

Figure 6.11 Tensile strength (Mpa) of initial paper and paper containing 20% of NanoCell [1, 18]. Reprinted from Ref. [1] with permission from *Engineering Journal of Don*.

Introduction of a low amount of the NanoCell into the paint composition permits replacing a significant amount of the TiO_2 pigment without changing in paint opacity value (Table 6.3 [1, 18]).

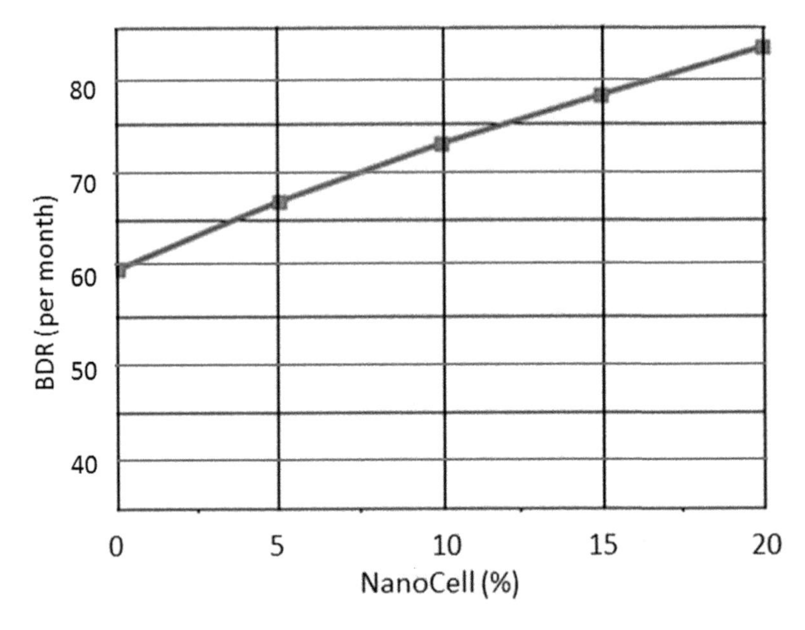

Figure 6.12 Biodecomposition rate (BDR) of PCL plastics filled with NanoCell [1, 18]. Reprinted from Ref. [1] with permission from *Engineering Journal of Don*.

Table 6.2 Properties of Wood's glues

Properties	Standard glue	NanoCell-based glue
Tackiness, Pa	70	120
Strength, Mpa	6.4	6.7

Source: [1, 18]. Reprinted from Ref. [1] with permission from *Engineering Journal of Don*.

6.3 Processing of Biodegradable Packaging Materials

6.3.1 Biodegradable Packaging Material GreenCoat™

Hydrophobic synthetic polymers are widely used for the production of packaging and other goods. These polymers have excellent protection properties, but they are highly resistant to biodegradation that causes pollution of the environment.

Table 6.3 Characteristics of interior paint compositions based on the UCAR binder (pigment volume concentration = 60%)

Components	Composition, %	
	No. 1	No. 2
TiO	20	13
GCC	30	30
UCAR binder	25	28
Medium	25	29
Characteristic		
TiO saving %	0	35
Opacity value	0.94	0.95

Source: [1, 18]. Reprinted from Ref. [1] with permission from *Engineering Journal of Don*.

To solve the burning ecological problem of biostable polymeric waste, various companies have been working out innovative polymers and polymeric materials that can be biodegradable: Biopol, Mater-Bi, Bioflex, polylactic acid (PLA), and others. But these polymeric materials have common disadvantages, like complicated technology, limited raw materials, high cost, and dissatisfaction in terms of some properties. World consumption of these biodegradable polymeric materials is only 0.1% of the consumption of polyolefins and other biostable hydrophobic synthetic polymers. Therefore, in principle, a new concept for creating hydrophobic biodegradable materials is needed.

The goal was to provide a composition for forming a protective coating layer on a biodegradable natural material that imparts to the material improved waterproofing and grease-resistant properties. The fundamental idea was based on a solution to the problem of susceptibility of natural materials to penetration by water and other liquids by filling the pores on the surface of the natural biodegradable packaging material with fine barrier particles.

Because paper and other natural materials contain micron-scale pores, filling of these micropores with protective biodegradable nanoscale particles closes the pores and thus makes the natural materials stable against penetration of water and other liquids.

Nanoparticles of cellulose, in combination with some other additives, appeared to be most suitable for the purpose.

Substrates for coating were paperboard of Weyerhaeuser Co. and starch-based trays of Hartmann Co. Bleached Kraft pulp (92% α-cellulose and DP = 1100) of Weyerhaeuser Co. was used as an initial material for the preparation of cellulose nanoparticles. Biodegradable PCL was delivered from Dow Plastics Co., while the natural hydrophobic agent (carnauba wax) was from Strohmeyer & Arpe Co. and other suitable chemicals (sulfuric acid, calcium oxide, organic solvent, and some others) were available from Sigma-Aldrich Co.

The initial cellulose sample was cut into pieces and mixed with water in a lab glass. Then 80 wt. % sulfuric acid was slowly added at cooling to obtain the required final concentration of sulfuric acid of 60 wt. % and an acid/cellulose ratio of 5. The glass was placed into a water bath having a temperature of 50°C and heated at stirring for 1 h. Hydrolyzed cellulose was separated from the acid by centrifugation at the acceleration of 4000 g for 15 min and washed and separated once again. Then calcium oxide was added to the acidic slurry to neutralize the acid and precipitate the inorganic nanofiller calcium sulfate.

The slurry containing agglomerates of nanoparticles was added to an acetone solution containing 30–35 wt. % PCL and 1.5–2 wt. % carnauba wax. Then this mixture was homogenized by means of a Gaulin-type homogenizer with 10 circulations at 100 Mpa to obtain the protective nanocomposition for coating of natural biodegradable substrates.

The substrates were coated with the liquid nanocomposition by means of a rod-type lab coater and dried at 100°C for 30 min. The weight of the dry coating on the surface of paperboard was 20 g/m^2 and on the surface of the starch-based substrate was 40 g/m^2.

The particle size of the nanocoating was studied by means of a Mastersizer-3000 apparatus of Malvern Instrument Ltd. The viscosity of the liquid compositions was measured by the Brookfield viscometer DVII at a rotation rate of 20 rpm. Water absorption of the coated substrates was determined by a Cobb test in accordance with ASTM D3285. The mechanical properties were tested by the

Instron 4201 test system. The biodegradability was evaluated by the weight loss of materials at composting in wet soil under conditions described in ASTM D2020-B and ASTM D5988.

After acid hydrolysis of the initial pulp and neutralization with calcium oxide about 15% wet cake was obtained containing 60–65% nanocellulose and 35–40% inorganic nanofiller ($CaSO_4$). About 100 g of the wet cake was mixed with a 200 g solution containing a biodegradable polymer and a natural hydrophobic agent and then homogenized. As a result the coating composition was obtained (Table 6.4 [1, 23]).

Table 6.4 Main ingredients of the coating composition

Ingredient	Content, wt. %
Solid nanocellulose	3.0–3.3
Solid inorganic nanofiller	1.8–2.0
Soluble polymer	20–23
Soluble wax	1.0–1.3
Organic solvent	42–45
Water	27–28

Source: [1, 23]. Reprinted from Ref. [1] with permission from *Engineering Journal of Don*.

Investigation showed that this composition has a viscosity of about 300 centipoise (cP) and an average size of the solid particles of about 170 nm (Fig. 6.13 [1, 23]).

After coating of the substrates, an increase of mechanical properties was observed (Fig. 6.14 [1, 23]). The initial substrate in a wet state loses about 90% of its strength, while the coated substrate after wetting, and vice versa, maintains up to 90% of its strength. Besides, introduction of the nanoparticles into the coating composition promotes obtaining higher mechanical properties of the coated substrate.

The initial substrates don't possess barrier properties to water and grease. In contrast, coating of the substrates with nanocomposition imparts to hydrophilic materials increased barrier properties (Table 6.5 [1, 23]).

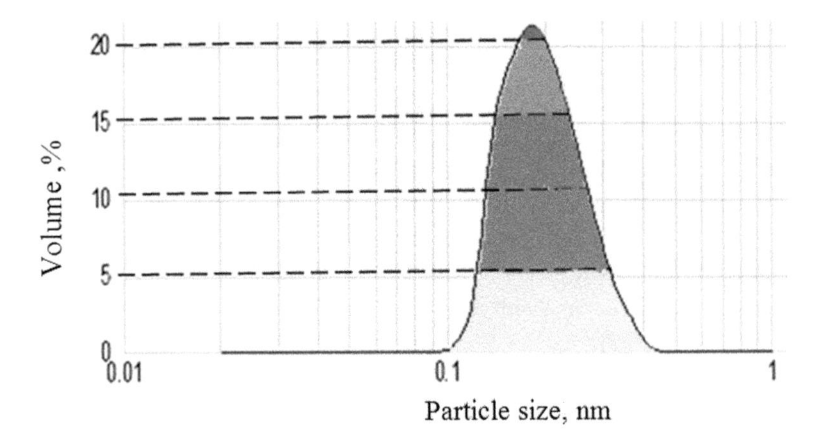

Figure 6.13 Size distribution of the solid nanoparticles in the coating composition [1, 23]. Reprinted from Ref. [1] with permission from *Engineering Journal of Don.*

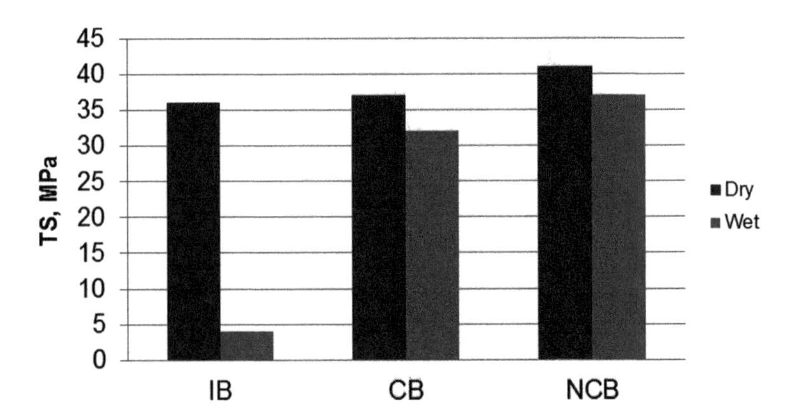

Figure 6.14 Tensile strength (TS) in the dry and wet states of the initial paperboard (IPB), board coated with composition without nanoparticles (CB), and board coated with composition containing nanoparticles (NCB) [1, 23]. Reprinted from Ref. [1] with permission from *Engineering Journal of Don.*

Due to special composition of the coating, the coated substrates decompose in the wet soil during a relative short time (Fig. 6.15 [1, 23]).

Table 6.5 Barrier properties of the initial (IB, IS) and nanocoated (NCB, NCS) materials

Substrate	Grease kit no.	Water $Cobb_{30}$, g/m^2
IB	1–2	170–200
NCB	10–11	10–15
IS	0	480–540
NCS	8–9	20–25

Source: [1, 23]. Reprinted from Ref. [1] with permission from *Engineering Journal of Don*.
Note: IB and IS are initial board and initial starch-based tray, respectively; NCB and NCS are nanocoated board and nanocoated starch-based tray, respectively.

Comparative properties of the GreenCoat™ composition with the similar biodegradable package materials are shown in Tables 6.6 and 6.7 [1, 23]. The patent in Ref. [11] presents the results of experimental investigation of some biodegradable nanocellulose composites samples (Table 6.8 [1, 23]).

As shown in Table 6.8, nanocompositions protect the natural packaging materials against water and grease, while microcompositions containing coarse micron-scale particles have poor barrier properties. Moreover, natural packaging materials coated with the proposed nanocompositions are biodegradable and decompose fully in 2–3 months when composting in wet soil.

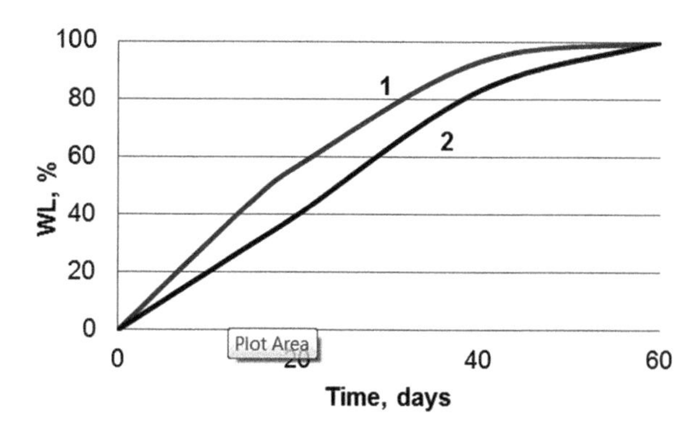

Figure 6.15 Weight loss (WL) of the initial (1) and nanocoated (2) paperboard due to biodestruction in wet soil [1, 23]. Reprinted from Ref. [1] with permission from *Engineering Journal of Don*.

Table 6.6 GreenCoat™ comparative table

Properties	Materials				
	LDPE	**Bioceta**	**Mater-Bi**	**Paper**	**GreenCoat™**
Producer	Borealis A/S, Denmark	Mazzucchelli 1849 S.p.A., Italy	Novamont S.p.A., Italy		Polymate Ltd., Israel
Tensile strength, (kg/mm^2)					
• Dry condition	2–3	3–4	2–3	2–4	**4.5–6**
• Wet condition	1.9–3	2.3–3.2	1.5–2.5	03–0.6	**3–3.5**
WVT (g/day m^2)	0.1–1	5–6	5–10	20–30	**4–5**
O_2 permeability $\times 10^{13}$ (cm^3 cm/cm^{2}s Pa)	0.2–0.5	0.5–0.7	0.5–0.8	5–10	**0.2–0.4**
Rate of biodegradation (%) in wet soil (25–30% moisture 28–30°C for 6 months 30°C) for 6 months	0–0.5	20–30	100	100	**100**
Time of biodegradation	>10 years	1 year	1–3 months	1–2 months	**1–2 months**

Source: [1, 23]. Reprinted from Ref. [1] with permission from *Engineering Journal of Don*.

Table 6.7 Properties of cardboard coated by synthetic rubber and GreenCoat™

Properties	Rubber coating	GreenCoat™
Water absorption, Cobb test for 30 min (%)	12	5
WVP (g/day m^2)	5	4
O_2 permeability $\times 10^{13}$ (cm^3 cm/cm^2 s Pa)	4	0.3
Resistance against oil and organic solvents (Kit number)	10	12

Source: [1, 23]. Reprinted from Ref. [1] with permission from *Engineering Journal of Don*.

6.3.2 Biodegradable Packaging Materials BHM

Kraft paper with a density of 50–70 g/m^2 and cardboard with a density of 300–400 g/m^2 were selected as cellulose substrates. Films of biodegradable plastics [2, 24] Mater-Bi (Novamont Co.), Natura (Natura Co.), and Biopol (Monsanto Co.) were used for protective coating of the cellulose substrates. The biodegradable plastics Mater-Bi and Natura contain PCL and starch; Biopol contains poly-hydroxy-butyrate-valerate (PHBV).

The novel biodegradable hydrophobic polymer materials (BHMs) in the form of aqueous dispersions were applied [2, 25]:

- **BHM-B** (the basic type of BHM) was synthesized by emulsion polymerization of vinylacetate with the addition of a small amount of acrylic acid and other needed reagents. The concentration of vinylacetate and acrylic acid copolymer (VAC) in BHM-B emulsion is 35–45%. BHM-B is intended to create a grease-repellent coating on the surface of a cellulose substrate. Other types of BHM composition were also developed.
- **BHM-GW**, intended to create a grease- and water-repellent coating, was prepared by the addition of a cross-linking agent (e.g., dimethylolurea) to the basic BHM emulsion.
- **BHM-W** is intended for waterproof coating; it was prepared by the addition to the basic BHM emulsion of a cross-linking agent and an emulsion of the natural carnauba wax.
- **BHM-U**, a two-layer-deep barrier coating of the type, was applied on the substrate surface: the first, primer, layer

Table 6.8 Results of water penetration (Cobb test for 30 min), grease resistance (3 Kit test number or degree), and biodegradability time (BDT) on full weight loss during composting in wet soil

Sample	Particle size (nm) in coating composition	Coating weight (g/m^3)	Cobb test ($g\ H_2O/m^2$)	Grease-resistance (degrees)	BDT (months)
1	240	20	18	12	2
2	180	15	21	12	3
3	200	40	27	10	2
4*	6000	20	62	8	2
5*	10,000	40	75	5	2

Source: [1, 23]. Reprinted from Ref. [1] with permission from *Engineering Journal of Don*.
*Comparative samples

consists of BHM-B and the second layer consists of a hot-melt glazing composition of natural wax, higher aliphatic acids, and natural fillers.

Polymer films were applied on the substrate surface by means of lamination technology. Coating of the cellulose substrate with BHM emulsions and compositions was carried out by a bar coater. The emulsion-coated substrate was dried at a temperature of 150–180°C.

The layer composites created have "bread-butter" (A) or "sandwich" (B) structures (Fig. 6.16 [2, 26]). These composites can consist of a thin (10–30 μm) protective coating layer (1) and a relatively thick (0.5–2 mm) layer of cellulose substrate (2).

The production scheme of the coating process is illustrated in Fig. 6.17.

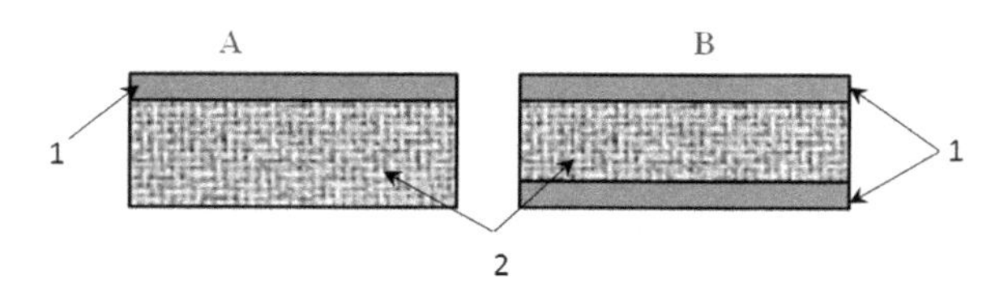

Figure 6.16 Structure types of the layer composites: 1, polymer coating; and 2, cellulose substrate [2]. Reprinted from Ref. [2] with permission from *Engineering Journal of Don*.

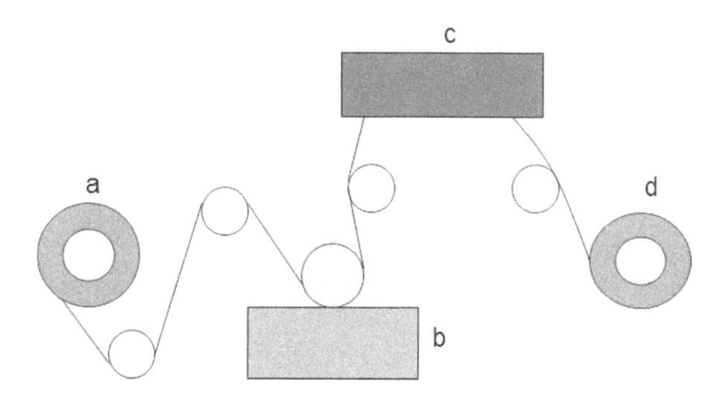

Figure 6.17 Scheme of a coater: (a) unwind station, (b) coating station, (c) drier, and (d) rewind station.

The effective size (diameter) of micropores in a protective coating layer was measured by the method of water vapor sorption. The effective diameter of micropores was calculated in accordance with the following equation [2, 26]:

$$d = 4V_p/S, \quad (6.4)$$

where S is the specific surface of a polymer material and V_p is the specific volume of micropores measured at a relative humidity close to 1.

For example, the Brunauer–Emmett–Teller (BET) isotherm of water vapor sorption for Coating A is shown in the Fig. 6.18 [2, 26]. The calculated specific surface (S) of the polymer coating is 35 m^2/g, and the specific volume of micropores (V_p) is 0.37 cm^3/g; therefore, the effective diameter of micropores (d) in Coating A is about 42 nm.

The following barrier properties of composite materials were studied:

- The water vapor transmission rate (WVT) through a sample tested at normal conditions (25°C, relative humidity (RH) = 0.85) [27].
- The absorption of liquid water for 30 min (A) was measured by the Cobb test method [28].
- The permeation of O_2 (P[O_2]) [29].
- Grease penetration (Kit number) was tested by the 3M Kit method [27].
- Time of biodegradation (t) and weight loss of the composites in composting conditions were measured by the methods [30, 31].

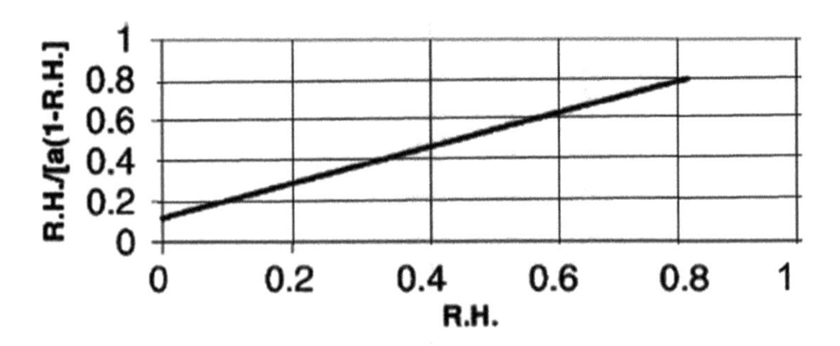

Figure 6.18 The BET isotherm of Coating A [2]. Reprinted from Ref. [2] with permission from *Engineering Journal of Don*.

- Repulping degree (RD) of the composite materials was checked by the method [32].

The barrier properties of the layer composite materials having various protective coatings are shown in Table 6.9 [2, 26].

As can be seen from the experimental results, the value of grease penetration is similar for all the composites studied. However, there is a considerable difference in other barrier properties of the various materials. The coatings of Mater-Bi and Natura do not ensure that the composite materials constitute a barrier against water penetration or moisture and oxygen permeation. The best barrier properties observed are for the composite having the BHM-U coating.

The effective diameter of the micropores (d) and the time of degradation (t) in the various coatings of the composite materials are shown in Table 6.10 [2, 26].

Table 6.9 Barrier properties of the various layer composites

Coating (thickness 25 μm)	Kit no.	A g/m^2	WVT g/day m^2	PO_2
Mater-Bi	12	13	330	16
Natura	12	11	270	14
Biopol	12	5	40	9
BHM-B	12	9	110	10
BHM-W	8	3	30	9
BHM-GW	12	7	100	10
BHM-U	12	1	4	7

Source: [2]. Reprinted from Ref. [2] with permission from *Engineering Journal of Don*.

The smallest micropores, with an effective size of about 5 nm, were found for the composite having the BHM-U coating; and the largest, about 60 nm, were found for the material having the Mater-Bi coating.

To prevent pollution of the environment waste materials should be broken down and destroyed. The waste of all advanced composite materials is biodegradable and can be decomposed in composting conditions. Decreasing the thickness of the protective coating leads to a reduction in the time taken for biodegradation. Another way of to get rid of the waste is to repulp it and then use the

Table 6.10 Sizes of micropores and time of degradation in the coatings

Coating (thickness 25 μm)	*D* nm	*t* months
Mater-Bi	60	2
Natura	42	2
Biopol	17	5
BHM-B	30	3
BHM-W	16	5
BHM-GW	20	3
BHM-U	5	6

Source: [2]. Reprinted from Ref. [2] with permission from *Engineering Journal of Don*.

recycled composites in the paper industry. This is preferable to biodegradation of the waste.

As can be seen from the results (Table 6.11 [2, 26]), the composites having coatings of the BHM type are repulpable, while the composites having coatings of Mater-Bi, Natura, or Biopol are nonrepulpable and that limits their application.

Table 6.11 Repulping degree (RD) for various types of layer composite materials

Coating (thickness 25 μm)	RD %	Note
Uncoated substrate	100	Repulpable
Mater-Bi	38	Nonrepulpable
Natura	36	
Biopol	31	
BHM-B	93	
BHM-W	90	Repulpable
BHM-GW	86	
BHM-U	83	

Source: [2]. Reprinted from Ref. [2] with permission from *Engineering Journal of Don*.

A correlation between the sizes of the coating micropores and the properties of the layer composite materials was found. Decreasing the micropore diameter of the protective coating reduced the water absorption and moisture and oxygen permeation, as well as retarded

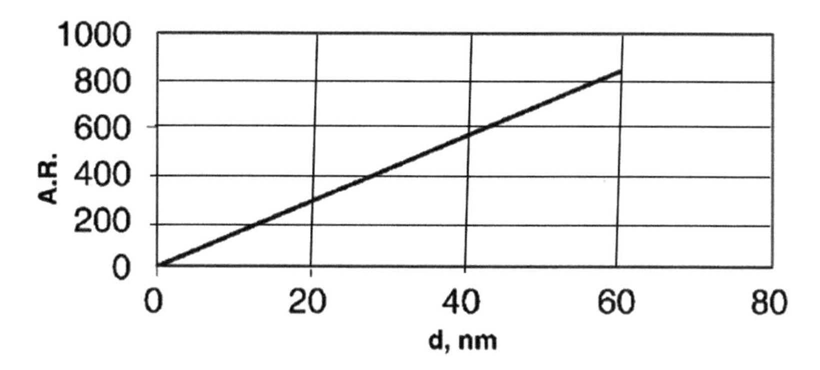

Figure 6.19 Correlation between micropore size (d) and rate of liquid water absorption (AR) [2]. Reprinted from Ref. [2] with permission from *Engineering Journal of Don*.

the biodegradation process of the composite materials (Table 6.11, Figs. 6.19–6.21 [2, 26]).

The larger micropores and poor barrier properties of the composites having PCL-based polymer coatings (Mater-Bi and Natura) can be explained by the loose packing of the supermolecular structure in these polymer layers. The denser supermolecular structure of PHBV-based Biopol coating and VAC-based BHM coating ensures better barrier properties in the composite materials.

Sizes of micropores and the barrier properties of the dense two-layer BHM-U coating are similar to that of a polyolefin coating. Therefore, the environment-friendly, recyclable composite material having this coating can be used instead of the environment-polluting, nonrecyclable polyolefin-based laminate.

Table 6.12 contains for example characteristics of the 1BT version of the BHM aqueous polymer emulsion.

Table 6.12 Characteristics of the 1BT version of the BHM aqueous polymer emulsion

Characteristics	Value
Solid content (%)	35–36
Density (g/cm^3)	1.03–1.05
Viscosity (cps)	100–150
pH	7–7.5

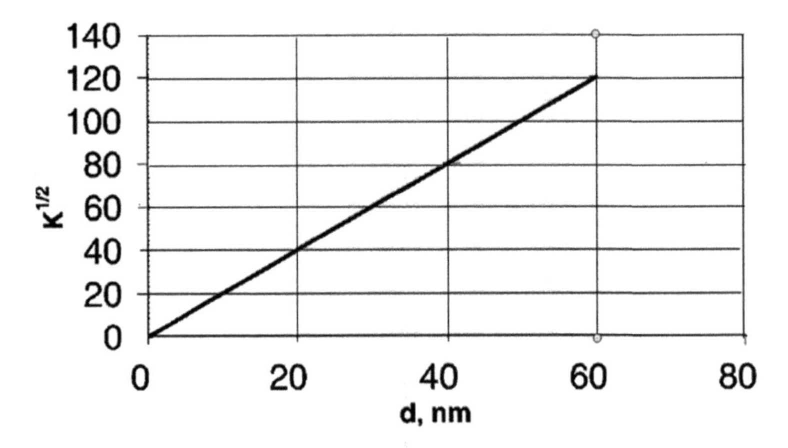

Figure 6.20 Correlation between micropore size (d) and the coefficient of water vapor transmission (K) [2]. Reprinted from Ref. [2] with permission from *Engineering Journal of Don*.

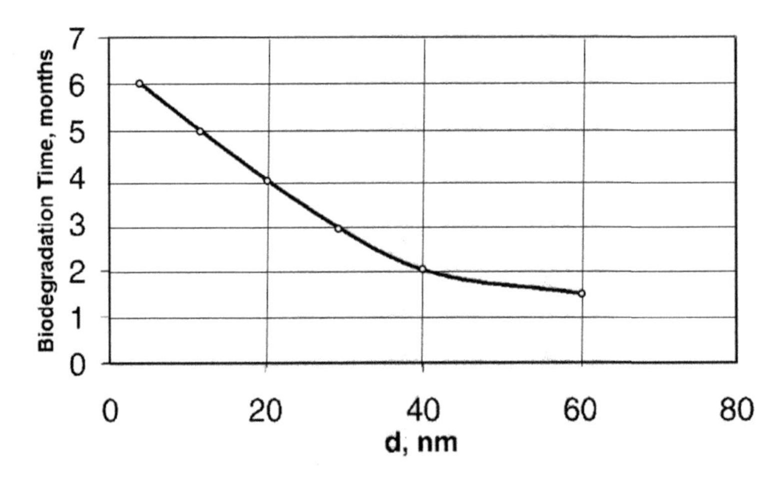

Figure 6.21 Dependence of the biodegradation time of composite materials on the size of micropores in protective coatings [2]. Reprinted from Ref. [2] with permission from *Engineering Journal of Don*.

Properties of cardboard coated with the 1BT version of BHM are the following (Table 6.13).

It should be noted that within the framework of the patent in Ref. [25] 18 different cellulose waterproof biodegradable composites are

Table 6.13 Characteristics of cardboard coated with the 1BT version of BHM

Characteristics	Value
Coating weight (g/m^2)	5–10
Tensile strength (Mpa)	40–50
Cobb value for 30 min (g/m^2)	10–20
WVT (g/day m^2 at RT)	50–60
3M Kit number	6–7
Time of biodegradation (months)	2–4
Repulpability	Repulpable

described in detail. These composites consist essentially of cellulose and hydrophobic polymer-based components.

The composition of the polymer-based component is:

- 45–94 wt. % of a first polymer, which is preferably polyvinyl acetate. Said polymer is capable of cross-linking and contains about 2 to 8 wt. % of free hydroxyl groups.
- 4–28 wt. % of a second polymer, which is not cross-linked and is capable of imparting to the first polymer improved elastic properties.
- 2–20 wt. % of a cross-linking agent having at least two functional hydroxyl, carboxyl, amine, and/or aldehyde groups.

Specific properties of the suggested cellulose composites are shown in Table 6.14 [2, 25] below.

Table 6.14 Properties of the cellulose composites

Cellulose composites, no.	1	2	3	4	5	6	7	8	9
Water absorption, wt. %	13	11	12	5	18	21	5	17	14
Water absorption after folding, wt. %	13	12	12	7	19	23	6	30	15
Weight loss % for 3 months due to biodegradation	93	90	93	88	95	100	72	100	100

As is shown in Ref. [25] the equivalent ratio of functional groups (FGs) in a cross-linking agent to the stoichiometric content of free OH-groups in the polyvinyl alcohol (PVA) should be 0.4 to 1.2. It

Table 6.14 (*Continued*)

Cellulose composites, no.	10	11	12	13	14	15	16	17	18
Water absorption, wt. %	4	3	26	10	5	7	6	3	2
Water absorption after folding, wt. %	6	4	30			N/A			
Weight loss % for 3 months due to biodegradation	70	55	95	93	90	88	92	84	80

Source: [2, 25]. Reprinted from Ref. [2] with permission from *Engineering Journal of Don*.

has been empirically revealed that if the above ratio is less than 0.4 the cross-linking process does not take place; and if the ratio is more than 1.2 the cross-linking degree is not increased and does not render the obtained composite material more waterproof. It is advantageous if the weight ratio plasticizer (PL)/PVA lies in the range 0.05–0.43 and the PL content in the polymer-based component lies in the range 4–28 wt. %. Decreasing the PL/PVA ratio below 0.05 or the PL content below 4 wt. % renders the obtained composite material more rigid (cracks after folding) and less waterproof (see example 8). Increasing the PL/PVA ratio above 0.43 or the PL content above 28 wt. % renders the obtained composite material less biodegradable.

6.3.3 Basic Principles for Biodegradation of Polymers

The decomposition process of the waste of the novel composite materials and articles in wet soil occurs under the effect of various microorganisms, fungi, and bacteria. These microorganisms exude hydrolases, esterases, peptidases, oxydases, reductases, and other enzymes that are biological catalysts of the degradation process.

An organic polymer is capable of biodegradation if it contains specific bonds sensitive to enzymatic destruction, for example, esteric (1), acetalic (2), and peptide (3) bonds [32]:

(1) -R-0—00
(2) -R-0C—0-
(3) -R-NH ~C=0

Natural hydrophilic polymers, such as cellulose starch and proteins, and also synthetic hydrophilic and hydrophobic polymers having the above-mentioned specific bonds are biodegradable.

Biodegradation of an elaborated hydrophobic polymer coating occurs in wet soil under normal enzymatic action of various microorganisms (fungi and bacteria) as follows:

- In aerobic conditions (top soil layers)
  $C_nH_mO_k$ (BHM) + O_2+ enzymes = CO_2+ H_2O + biomass (humus)
- In anaerobic conditions (lower soil levels)
  $C_nH_mO_k$ (BHM) + enzymes = CH_4+ biomass (humus)

The decomposition of the biodegradable composite occurs in three basic steps [32]:

(1) The process of the enzymatic detachment of lateral groups occurs, for example:

```
– A – CO -- O – A –  + 2H2O  ——→ – A – CO -- O – A–  + 2RCOOH
  |             |                  |             |
  O             O                  OH            OH
  |             |
 RC= O         RC = O
```

(2) The process of the enzymatic depolymerization occurs, for example:

```
– (ACO – OA)n u + 2nH2O——→   n(HO2) ACOOH +n A(OH)3
    |        |
    OH       OH
```

(3) The metabolic conversion of monomeric products into water, carbon dioxide (in aerobic conditions), or methane (in anaerobic conditions) occurs, for example:

$$(HO_2)\ ACOOH + A(OH)_3 + O_2 \longrightarrow CO_2 + H_2O$$

Microorganisms of soil after the finish of their vital function are converted into humus. Thus, only the environment-friendly substances are formed as a result of an organic polymer biodegradation in aerobic conditions: water, carbon dioxide, and humus. There

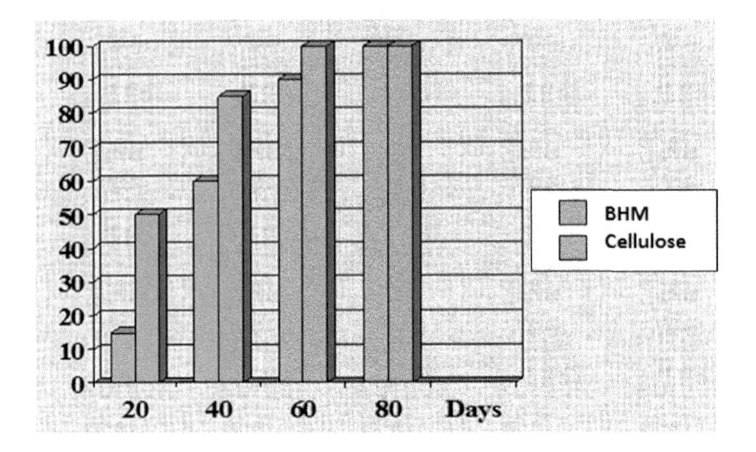

Figure 6.22 Weight loss percentage of materials in wet soil.

are additional factors that are influenced on the decomposition of biodegradable polymers. Increase in the degrees of polymerization and crystallinity delays the biodegradation process of a polymer. Porous polymer decomposes faster than monolythic and glass-type polymer decomposes more slowly than elastic.

The decomposition of a cellulose base coated with BHM at composting conditions is only slightly longer than the basic cellulose material (Fig. 6.22). Moreover the coated cellulose substrate is repulpable and can be utilized together with uncoated cellulose materials.

Soil microorganisms, after their vital functions are finished, are converted into humus. Thus, in aerobic conditions only environment-friendly substances are formed as a result of the decomposition of biodegradable polymer materials.

The coated substrates—novel layer composites—are produced on the basis of plant raw materials. The end products (water, carbon dioxide, and humus) formed as a result of biodegradation of the composite materials are returned to their natural sources (Fig. 6.23 [2, 26]).

The layer composites made by lamination with films of Mater-Bi, Natura, Biopol, etc., are biodegradable but nonrepulpable. Only composite materials having BHM-type coatings are both biodegradable

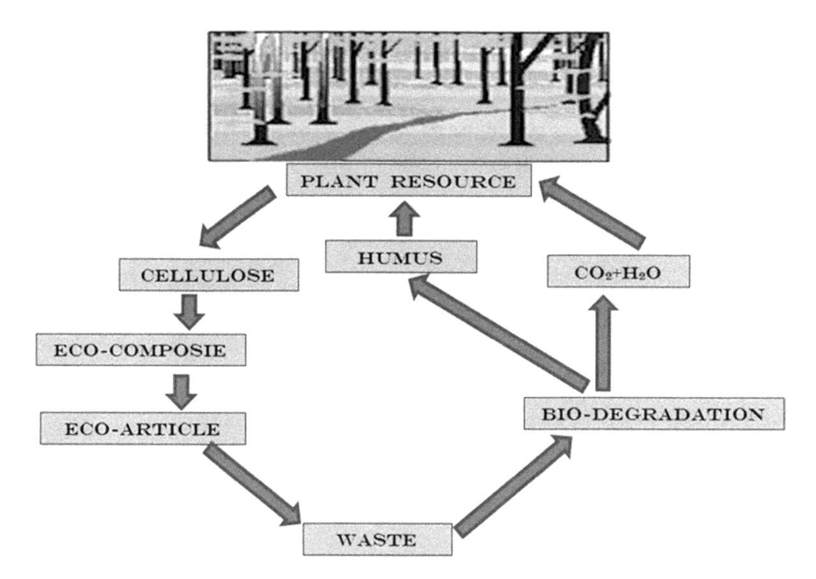

Figure 6.23 Biodegradable cycle of the novel composite material in nature [2]. Reprinted from Ref. [2] with permission from *Engineering Journal of Don*.

and repulpable. Widespread use of the novel composite materials will improve the ecological state of the environment.

It should be noted that within the framework of the patent [25] eighteen different cellulose waterproof biodegradable composites are described in detail. These composites consist essentially of cellulose and hydrophobic polymer-based components.

The composition of the polymer-based component is:

- 45–94 wt. % of a first polymer, which is preferably polyvinyl acetate. Said polymer is capable of cross-linking and contains about 2 to 8 wt. % of free hydroxyl groups.
- 4–28 wt. % of a second polymer, which is not cross-linked and is capable of imparting to the first polymer improved elastic properties.
- 2–20 wt. % of a cross-linking agent having at least two functional hydroxyl, carboxyl, amine, and/or aldehyde groups.

6.3.4 Application of the Novel Biodegradable Packaging Materials

- Everyday items: Trash bags, grocery bags, cups, plates, tablecloths and other household goods, etc.

- Packaging materials: Carton boxes, disposable containers for food processing, bags, boxes and containers for industrial products, building materials and chemicals, etc.

- Agricultural use: Composting bags for agriculture waste, bags for fertilizers, mulch sheets, flowerpots, seeding planter trays, etc.
- Textile and polymer industry: Hydrophobizing of natural textile materials, production of synthetic leather, biodegradable membranes, etc.
- Sanitary products: Protection layer for disposable diapers, sanitary napkins, panties, towels, etc.

- Other applications: Water- and grease-resistant paper/board, filler binding, paper sizing, printing compositions, etc.

6.4 Nanocellulose as a Promising Biocarrier

Pure cotton cellulose (degree of polymerization [DP] = 3000 98.7%, α-cellulose = 98.7%) was used as initial cellulose materials. The initial cellulose was treated with a ellulose complex of *Trichoderma reesei* in a bioreactor. To prevent aggregation of the fine particles, all experiments were carried out using never-dried or non-water-dried samples. The biohydrolyzed cellulose was washed and squeezed on vacuum filter, and the wet cake was chemically modified to introduce form carbonyl and some other types of specific FGs. Then, a biologically active substance (BAS), for example, an enzyme, was coupled to the specific groups. The semiproduct was diluted with water and dispersed by the high-pressure homogenizer APV-2000 [34].

The average DP of the cellulose samples was measured by the Cuen viscosity method. The diffract meter Rigaku-Ultima Plus (CuK – radiation, $\lambda = 0.15418$ nm) was used for x-ray investigations. The degree of cellulose crystallinity and average lateral size of crystallites were calculated according to improved methods [35, 36]. The particle size distribution (PSD) and the average particle size of aqueous suspensions were tested by a method of laser-light scattering (LLS) using Malvern's Mastersizer-2000 apparatus. Scanning electron micrographs were obtained with a Hitachi S-430 apparatus.

The activity of free and coupled drugs was tested by standard biological methods.

The process for the preparation of the biocarrier includes the following main steps:

(1) Controlled biodepolymerization of the initial cellulose up to the minimal "level-off DP" (LODP); washing and squeezing to obtaining a wet cake.
(2) Modification in order to introduce specific FGs and join various BASs to fine cellulose particles.

(3) High-power mechanical disintegration of the modified cellulose particles in liquid media to produce dispersions of the bioactive nanocellulose.

This process can be represented by means of the following scheme:

Step 1: Enzymatic hydrolysis

The initial cellulose is treated with the ellulose complex, allowing selective cleavage of the macromolecular chains in the poorly ordered noncrystalline domains. The catalytic splitting of the poorly ordered domains promotes mechanical disintegration, forming superfine particles uniformly distributed in a liquid medium (Fig. 6.24 [2, 37]):

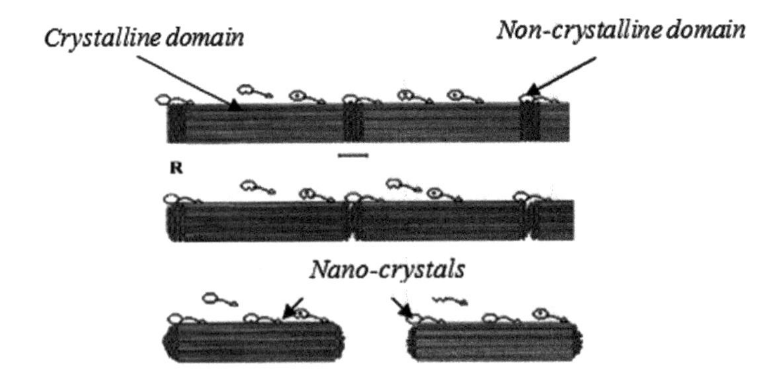

Figure 6.24 Scheme of enzymatic hydrolysis [2, 37]. Reprinted from Ref. [2] with permission from *Engineering Journal of Don*.

Step 2: Modification

To provide chemical coupling between the carrier and BAS, the specific FG should be introduced into the particles, for example, carboxyl, carbonyl, amine, and epoxy (Fig. 6.25 [2, 37]).

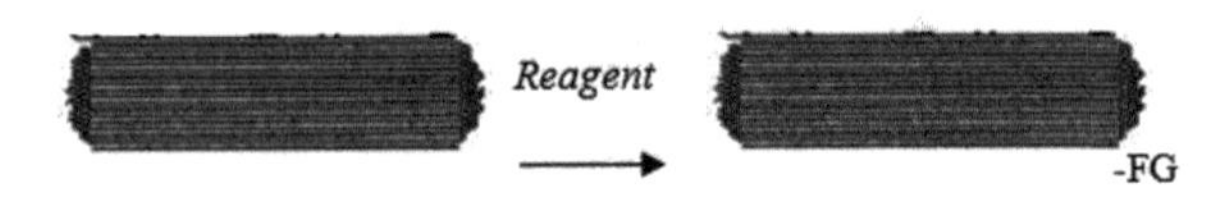

Figure 6.25 Scheme of modification [2, 37]. Reprinted from Ref. [2] with permission from *Engineering Journal of Don*.

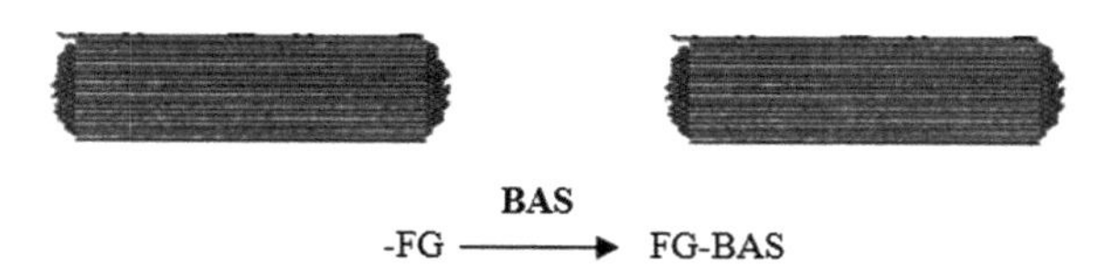

Figure 6.26 Scheme of attachment of a BAS to a FG of the carrier [2, 37]. Reprinted from Ref. [2] with permission from *Engineering Journal of Don*.

The nondried particulate carrier having specific reactive groups are able to quickly interact with various BASs at optimal conditions, allowing chemical attachment and forming of a carrier-BAS complex (Fig. 6.26 [2, 37]):

Step 3: High-power mechanical disintegration

The uniform length distribution of the nanocarrier particles in the water dispersion is shown in Fig. 6.27 [2, 37].

The structural characteristics of the nanocarrier having a low amount of carbonyl groups (0.3–0.5%) are presented in Table 6.15 [2, 37].

Electron microscopic investigations showed the suspension of the nanocellulose carrier contains rod-like particles having lengths of 150–200 nm and lateral sizes of 20–40 nm (Fig. 6.28 [2, 37]).

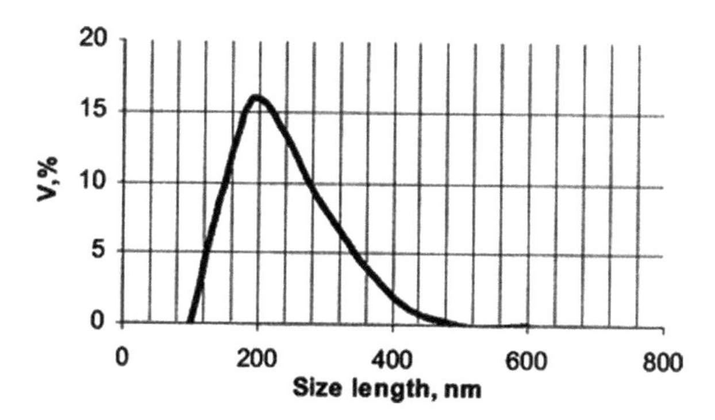

Figure 6.27 Distribution of the nanocarrier particles in the water dispersion [2, 37]. Reprinted from Ref. [2] with permission from *Engineering Journal of Don*.

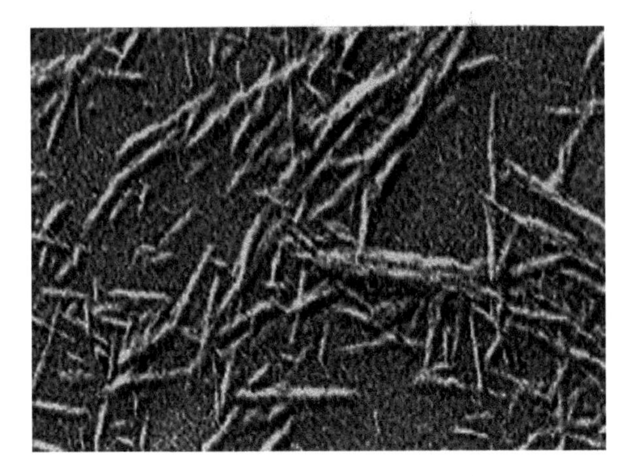

Figure 6.28 SEM of rod-like particles of the nanocellulose carrier [2, 37]. Reprinted from Ref. [2] with permission from *Engineering Journal of Don*.

Table 6.15 Main characteristics of the nanocellulose carrier

Characteristics	Value
Crystalline modification	CI
Degree of crystallinity, %	78–80
Degree of polymerization	100–120
Average particle length, nm	150–200
Lateral particle size, nm	20–40
Length of crystallites, nm	40–60
Lateral size of crystallites	10–12

Source: [2, 37]. Reprinted from Ref. [2] with permission from *Engineering Journal of Don*.

Since individual nanocrystallites have diameters of 10–12 nm and length of 40–60 nm, each such nanoparticle is built from aggregates comprising about 2–4 of the crystallites.

6.4.1 Activity of the Nanocarrier

To check the activity of the coupled BAS the standard testing methods were used. For example, the activity of some enzymes was investigated by UV/VIS-spectroscopic method using specific substrates.

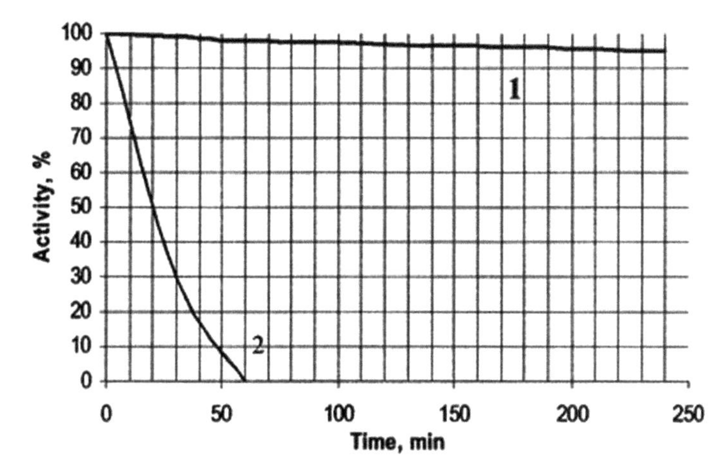

Figure 6.29 Activity of the coupled (1) and free (2) trypsine in aqueous media at 60°C [2, 37]. Reprinted from Ref. [2] with permission from *Engineering Journal of Don*.

The free enzyme has low stability in the aqueous solutions, and it quickly deactivates, particularly at increased temperatures. In contrast to the free enzyme, the enzyme coupled to the nanocellulose carrier is stable in the aqueous medium even at heating (Fig. 6.29 [1, 37]).

6.4.2 Applications of Cellulose Nanocarriers

The application potential of cellulose nanocarriers can be immense. Due to these unique properties and unlimited sources of the raw material, the biocarrier can find wide application in various health-care branches, such as cosmetology, personal care, dermatology, otolaryngology, dietary food, and biotechnology.

The cellulose nanocarrier containing attached proteolytic enzymes can be used in cosmetology as a gentle skin peeler, while containing bound aminoacids—an excellent nutrient agent for the skin. The carrier containing bound lipases can be used for selective degreasing of the skin. Nanocarriers with coupled proteolytic enzymes can be used in medicine for treatment of wounds and burns and also postoperating scars, while containing attached anesthetics—antipain agents.

6.5 Conclusion

The organic nanocellulose products have an average particle size of 100–300 nm in the form of water-based dispersion, paste, and dry powder. Introducing of NanoCell to composite materials (plastics, paper compositions, adhesives, etc.) imparts them some peculiar properties, such as a light weight, increased strength, raised biodegradability, and low cost. Moreover, NanoCell improves properties of water-based paints and coatings. A wide application of the organic NanoCell products can create new nanomaterials having unique characteristics and properties.

The main problem of natural biodegradable packaging materials is their susceptibility to water, grease, and various other liquids. To solve this problem, a special protective nanocoating was developed and applied onto the surface of these biodegradable substrates.

The green technology for the preparation of a nanocoating containing nanoparticles of cellulose and inorganic filler, as well as a biodegradable hydrophobic polymer, was proposed. This technology is zero discharge because it provides a full utilization of the raw materials. The initial cellulose pulp hydrolyzes completely and turns into solid nanocellulose; sulfuric acid turns into solid inorganic nanofiller. The organic solvent after coating of the substrates is condensed to utilize again.

As known, paper, starch, and other natural packaging materials contain micron-scale pores. Filling of these micropores with nanoscale particles in combination with a polymer binder and some other additives closes the pores and thus makes the natural packaging materials more stable against the effect of water, grease, and other penetrable liquids. Therefore, nanocoating containing solid nanoparticles with an average size of about 200 nm imparts to natural packaging materials increased dry and wet strength, an effective barrier against water and grease. Since the coating composition contains mainly biodegradable ingredients, coated substrates can decompose in nature in a relative short time.

Novel package materials consist of cellulose, biodegradable polymers, and other biodegradable organic additives. Proposed nanostructured polymer coating (NPC) is hydrophobic; oil-, fat-, and gasoline resistant; strong; inexpensive; recyclable; and biodegradable. Due to a biodegradable nature the package materials decompose in wet soil in 2–3 months by the enzymatic action of various microorganisms, such as fungi and bacteria, pretty much similar to ordinary paper, forming a biomass. All the components of the materials have Food and Drug Administration (FDA) approval as food packaging materials.

The main advantages of various NPC versions are followings:

- Very strong
  The tensile strength is 40–60 Mpa. Such strength characteristics, especially combined with low elongation and acquired water resistance of the material, make NPC unique and highly desirable for packaging applications.
- Water resistant
  NPC has good water resistance to the cellulose basic material. Thus, it has excellent prospects for many packaging applications. Most of the existing biodegradable packaging is not hydrophobic and fails in wetting conditions.
- Grease and oil resistant
  NPC is stable against greases, oils, and nonpolar organic solvents. Common paper and cardboard packaging is grease and oil penetrable.
- Recyclable
  NPC can be recycled and repulped like basic material (cellulose, paper, board, etc.).
- Degradable in the environment
  Natural microorganisms begin breaking down NPC and NPC cellulose substrate in water presence. Then microorganisms decompose the material with metabolic reactions. The material is converted into carbon dioxide, water, and biomass at composting in wet soil in 2–3 months. Thus, this process completely coincides with the definition of biodegradability given by most experts.
- Relatively low in cost
  The main obstacle to widespread use of biodegradable polymers is high cost. New biodegradable polymers, Mater-Bi, polylactic acid (PLA), Biopol, etc., are significantly more expensive than common biostable polymers, like polyethylene

(PE), polypropylene (PP), polyethylene terephthalate (PET), and polyvinyl chloride (PVC). This expensiveness blocked the widespread adoption of biodegradable plastics in major consumer applications. The high costs involved in the production of biodegradable polymers means that they cannot compete favorably with conventional polymers. NPC can be manufactured on the basis of relative cheap components, with the existing industry equipment using existing technologies. Paper or board with NPC is only insignificantly more expensive to produce than to produce paper or board itself. Currently available degradable materials on the other hand can cost twice as much.

- Harmless
  NPC does not contain harmful components. It is environment-friendly and FDA approved.

The cellulose nanocarrier has the following general features:

- It is natural, biocompatible, and harmless, permitting its use as a health-care agent.
- Specific FGs and developed surface promote expressed reactivity.
- Superfine and soft particles contribute to a gentle sensation.
- It has excellent compatibility with various organic ingredients.
- It is insoluble in water, oils, and organic solvents.
- It permits increasing the local concentration of the attached BAS.
- It ensure a slow-release effect of the attached BAS.
- It increases the stability of the attached BAS to decomposition and deactivation.
- Due to expressed thickening effect, it imparts rheological properties.
- A settling stability enables homogeneity of liquid-based systems.

The carrier-BAS complex has increased stability against external factors. Moreover, due to a developed surface, this product can form high viscous suspensions and pastes that are convenient for applications. Various BAS types, such as enzymes, biocides, anesthetic, cosmetic, and active health-care agents, can be attached simultaneously to the reactive nanocarrier.

Due to a superfine nanosize, the particles of cellulose nanocarriers having coupled BAS are capable of cleaning the skin's micron-scale pores, opening them, and penetrating through the epidermis within the skin strata. This effect of the biocarrier can be employed to develop advanced types of biomedical and cosmetic remedies used for gentle care and effective treatment of the skin.

References

1. Figovsky O., Beilin D. (2016). Green nanostructural biodegradable materials - short review, Part I, *Electronic Scientific Journal of Don*, no. 1. http://www.ivdon.ru/uploads/article/pdf/IVD_37_Figovsky.pdf_36831579c4.pdf
2. Figovsky O., Beilin D. (2016). Green nanostructural biodegradable materials - short review, Part II, *Electronic Scientific Journal of Don*, no. 2. http://www.ivdon.ru/uploads/article/pdf/IVD_38_Figovsky.pdf_dae12dada8.pdf
3. Swift G. (1997). Non-medical biodegradable polymers: environmentally degradable polymers, in: *Handbook of Biodegradable Polymers*, Harwood Academic: Amsterdam.
4. Griffin G. J. L. (1980). Synthetic polymers and the living environment, *Pure and Applied Chemistry*, **52**, 399–407.
5. Gautam R., Bassi A. S., Yanful E. K. (2007). A review of biodegradation of synthetic plastic and foams, *Applied Biochemistry and Biotechnology*, **141**(1), 85–108.
6. Ioelovich M. (1999). Structure and properties of cellulose-based biodegradable hydrophobic materials, *Journal Scientific Israel - Technological Advantages*, **1**(2), 75–81.
7. http://www.marketsandmarkets.com/PressReleases/paper-paperboard-packaging.asp
8. http://www.businesswire.com/news/home/20151216005732/en/Research-Markets-Global-Paper-Packaging-Market-Worth
9. http://www.envsec.org/publications/vitalwaste3_rus_1.pdf
10. http://ec.europa.eu/environment/waste/studies/pdf/plastics.pdf
11. Ioelovich M., Figovsky O., Leykin A. (2012). Biodegradable nanocomposition for application of protective coatings onto natural materials, US Patent 8268391.

12. Perez J., Munoz-Dorado J., de la Rubia T., Martınez J. (2005). Biodegradation and biological treatments of cellulose, hemicellulose and lignin: an overview, *International Microbiology*, **5**, 53–63.
13. Degli-Innocenti F., Tosin M., Bastiol C. (1998). Evaluation of the biodegradation of starch and cellulose under controlled composting conditions, *Journal of Polymers and the Environment*, **6**, 197–202.
14. Schwach E., Averous L. (2004). Starch-based biodegradable blends: morphology and interface properties, *Polymer International*, **53**, 2115–2124.
15. Lu D. R., Xiao C. M., Xu S. J. (2009). Starch-based completely biodegradable polymer materials, *Express Polymer Letters*, **3**(6), 366–375.
16. Zhang X. Y., Niu J. N., Tian Q. L. (2012). Study on the biodegradability of protein fibers and their products in activated sludge, *Advances in Materials Research*, **602–604**, 1227–1230.
17. Ioelovich M., Leykin A. (2008). Structure of cotton, *BioResources*, **3**, 170–177.
18. Ioelovich M., Leykin A. (2004). Nano-cellulose and its applications, *Journal Scientific Israel - Technological Advantages*, **6**(3–4), 17–24.
19. Ioelovich M., Dizhbite T., Plotnikov O. (1997). Effect of parameters of cellulose supermolecular structure on stable free radicals upon γ-radiation, *Cellulose Chemistry and Technology*, **31**, 291–295.
20. Ioelovich M. (1999). Concept of the native cellulose structural organization, *Journal Scientific Israel - Technological Advantages*, **1**(1), 68–77.
21. Battista O. A. (1971). Mtcrocn stalline cellulose, in: *Cellulose and Cellulose Derivatives*, Parts 4–5 (N. M. Bikaies and L. Segal, eds.), Wiley Interscience: New York, London, Sydney, Toronto.
22. loelovich M., Gordeev M. (1994). Crystallinity of cellulose and its accessibility during deuteralion, *Acta Polvmerica*, **45**, 121–123.
23. Ioelovich M., Figovsky O. (2013). Green nano-protective coating, *Journal Scientific Israel - Technological Advantages*, **15**(2), 63–67.
24. Witt U., Muller R., Klein J. (1997). *Biologisch Abbaubare Polymere: Status und Perspektiven*, FPZ: Braunschweig.
25. Ioelovich M., Figovsky O. (2001). Hydrophobic biodegradable cellulose containing composite materials, US Patent 6294265.
26. Ioelovich M., Figovsky O. (2002). Advanced environment-friendly polymer materials, *Polymers for Advanced Technologies*, **13**, 1112–1115.

27. Standard Test Method for Water Vapor Transmission of Materials, ASTM E 96.
28. Standard Test Method for Water Absorptiveness of Nonbibulous Paper and Board (Cobb Test), ASTM D 3285.
29. Standard Test Method for Gas and Vapor Permeation through Polymers, ASTM D 1434.
30. Quality Control of Treaded Paper and Board, 3M Kit Test.
31. Standard Test Method for Mildew (Fungus) Resistance of Paper and Paperboard, ASTM D 2020 B.
32. Determination of Weight Loss of Plastic Materials Exposed to Aerobic Compost, ASTM D 6003.
33. Ioelovich M. (2001). Engineering of environmental friendly polymer materials, *Journal Scientific Israel - Technological Advantages*, **3**(1–2), 152–157.
34. Ioelovich M., Leykin A. (2006). Microcrystalline cellulose: nano-structure formation, *Cellulose Chemistry and Technology*, **40**(5), 313–317.
35. Ioelovich M. (1992). Supermolecular structure of native and isolated cellulose, *Acta Polymerica*, **43**, 110–113.
36. Ioelovich M. (1999). Concept of the native cellulose structural organization, *Journal Scientific Israel - Technological Advantages*, **1**(1), 68–77.
37. Ioelovich M., Figovsky O. (2008). Nano-cellulose as promising biocarrier, *Journal Scientific Israel - Technological Advantages*, **10**(1).

Chapter 7

Nanotechnology in Agriculture

7.1 Introduction

Contemporary agricultural production is conducted while being exposed to global natural and anthropogenic challenges. Climate changes, environmental pollution with ecotoxicants, emergence of large arid areas, substrate salinization, and water shortage result in a reduction in the agricultural planting footprint, lower plant tolerance to adverse environmental factors, and the emergence of new populations of pathogenic microorganisms and cultural plant pests, along with their rising aggressiveness. All of the above factors result in decreased agricultural yields, lower-quality produce, seeds with a short shelf life and low germination, and price increases in consumer markets. Therefore, the agricultural sector utilizes environmentally friendly technologies, including nanotechnologies.

Heretofore, methods and compositions have been known for treating seeds to improve productivity of high crops, tolerance to changes in weather conditions, resistance to pathogenic microorganisms and cultural plant pests, etc. [1].

However, these compositions and methods are insufficient when considering harshly changing environmental exposures and the fact

Green Nanotechnology
Oleg Figovsky and Dmitry Beilin

ISBN 978-981-4774-10-9 (Hardcover), 978-1-315-22928-7 (eBook)
www.panstanford.com

that subsequent seed generations do not always meet the required planting seed preservation criteria and demonstrate poor field-germination performance.

As an alternative to chemical drugs, new farming practices and processing methods come forward: specialized crop rotation, newly raised cultures in a given region that are resistant to pathogens and pests, optimal schemes of planting, bio-methods, microbiological agents, phyto-regulators of plant growth and development in order to enhance natural protective reactions, biologically active substances of natural origin and their derivatives, etc., all of which induce plant resistance to pests and diseases [2, 3].

The concept of structured systematic approaches to dealing with the nutrition issues, together with the plant growth and development stimulation problems using novel nanotechnologies and nanotechnological methods, is currently the basis for the development of new types of fertilizers and new plant protection means in the sphere of plant breeding and in the agro-industrial complex. Considering the fact that a lot of developed nanotechnologies are unique, it is sensible to transfer the practical professional experience into the agro-industrial sector to resolve the crucial problems in this sphere. As the performed literature analysis indicates, a successive, gradual implementation of certain techniques in the area of plant breeding and farming independently from agricultural scientific methods is the basis of the approach suggested. Finally, the concept is now being developed of applying common agricultural methods for presowing seed treatment with nonroot and root fertilizers on various plant development stages [3].

Furthermore, a goal of increasing crop yield, natural raw material quality, and competitive ability through the interaction and reciprocal influence of certain factors is being pursued. Using nano-objects commensurable to the action radius of intermolecular forces allows the implementation of new inventions, stimulation of various processes, and application of novel nanomaterials. The tendency to reduce the market price of raw material (nano)components used is also important.

To improve adaptability of plants to adverse factors in storage, to obtain full-value and healthy sprouts and good plant development, and to increase seed productivity and quality in subsequent generations, a new agro-biological nanotechnology has been developed that features a composition with properties such as lability and mobility, which can be modified on the basis of predictions to ensure steady seed production and plant growth and to improve the agricultural industry in general. The proposed method relates to natural nanodevices, such as seeds of various plants irrespective of species, varieties, and geographical spread.

It is an object of the new technology:

- To provide physiologically active multifunctional nanochips and a method of application for the production of high-quality seed
- To provide the aforementioned nanochips to be pretreated for sowing on the basis of a nanotechnology that enhances seed and plant adaptability to real-life adverse environmental conditions and to be constructed as multifunctional nanochips that are integrated in the nanopores of the seed cover
- To provide a method for presowing treatment wherein on the basis of prediction of adverse effects on plant growing, the composition and properties of the biologically active nanochips (BANs) can be modified by populating pores of carriers with appropriate biologically active nanoparticles and phytosanitary nanoparticles that enhance plant tolerance to new adverse environmental factors, improve germination properties, and increase yield and productivity

Further objects are to provide the aforementioned chips and method of application that will:

- Extend seed dormancy
- Allow the planting seeds to be stored for a long time without compromising quality
- Initialize termination of seed dormancy under changing environmental conditions by using variously composed and structured BANs for seed preparation before planting
- Enhance seed germination
- Enhance seed tolerance to pathogens, salinization, drought, frost, and other adverse environmental effects

- Increase yield
- Improve produce quality
- Reduce the rate of consumption of physiologically active and phytosanitary components
- Easily adapt to currently existing technologies of seed preparation for planting

Soil protection from various pesticides will become even more urgent in the near future because of the constantly growing world population and low artificial food development tempos. That is why fertile soil protection, which is only 6% of the total land area, is becoming urgent in the twenty-first century. One should note the importance of agricultural land protection in Russia, which has 38% of world chernozem—the most fertile soils for agricultural cultivation.

Precisely soil detoxication technology will not only allow yield increase in the areas of intensive cultivation (China, southeastern Asia, the Middle East, the African oases) but also facilitate the essential improvement of the quality of agricultural products and, as a result, will lead to healthier nations [4].

7.2 Biologically Active Multifunctional Nanochips and Method of Application for the Production of High-Quality Seed

Given the fact that seeds are now produced under harshly changing environmental exposures, with the prevalence of negative exposures, subsequent seed generations do not always meet the required planting seed preservation criteria and demonstrate poor field germination performance. Other conditions adverse for normal seed germination and development of agricultural plants comprise diseases of agricultural plants, attacks from various insects, violation of balance between useful and "harmful" microflora and insects, as well as variations in environmental temperature and humidity that often do not coincide with optimal environmental conditions, which in turn weaken early development of plants.

Since it is impossible to predict or forecast all such unfavorable conditions with high accuracy for massive growing of agricultural plants in actual field conditions rather than in laboratories, nanochip compositions disclosed herein and in the attached claims are designed for anticipated and averaged adverse conditions.

The nanochip compositions presented herein are based on the applicants' experiences and are most optimal for treating seeds of specific agricultural plants mentioned in the examples and claims and for growing plants under anticipated and averaged adverse conditions.

To improve performance and enhance yield and quality, seeds should be additionally provided with physiologically active compounds and phytosanitary substances that improve seed tolerance to adverse exposures.

Phytosanitary substances comprise insectofungicides, bactericides, herbicides, nematocides, acaricides, antiviral preparations, and substances that induce protective functions in plants, immunomodulators, elicitors, desiccants, etc. Phytosanitary measures are aimed at revealing and eliminating contamination of soil with weeds, as well as treating the soil affected by "diseases" and pests.

Nanotechnology development, which allows use of natural seed adaptation systems and BANs in seed cover pores, is the most effective way to enhance seed reliability and resistance to adverse environmental factors.

Nanotechnology of the above-described type allows for modifying the composition of physiologically active substances, including phytosanitary substances, and altering their character based on specifics of nanosystem formation and on interactions of components (nanoparticles) on molecular and supramolecular levels within BANs, depending on specific soil/climatic conditions of cultivation of various plants and specifics of diseases caused by microorganisms and soil-based and other pests and extending planting seed shelf life without compromising planting properties.

One of the novel concepts of nanotechnology implementation is creating polyfunctional, multicomponent, BANs for presowing seed treatment.

In the context of the present technology application, the term "biologically active nanochip" designates a system that comprises a carrier based on natural mineral derivatives (such as mineral, clay, turf, or polymer) having nanopore-filling molecules of physiologically active substances with biopesticides of natural origin and other low-toxic and nontoxic ingredients. These substances may be in the form of water solutions, colloidal systems, and fine-dispersed suspensions and emulsions for filling carriers' nanopores (such as plant development and growth-control components, micro- and macroelements of plant nutrition, and phytosanitary substances).

Depending on the nature and structure of the carrier, the dimensions of BANs range from several microns to 1 to 2 mm, while pores of the carrier range from less than 2 nm (micropores) to 2 to 50 nm (mesopores) or 50 nm and greater (macropores).

BANs contain biologically active components that protect the plants from unfavorable factors and increase production efficiency of agricultural goods. BANs are applied to the surface of a seed and adhere to it, thus forming a film for protection of the seed from unfavorable environmental conditions. Each BAN has a carrier with nanopores penetrable by the aforementioned biologically active substances.

The carrier, which is preloaded with respective physiologically active and phytosanitary substances, is ground to the dimension of the carrier pores, and then the finely ground carriers with physiologically active components are incorporated into the nanopores of the seed cover by means of any conventional method of presowing treatment of seeds (wetting, spraying, blowing, powdering, encapsulating, incrusting, etc.).

When, after sowing, the seeds come into contact with moisture, the physiologically active substances that fill the pores of the carrier are "sucked" through the pores of the seeds into a space between the seed coat and seed embryo, where they fulfill their functions.

The method and BANs apply to seed of various types, such as cotton seed, sugar beet seed, and rice seed.

7.2.1 Biologically Active Nanochips: Species and Compositions

As mentioned above, BANs are based on the use of carriers such as minerals, clay, peat, and soot; products of modification thereof; and other systems that additionally contain stabilizers, ionogenic and nonionogenic surfactants, emulsifiers, various natural and synthetic oligomers and polymers, and homopolymers and copolymers and their derivatives, as well as mixtures thereof in various proportions. In addition, BANs incorporate molecules of physiologically active substances and phytosanitary substances, which ensure preservation of the planting seeds and their properties for a long time, seed and plant resistance to pathogens, tolerance to salinization and other adverse environmental factors, activation of growth processes, immunity enhancement, yield increase, and improved quality of produce.

BANs are constructed according to several methods (synthesis and modification) on the basis of the following [2]:

- Character of nanochip components
- Functional tasks of the chip, such as the shelf life
- Seed and plant protection against phytopathogens and pests
- Enhanced tolerance to adverse environmental effects (plant salinity, drought, and the like)
- Enhancement of growth processes and the like
- Production of subsequent seed generations that have high planting properties
- Increased yield
- Conditions under which seeds are produced and used

BANs may include ions of zinc, copper, cobalt, iron, lithium, manganese, molybdenum, and other trace elements, which function as enzyme activators and cofactors. A shortage of these substances may result in plant metabolism disorders, lower yields, and impaired quality of produce. The above list of elements does not rule out the use of other plant nutrient microelements and mesoelements.

Depending on the level of certain microelements and mesoelements in the soil and on the physiological response of certain plants to the effect of such nutrient elements, BANs include various quantities and combinations of meso- and microfertilizers.

According to our researchers BANs may also contain nutrient trace elements such as nitrogen, phosphorus, potassium, and other fertilizers in the form of various salts (mono-, di-, and triphosphates and the like) and bio-organic compounds that contribute to the intensification of all vital processes that occur in a vegetable organism and constitute the basis of its functioning, growth, development, and productivity when used in various combinations and proportions based on their level of nitrogen, phosphorus, and potassium forms that are available for the plants in the soil and on plant demand for such nutrients. They can be used individually or in combination with various nutrient micro- and mesoelements, various sources of amino acids, proteins (casein, sericin, and the like), biological humus, and other nutrients.

BANs may also contain organic acids, which constitute a substrate for respiratory metabolism, as well as endogenous and exogenous plant growth regulators (both low and high molecular), such as auxins, gibberellins, and cytotoxins and their derivatives, as well as certain metabolites (maleic acid hydrazide) and other substances that are capable of controlling both individual metabolic elements and metabolism in general, which, depending on the seed type, plant cultivation conditions, and the need to accelerate or decelerate growth processes, will be used in various concentrations and combinations.

BANs contain immunity enhancers such as natural phytoalexins and their derivatives, as well as elicitor molecules: oligoamino-polysaccharides and chitooligosaccharides and other substances, which will be used in various combinations to contribute to the improvement of plant protective functions. They may contain vitamins, adaptogenes, antibiotics, and other substances that improve plant immunity and resistance to viral diseases and other stress factors and also the use of microorganisms that produce antibiotics and enzymes and the use of phenolic compounds, nitrogen fixators, and other rhizospheric microorganisms. These ingredients are adaptable to higher levels of salts, ecotoxicants, and other agents that enhance plant tolerance to pathogens, contribute to providing an extended period before pathogenic microorganisms become resistant, and supply the plants with accessible and readily digestible forms of fertilizers, ensuring remediation (restoration) of soil.

It is worth noting that BANs may contain fungal disease inhibitors; microorganisms that are antagonists of plant pathogens; fungicides such as cyprocanazole, propiconazole, triadiaphenone, bromiconozol, tebuconazole, triforin, thiophante-methyl, and sulfur; and other phytosanitary agents in various combinations and concentrations depending on pathogens, morbidity rate, plant cultivation conditions, and plant varieties as also other topical and systemic fungicides in combination with physiologically active agents and phytosanitary agents.

According to one aspect of the proposed method, BANs may also contain inhibitors of bacterial diseases or bactericides such as bronotak, bronopol, vitawax, carboxin, thyram, 2-(thiocyanomethylo)benzothia-zole, dimethylol carbamide, and propamocarb hydrochloride, which, depending on pathogen species, degree of seed contamination, plant cultivation conditions, and plant species and variety, will be used in various quantities and different ratios to each other. The above-listed ingredients may be used in combinations with plant protection agents, plant growth regulators, fertilizers, and other physiologically active compounds in various concentrations and proportions.

BANs may contain insecticides (including thiamethoxam, acephate, and imidachloprid) as well as nematocides (such as oxamyl and chitosan), which are used to control pests such as insects and nematodes in various quantities and combinations, depending on the species of pests and plants that are being attacked as well as their combinations with each other and with plant growth regulators; nutrient micro-, meso-, and macroelements; fungicides; herbicides; and other physiologically active compounds.

At last we obtain BANs containing herbicides or weed-control agents (under the chemical names or trademarks) such as: Fluortamon, Bispiribac sodium™, Tribenuron-methyl, Dicamba™, Chlorsulfuron, Prometrin, Fuazifop-metiryl, Haloaxyfop-R-methyl, Glyphosate™, Azimesulfuron™, Fumetsulam™, Forasulam, Bensulfuron-methyl, and Rimsulfuron, which are used in various quantities and combinations, depending on the species of weed and of the basic plant whose seeds are subjected to the preparation before sowing. These agents and future herbicides, which will be synthesized by manufacturers, are used with plant growth regulators; nutrient micro-, meso-, and macroelements; insecticides and fungicides; bactericides; immunoregulators; and other physiologically active substances and phytosanitary compounds.

7.2.2 Carriers, Stabilizers, and Solvents as the Components of BANs

Carriers for physiologically active agents and phytosanitary agents that are used in BANs contain various natural sorbents and their modified forms, such as natural minerals, lignin, peat, soot, cyclone fluff, clays, organoclays, and schistose silicates; other substances such as montmorillonite, hectorite, vermiculite, kaolin, and saponite and products of modification thereof; and other matrices that are capable of retaining nanoparticles of biologically active agents and phytosanitary agents on a surface or within pores.

Vermiculite as a carrier of BAN occupies a special place. As the analysis of modern research information sources has shown, 55% of world vermiculite resources are applied in the sphere of agriculture [5].

Vermiculite slows down permanent soil salinization process, retains soil moisture, and decreases root rot disease. At the same time, this natural mineral is used as a source of and also a carrier of plant nutrition microelements for it contains silicon, magnesium, copper, calcium, and so on. Thus, this mineral increases the yield of various cultures by 12–17%; decreases the amount of soil ecotoxins; prolongates the period of exchangeable peaty soil usage; and is the medium for an extended vegetables, fruits, and flower bulbs storage. Prepared in accordance with a specific technology vermiculite has a high moisture capacity and can be used as a soil "conditioner" due to its ability to retain not only water but also air. It improves the soil structure and controls water, air, and thermal conditions, which allows its wide usage within agricultural cultivation technologies.

Vermiculite's ability to absorb and retain liquids up to 400% of its size permits using it together with various organic fertilizers (peat, biohumus, manure, and so on). Vermiculite peat is frequently used in gardening. Sterility and inactivity are also its advantages. Adding 25–75% of vermiculite peat mass maintains almost a stable moisture rate, even in drought conditions. All these vermiculite features permit using it on open grounds to prevent moisture evaporation from soil surface and for maintaining the soil level. The same vermiculite characteristics allow its wide application as the carrier of phosphoric potash, nitrogen, and other fertilizers. Having an ability to retain fertilizers, porous vermiculite granules also provide their activity prolongation and create more favorable conditions for strengthening the root system of a cultivar.

It is also known that vermiculite can be used as a carrier for insecticides, herbicides, and so on. "Vermiculitoponics" has also come into extensive use in various countries. Vermiculite is considered to be a perfect medium for plant growing due to 10–15 times reduction of plant nutrients expenditure, easy treatment process, purity, sterility, its reutilization ability, and yield increase (up to 30 times) of various vegetable cultures (tomatoes, cucumbers, potatoes, onions, lettuce, etc.), legumes, and other cultures. The efficiency of vermiculitoponics is high for the fact that modified vermiculite is an active biogenic stimulator of crop yields.

Nanochip stabilizers are represented by:

- various low- and high-molecular ionogenic and nonionogenic surfactants;
- various natural and synthetic oligomers and polymers;
- their homo- and copolymers (both ionogenic and nonionogenic), including vinyl-series polymers (such as polyvinyl lactams and polyvinyl acetates);
- polyacrylonitrile;
- polyacrylic acid;
- urea formaldehyde resin and the like;
- oligo- and polysaccharides (pectins, starch and its copolymers, carboxymethyl cellulose, and the like);
- amino polysaccharides (chitosan and others);
- products of modification thereof as well as their derivatives;
- proteins;
- lipids;
- mixtures of the above-mentioned substances in different ratios, concentrations, and combinations, etc.

The above ingredients do not rule out the use of newly synthesized low- and high-molecular substances that exhibit surfactant properties and are capable of stabilizing nanoparticles and supporting their functionality. In addition, polymeric materials form a matrix that fixates multifunctional physiologically active and phytosanitary nanoclusters and nanochips.

Examples of solvents to be used for BAN components are water, various acids, organic solvents, and other substances, as well as multiple-component compositions in different ratios and combinations. The pH values of the above-mentioned agents and substances should range from 5.0 to 8.0 and specified on the basis of the acid/alkali balance of the soil in which the plants are cultivated.

7.2.3 Use of BANs for Treating Plant Seeds

BANs comprise all necessary components for seed germination and seed protection against anticipated factors adverse to seed germination and plant growth. The components are used optionally in various combinations, with at least one carrier and at least one physiologically active component being indispensable, and the components are used in the proportions (percentage by mass) shown below and comprising the following:

- Physiologically active substances: $(10^{-10} \div 100)\%$
- Carrier for physiologically active substances: $(10^{-2} \div 10)\%$
- Plant growth regulators: $(10^{-10} \div 10)\%$
- Phytosanitary agents: $(10^{-5} \div 10)\%$
- Nutrient element: $(10^{-2} \div 90)\%$
- Solvent: the balance

It follows from the above that a BAN in its simplest form comprises only two components, that is, at least one carrier and at least one biologically active substance, both selected with reference to anticipated adverse factors such as cold weather, salinization of soil, and emergence of new populations of pathogenic microorganisms and cultural plant pests, along with their rising aggressiveness, etc.

BANs may be produced either as vendible products as agents, as dry or liquid substances, or in the form of a preparation.

The composition and quantity of BANs to be applied to seeds depend on the results of the monitoring of agricultural plant cultivation conditions, environmental statistics, and also predictions of the following indicators for the coming year: soil and ambient temperatures, humidity, attacks of pathogenic microorganisms, nature of diseases, seed types, true or light dormancy, as well as seed size and seed potentials such as germination energy and germinating capacity. In addition, the BAN composition is defined with consideration of the availability of digestible forms of potassium, phosphorus, nitrogen, and various nutrient trace elements, such as zinc, copper, cobalt, iron, lithium, manganese, molybdenum, and other nutrient micro- and mesoelements in the soil. For this reason, nanochip components vary within the very broad range of $1–10^{-10}$% to 100%. Trace quantities of nanochip components are used for steeping plant seeds, macro quantities are used for dusting seeds, and intermediate quantities are used for pelleting.

From the processing point of view, the difference in the use of BANs for treating plant seeds having different dormancy types consists of the fact that seeds with light dormancy are treated without using additional steps, whereas seeds that have true dormancy are subjected to scarification, that is, mechanical damage to seed cover. Scarification allows BANs to penetrate deep into the seed cover pores so as to effect growth activation and to contribute to and induce the protective response of the plant to phytopathogens that cause diseases, as well as to stress conditions caused by soil salinization, ecotoxicants, and shortage of molecules providing nutrition for the plants at the earliest stages of development (nutrient macro-, meso-, and microelements).

7.2.4 Practical Preparation of Biologically Active Nanochips for Seed Germination

The following examples demonstrate practical preparation of BANs for seed germination. In these examples, BAN compositions vary, depending on the type of culture to be grown and anticipated and averaged adverse conditions that will affect the germination capacity and growth of the plant.

Since it is impossible to precisely forecast all specific external factors that may simultaneously affect seed germination and plant growth, as well as all attacks from the side of the pathogens along with variations in environmental parameters that may be closely associated with activation or suppression of pathogenic activity, the examples that follow disclose BANs maximally filled with substances that cover a wide range of different biological activities for specific agricultural plants.

We will do well to bear in mind that these examples should not be construed as limiting the fields of practical application and that any changes and modifications are possible without departure from the scope of the proposed method, for example, the principle of the method applicable to treating seeds of other agricultural plants with other nanochip compositions specifically selected for those specific agricultural plants and growing conditions.

7.2.4.1 Composition of BANs for rice seed preparation for planting

Rice seeds (Tables 7.1 and 7.2 [1, 2]) precalibrated and presorted by passing through Petkus sieves of different cell diameters (depending on seed dimensions) were fed to an accumulation hopper in the amount of 100 kg, from where the seeds were periodically unloaded in small portions under gravity onto a rotary pelletizer drum. The surfaces of the seeds were coated in the pelletizer drum by spraying the nanochip-containing finely dispersed homogeneous colloidal system (or solution) with a dosing device (based on 20 mL of solution per 1 kg of seeds) for 2 to 3 s. The nanochip compositions were prepared in five different variants, as shown in Table 7.1.

To form and fix the BANs on the surfaces of the seeds, the treated seeds were tumbled and mixed for 5 min in the pelletizer and were then unloaded to a feed screw, where during transportation the treated seeds were dried in a flow of air heated to 20 to 30°C. Following this, the seeds were fed to a receiving hopper, packaged, and sent to storage until sowing.

Here and hereinafter, the term "control" means untreated seeds, that is, seeds that have not been pretreated with BANs.

Table 7.1 Concentration of nanochip components and their composition for presowing processing of rice seeds

		Content of components in various preparations				
		Preparation variant				
Component	**Units**	**1**	**2**	**3**	**4**	**5**
Vermiculite[a]	%	25	50	75	100	125
	kg/ton of seed	5.0	10.0	15.0	20.0	25.0
Sodium salt of carboxymethyl cellulose[b]	%	0.5	1.0	2.0	3.0	3.5
	kg/ton of seed	0.10	0.20	0.40	0.60	0.70
Chitosan[c]	%	0.0005	0.005	0.010	0.025	0.500
	kg/ton of seed	0.0001	0.001	0.002	0.005	0.100
Roslin[d]	%	1.25	2.50	5.00	7.50	10.00
	kg/ton of seed	0.25	0.50	1.00	1.50	2.00
Topsin M[e]	%	0.50	2.50	5.00	7.50	10.00
	kg/ton of seed	0.10	0.50	1.00	1.50	2.00
Molybdenum salts[f]	%	0.50	1.25	1.75	2.50	5.00
	kg/ton of seed	0.10	0.25	0.35	0.50	1.00
Manganese salts[f]	%	0.50	1.75	2.25	3.00	5.00
	kg/ton of seed	0.10	0.35	0.45	0.60	1.00
Zinc salts[f]	%	0.50	1.00	2.00	3.75	5.00
	kg/ton of seed	0.10	0.20	0.40	0.75	1.00
Gulliver®[g]	%	0.050	0.200	0.275	0.375	0.500
	kg/ton of seed	0.05	0.09	0.11	0.15	0.30
Water	%	balance	balance	balance	balance	balance
	l/ton of seed	15	17	20	22	25

Source: [1, 2]

[a] Carrier for biologically active components; comprises a natural mineral that expands with heat application

[b] Polymeric binder; water-soluble polymer

[c] Linear polysaccharide composed of randomly distributed β-(1-4)-linked D-glucosamine (deacetylated unit) and N-acetyl-D-glucosamine (acetylated unit); in agriculture, chitosan is used primarily as a natural seed treatment and plant growth enhancer and as a substance that boosts the ability of plants to defend against fungal infections

[d] Plant growth regulator—copolymer of nitron fibers with nitrolignin

[e] Fungicide—dimethyl 4,4'-o-phenylenebis[3-thioallopahnate]

[f] Nutritive microelement

[g] Herbicide produced by DuPont Company; contains active constituent of sulfonylurea compound

The effect of treating rice seeds with BANs of different compositions on sowing properties and yield is shown in Table 7.2.

Thus, as can be seen from Table 7.2, the following component contents (kg/ton of seed) can be recommended for BANs for presowing treatment of rice seed:

Table 7.2 Characteristics of rice seeds treated with biologically active nanochips, weed-suppressing capacity, and effect on rice yield

Characteristic	Preparation variant					
	1	2	3	4	5	Control
Germination capacity, %	37.4	42.6	50.5	45.2	34.6	30.2
Suppression of pathogens, %	61.0	75.1	81.5	80.4	84.3	–
Suppression of weeds:						
• *Echinochloa*	30.0	93.4	95.1	96.3	97.2	–
• *Bolboschoenus*	89.6	95.0	96.0	96.0	97.0	–
• Yield (100 kg/ha)	34.3	41.2	43.3	42.1	30.4	36.3

Source: [1, 2]

- Vermiculite: 10 to 20 kg
- Sodium salt of carboxymethyl cellulose: 0.20 to 0.60 kg
- Chitosan: 0.001 to 0.005 kg
- Roslin: 0.50 to 1.5 kg
- Topsin M: 0.50 to 1.5 kg
- Molybdenum salts: 0.25 to 0.50 kg
- Manganese salts: 0.35 to 0.60 kg
- Zinc salts: 0.20 to 0.75 kg
- Gulliver®: 0.09 to 0.15 kg
- Water: 15 to 25 l

Compared with the control group, variants 1 and 5 did not provide optimal contents of the BAN components with regard to seed germination, suppression of 30 pathogens, suppression of weeds, and improvement of yield.

In the research in Ref. [6] natural polysaccharide derivatives (cellulose and chitosan) and modified minerals (modified vermiculite) as the carriers of physiologically active substances were used. Oligoaminosaccharide AgroHit was applied as the elicitor that increases plant adaptability to adverse environmental aspects with growth-regulating activity. As a weed control means the herbicide Rainbow was used.

Studying the influence of developed polyfunctional nanosystems with herbicide activity on weed control has shown that on the 15th day the number of weeds (barnyard grass and sedge species) in the experiment variants was the least and depended on the structure of the systems with different components in various combinations (sodium salt of carboxymethyl cellulose [NaKMC], modified vermiculite elicitor AgroHit, and Magnacide^{TM} H herbicide).

On the 30th day under the influence of the developed nanosystems, the weed control index was slightly lower in comparison with the indices obtained on the 15th day but was significantly higher than the control values.

On the 60th day developed polyfunctional nanochips with herbicide activity were also efficient at weed control in rice. The peak values of weed control indices were 87.0% for sedge species and 84.9% for barnyard grass species.

Examining rice development biometrical data allowed us to conclude that plant stand density and plant height indices were higher than the control values by the end of the vegetation period due to the influence of our developed multicomponent, polyfunctional nanochips with herbicide activity.

It has been also determined that the major yield increase to 19.0 dt/ha relative to the control variant (where the seeds were not treated) and to 10.8 dt/ha relative to the etalon variant (where the plants were treated with herbicide at the 3–4-leaf stage according to the manufacturing company recommendations) occurred at presowing treatment with environmentally sound, polyfunctional, multicomponent nanochips with herbicide activity.

7.2.4.2 Composition of BANs for wheat seed preparation for planting

Wheat seeds (Tables 7.3 and 7.4 [1, 2]) precalibrated and presorted by passing through Petkus sieves of different cell diameters (depending on seed dimensions) were fed to an accumulation hopper in the amount of 100 kg, from where the seeds were periodically unloaded in small portions under gravity onto a rotary pelletizer drum. The surfaces of the seeds were coated in the pelleting drum by spraying nanochip-containing finely dispersed homogeneous colloidal system (or solution) with a dosing device (based on 10 mL of solution per 1 kg of seeds) for 2 to 3 s. The nanochip compositions were prepared in five different variants, as shown in Table 7.3.

Table 7.3 Concentration of nanochip components and their composition for presowing processing of wheat seeds

		Content of components in various preparations Preparation variant				
Component	**Units**	**1**	**2**	**3**	**4**	**5**
Kaolon[a]	%	50	100	200	250	350
	kg/ton of seed	5.0	10.0	20.0	25.0	35.0
Polyvinyl alcohol[b]	%	0.75	1.00	1.50	2.00	3.00
	kg/ton of seed	0.075	0.10	0.15	0.20	0.30
Pectin[c]	%	0.5	1.0	1.5	2.5	4.0
	kg/ton of seed	0.05	0.10	0.15	0.25	0.40
Sodium gummate[d]	%	2.5	5.0	7.0	9.0	12.0
	kg/ton of seed	0.25	0.50	0.70	0.9	1.2
Lamardor®[e]	%	0.25	1.00	1.50	2.00	2.50
	kg/ton of seed	0.025	0.10	0.15	0.20	0.25
Copper salts[f]	%	0.050	0.075	0.100	0.300	0.500
	kg/ton of seed	0.0050	0.0075	0.0100	0.0300	0.0500
Boron[f]	%	0.005	0.009	0.010	0.050	0.100
	kg/ton of seed	0.0005	0.0009	0.0010	0.0050	0.0100
Sericine[g]	%	0.01	0.10	0.20	0.50	1.00
	kg/ton of seed	0.001	0.01	0.02	0.05	0.100
Granstar®[h]	%	0.5	0.9	1.1	1.5	3.0
	kg/ton of seed	0.010	0.040	0.055	0.750	1.000
Water	%	balance	balance	balance	balance	balance
	l/ton of seed	7	8	10	12	14

Source: [1, 2]

[a]Carrier

[b]Polymeric binder

[c]Structural biodegradable heteropolysaccharide contained in the primary cell walls of terrestrial plants

[d]Plant growth regulator

[e]Treatment fungicide; prothioconazole plus tebuconazole

[f]Nutritive microelement

[g]Natural water-soluble biopolymer having high content of oxyamino acids

[h]Herbicide used to control broad-leaved weeds in wheat and barley

To form and fix the BANs on the surfaces of seeds, the treated seeds were tumbled and mixed for 5 min in the pelletizer and were then unloaded to a feed screw, where during transportation the treated seeds were dried in a flow of air heated to 20 to 30°C. Following this, the seeds were fed to a receiving hopper, packaged, and sent to storage until sowing.

Table 7.4 Characteristics of wheat seeds treated with biologically active nanochips; effect on productivity and biochemical characteristics of food products obtained from treated wheat seeds

Characteristic	Preparation variant					
	1	2	3	4	5	Control
Laboratory germination, %	80.5	92.5	95.5	90.3	85.0	78.9
Yield, ton/ha	2.65	2.85	3.00	2.90	2.76	2.70
Protein content in:						
• Seeds, %	12.4	12.8	13.4	13.0	12.8	12.0
• Gluten, %	24.0	24.1	24.5	24.4	24.3	24.0

Source: [1, 2]

The effect of treating the seeds with BANs of different compositions on sowing properties and yield is shown in Table 7.4.

As seen in Table 7.4 in variants 1 to 5, wheat seeds treated with BANs of the invention showed improvement in germination and content of protein and gluten.

The following contents (kg/ton of seeds) of the nanochip components can be recommended for presowing treatment of wheat seeds:

- Kaolin: 5.0 to 35.0 kg
- Polyvinyl alcohol: 0.075 to 0.30 kg
- Pectin: 0.05 to 0.40 kg
- Sodium gummate: 0.25 to 1.2 kg
- Lamardor®: 0.025 to 0.25 kg
- Sericine: 0.005 to 0.05 kg
- Boron: 0.0005 to 0.01 kg
- Copper salts: 0.001 to 0.1 kg
- Granstar®: 0.01 to 1.0 kg
- Water: 7 to 14 l

7.2.4.3 Composition of BANs for cotton seed preparation for planting

Cotton seeds (Tables 7.5 and 7.6 [1, 2]) precalibrated and presorted by passing through Petkus sieves of different cell diameters (depending on seed dimensions) were fed to an accumulation hopper in the amount of 100 kg, from where the seeds were periodically

Table 7.5 Concentration of nanochip components and their composition for presowing processing of cotton seeds

		Content of components in various preparations Preparation variant				
Component	**Units**	**1**	**2**	**3**	**4**	**5**
Lignin[a]	%	16.7	33.3	50.0	66.7	166.7
	kg/ton of seed	5.0	10.0	15.0	20.0	50.0
Oxyethyl cellulose[b]	%	0.33	1.67	2.00	2.67	3.33
	kg/ton of seed	0.10	0.50	0.60	0.80	1.00
Cruiser®[c]	%	3.3	10.0	13.3	16.7	33.3
	kg/ton of seed	1.0	3.0	4.0	5.0	10.0
Panoctine®[d]	%	1.67	6.67	13.33	23.33	33.33
	kg/ton of seed	0.5	3.0	4.0	7.0	10.0
Extrasol[e]	%	0.53	1.67	3.33	6.67	16.67
	kg/ton of seed	0.10	0.50	1.00	2.00	5.00
Iron Hydro-xyacetate[f]	%	0.0033	0.0033	0.0100	0.0167	0.0267
	kg/ton of seed	0.0001	0.0010	0.0030	0.0050	0.0080
Vitawax®[g]	%	3.33	6.67	10.00	16.67	33.33
	kg/ton of seed	1.0	2.0	3.0	5.0	10.0
Water	%	balance	balance	balance	balance	balance
	l/ton of seed	20	22	25	30	35

Source: [1, 2]

[a]Carrier for biologically active component; a complex chemical compound most commonly derived from wood; organic polymer

[b]Polymeric binder; derivative of natural polysaccharide

[c]Insecticide; active ingredient in cruiser, thiamethoxam, a systemic insecticide in the neonicotinoid class of chemicals

[d]Nonvolatile liquid seed treatment for control of certain seed-borne diseases; used as solution containing guazatine

[e]Plant extract; nitrogen-fixing fertilizer; increases the germinating power of seeds; improves absorption of nutrient elements by plants

[f]Growth stimulator

[g]Seed-treatment fungicide effective against early season diseases; contains carboxyn, a systemic fungicide

unloaded in small portions under gravity onto a rotary pelletizer drum. The surfaces of the seeds were coated in the pelletizer drum by spraying the nanochip-containing finely dispersed homogeneous colloidal system (or solution) with a dosing device (based on 30 mL of solution per 1 kg of seeds) for 2 to 3 s. The nanochip compositions were prepared in five different variants, as shown in Table 7.5.

To form and fix the BANs on the surfaces of the seeds, the treated seeds were tumbled and mixed for 5 min in the pelletizer and were

Table 7.6 Characteristics of cotton seeds treated with biologically active nanochips; effect of nanochip composition on the yield of cotton seeds

Characteristic	Preparation variant					
	1	2	3	4	5	Control
Germination energy, %	85.0	90.3	92.8	90.0	80.6	80.1
Laboratory germination, %	93.0	95.1	96.9	92.4	88.0	85.0
Yield, ton/ha	32.7	33.0	35.2	32.0	31.5	30.0

Source: [1, 2]

then unloaded to a feed screw, where during transportation the treated seeds were dried in a flow of air heated to 20 to 30°C. Following this, the seeds were fed to a receiving hopper, packaged, and sent to storage until sowing.

The effect of treating seeds with BANs of different compositions on sowing properties and yield is shown in Table 7.6.

As seen in Table 7.6 in variants 1 to 5, cotton seeds treated with BANs of the invention showed improvement in germination and yield.

The following ranges of nanochip components (kg/ton of 65 seeds) can be recommended for presowing treatment of cotton seed:

- Lignin: 5.0 to 50.0 kg
- Oxyethyl cellulose: 0.1 to 1.0 kg
- Iron hydroxyacetate: 0.0001 to 0.008 kg
- Cruiser®: 1.0 to 10.0 kg
- Panoctine®: 0.5 to 10.0 kg
- Extrasol: 0.1 to 5.0 kg
- Vitawax®: 1.0 to 10.0 kg
- Water: 25 to 30 l

7.2.4.4 Composition of BANs for sugar beet seed preparation for planting

Sugar beet seeds (Tables 7.7 and 7.8 [1, 2]) precalibrated and presorted by passing through Petkus sieves of different cell diameters (depending on seed dimensions) were fed to an accumulation hopper in the amount of 100 kg, from where the seeds were periodically unloaded in small portions under gravity onto a

Table 7.7 Concentration of nanochip components and their composition for presowing processing of sugar beet seeds

		Content of components in various preparations				
		Preparation variant				
Component	**Units**	**1**	**2**	**3**	**4**	**5**
Vermiculite[a]	%	12.5	25.0	75.0	125.0	250.0
	kg/ton of seed	5.0	10.0	30.0	50.0	100.0
Polyethylene glycol[b]	%	0.25	1.25	2.00	2.50	5.00
	kg/ton of seed	0.10	0.50	0.80	1.00	2.00
Heteroauxin[c]	%	0.0025	0.0050	0.0075	0.0125	0.0250
	kg/ton of seed	0.001	0.002	0.003	0.005	0.010
Unigol®[d]	%	0.125	0.250	0.500	1.250	2.500
	kg/ton of seed	0.05	0.10	0.20	0.50	1.00
Impact®[e]	%	0.125	0.250	0.500	1.250	2.500
	kg/ton of seed	0.05	0.10	0.20	0.50	1.00
Fury®[f]	%	0.025	0.062	0.175	0.225	0.250
	kg/ton of seed	0.010	0.025	0.070	0.090	0.100
Caribou®[g]	%	0.0050	0.0075	0.0100	0.0125	0.0150
	kg/ton of seed	0.125	0.250	0.375	0.625	1.250
Water	%	balance	balance	balance	balance	balance
	l/ton of seed	35	37	40	42	45

Source: [1, 2]
[a]Carrier for biologically active component; comprises a natural mineral that expands with the application of heat
[b]Polymeric binder
[c]Growth-promoting hormone, 3-indoleacetic acid, occurring in some plants
[d]Nutritive fertilizer; comprises salts of nitrogen, phosphorus, and potassium enriched with nutritive microelements
[e]Flutriafol—one of the very few compounds to reach and disinfect the embryo of a seed; systemic fungicide used against fungal pathogens in beets, cereals, etc.
[f]Insecticide; zeta-cypermethrin
[g]Herbicide of DuPont Co; contains triflusulfuron

rotary pelletizer drum. The surfaces of the seeds were coated in the pelletizer drum by spraying the nanochip-containing finely dispersed homogeneous colloidal system (or solution) with a dosing device (based on 40 mL of solution per 1 kg of seeds) for 2 to 3 s. The BAN compositions were prepared in five different variants, as shown in Table 7.8.

To form and fix the BANs on the surfaces of the seeds, the treated seeds were tumbled and mixed for 5 min in the pelletizer and were then unloaded to a feed screw, where during transportation the

Table 7.8 Characteristics of sugar beet seeds treated with biologically active nanochips: yield, productivity, and effect of nanochips' composition on the content of sugar in the sugar beet

Characteristic	Preparation variant					
	1	2	3	4	5	Control
Germination energy, %	68.9	69.1	71.5	69.5	68.3	70.7
Laboratory germination, %	86.3	87.51	89.8	84.5	81.9	75.5
Field germination, %	64.6	65.6	74.5	68.4	67.3	50.0
Yield, 100 kg/ha	48.3	51.7	55.5	53.2	50.4	48.3
Sugar recovery, ton/ha	7.47	7.59	8.41	7.97	7.50	7.28

Source: [1, 2]

treated seeds were dried in a flow of air heated to 20 to 30°C. Following this, the seeds were fed to a receiving hopper, packaged, and sent to storage until sowing.

The effect of treating the seeds with the BANs of different compositions on sowing properties and yield is shown in Table 7.8.

It can be seen in Table 7.8 that the contents of BANs as shown in variants 1 to 5 of Table 7.7 provide higher sugar recovery than the control group of untreated seeds. Therefore, the following contents (kg/ton of seeds) can be recommended for BANs intended for presowing treatment of sugar beet seeds:

- Vermiculite: 5.0 to 100 kg
- Polyethylene glycol: 0.1 to 2.0 kg
- Heteroauxin: 0.001 to 0.01 kg
- Impact®: 0.05 to 1.0 kg
- Fury®: 0.01 to 0.1 kg
- Caribou®: 0.125 to 0.250 kg
- Water: 35 to 45 l

7.2.4.5 Composition of BANs for soybean seed preparation for planting

Soybean seeds (Tables 7.9 and 7.10 [1, 2]) precalibrated and presorted by passing through Petkus sieves of different cell diameters (depending on seed dimensions) were fed to an accumulation hopper in the amount of 100 kg, from where the seeds were periodically unloaded in small portions under gravity onto a

Table 7.9 Concentration of nanochip components and their composition for presowing processing of soybean seeds

		Content of components in various preparations Preparation variant				
Component	**Units**	**1**	**2**	**3**	**4**	**5**
Perlite[a]	%	20.0	33.3	66.7	100.0	166.7
	kg/ton of seed	3.0	5.0	10.0	5.0	25.0
Polyvinyl alcohol[b]	%	0.67	1.67	3.00	4.00	2.50
	kg/ton of seed	0.10	0.25	0.45	0.60	1.00
Albit[c]	%	0.067	0.167	0.267	0.600	1.000
	kg/ton of seed	0.010	0.025	0.04	0.09	0.15
Baikal EM-1[d]	%	0.013	0.13	0.267	0.400	0.667
	kg/ton of seed	0.002	0.02	0.04	0.06	0.100
Terpenol[e]	%	0.0067	0.0167	0.0333	0.0667	0.3333
	kg/ton of seed	0.0010	0.0025	0.0050	0.010	0.050
Boron salts[f]	%	0.067	0.667	1.333	2.000	6.667
	kg/ton of seed	0.01	0.1	0.2	0.3	1.0
Molybdenum salts[g]	%	0.67	1.33	2.00	3.33	6.67
	kg/ton of seed	0.1	0.2	0.3	0.5	1.0
Frontier®[h]	%	0.67	1.67	3.67	5.00	6.67
	kg/ton of seed	0.10	0.25	0.55	0.75	1.0
Water	%	balance	balance	balance	balance	balance
	l/ton of seed	10	12	15	17	20

Source: [1, 2]
[a] Carrier for BANs
[b] Binder
[c] Growth-control regulator
[d] Biofertilizer
[e] Plant growth regulator
[f] Boron salts (nutritive component)
[g] Nutritive component
[h] Herbicide

rotary pelletizer drum. The surfaces of the seeds were coated in the pelletizer drum by spraying the nanochip-containing finely dispersed homogeneous colloidal system (or solution) with a dosing device (based on 15 mL of solution per 1 kg of seeds) for 2 to 3 s. The nanochips compositions were prepared in five different variants, as shown in Table 7.9.

For forming and fixing the BANs on the surfaces of seeds, the treated seeds were tumbled and mixed for 5 min in the pelletizer and then unloaded to a feed screw, where during transportation

Table 7.10 Sowing characteristics, productivity, and biochemical characteristics of soybean seeds treated with biologically active nanochips and the effect of the treatment on the productivity and nutritive characteristics of food products obtained from the treated soybean

Characteristic	Preparation variant					
	1	2	3	4	5	Control
Laboratory germination, %	1.40	1.45	1.55	1.50	1.42	1.10
Yield, 100 kg/ha	40.0	41.5	43.0	42.5	41.8	39.5

Source: [1, 2]

the treated seeds were dried in a flow of air heated to 20–30°C. Following this, the seeds were fed to a receiving hopper, packaged, and sent to storage until sowing.

The effect of treating seeds with BANs of different compositions on sowing properties and yield is shown in Table 7.10.

It can be seen in Table 7.10 that the contents of BANs as shown invariants 1 to 5 in Table 7.9 provide higher yield and protein content than the control group of untreated seeds. Therefore, the following contents (kg/ton of seeds) can be recommended for BANs intended for presowing treatment of soybean seeds:

- Pearlite: 3.0 to 25.0 kg
- Polyvinyl alcohol: 0.10 to 1.0 kg
- Albit: 0.01 to 0.15 kg
- Terpenol: 0.005 to 0.01 kg
- Boron salt: 0.01 to 1.0 kg
- Molybdenum salts: 0.1 to 1.0 kg
- Frontier®: 0.1 to 1.0 kg
- Baikal EM-1: 1 kg
- Water: 10 to 20 l

In the course of laboratory research the authors [5] have studied the influence of polyfunctional, polycomplex, multicomponent nanochips on soya germination energy and capacity and soya germ's growth and development data on the earliest stages of development under the influence of different (nano)coatings. It has been noted that the presowing seed treatment with the developed efficient nanochips containing

- NaKMC,
- modified vermiculite,
- Topsin-M, and
- AgroHit

led to the following:

- Germination energy increase from 72.1% (control) to 81.0%
- Laboratory germination capacity expansion from 92.0% (control) to 99.1%
- Increase in stem lengths by 0.9–8.1 cm and germ roots' lengths by 0.9–1.7 cm, depending on developed nanochips' composition and germ's wet weight
- Stem height increase from 6.2 cm of the control variant to 9.6–11.0 cm
- Yield index increase from 47.1 (100 kg/ha) in the control to 54.8 (100 kg/ha)

Thus, the conducted research allowed detecting the most efficient polyfunctional BANs for presowing soya seed treatment. Application of these BANs within the technology for presowing seed treatment promoted higher field germination, branches, beans' number increase, and soya yield increase in comparison with the control variant.

7.2.4.6 Composition of BANs for corn seed preparation for planting

It is well known that corn (sort Zea mays L.) is one of the major crops of the modern agricultural world, which possesses high crop capacity and is widely used in various spheres of human activities. In the global grain balance corn ranks third (after rice and wheat) and is cultivated mainly as a cereal.

Worldwide corn is used for food supply (20%) and technical needs (15–20%) and approximately two-thirds goes to forage. Corn grain contains carbohydrates (65–70%), protein (9–12%), fat (4–8%), mineral salts, vitamins, and other important components, which represent nutritional value of this crop.

The following are being produced from the corn grain: flour, grits, cereal, canned food (corn sugar), starch, alcohol, dextrin, beer, glucose, sugar, molasses, syrups, honey, butter, vitamin E, ascorbic, glutamic acid, and some other substances. Stigma pistils are used in medicine. Out of stems, leaves, and cobs we make paper, linoleum, rayon, activated charcoal, artificial cork, plastics, anesthetics, etc. In addition, corn is excellent forage for livestock.

A kilogram of grain contains 1.34 feed units and 78 grams of digestible protein, which makes corn a valuable component of fodder. However, the protein of corn grain is poor in essential amino acids (lysine and tryptophan) and is rich in low-value forage of protein zein. Corn-based silage is easy to digest and has dietary properties. A hundred kilograms of silage made from corn during the phase of milky ripeness contains about 21 food items and is up to 1800 grams of digestible protein. Corn is used as green fodder, which is rich in carotene.

Corn's dry leaves, stems, and cobs, left after harvest, are used in fodder. A hundred kilograms of maize straw contains 37 food units, and 100 kg of milled rods contain 35 food units. As a tilled crop, corn is a good precursor in rotation cycles and helps to get rid of weeds from fields. In addition it does not have pests and diseases common to most grain crops. When harvesting grain, corn is a good precursor of crops and during cultivation of green fodder it's an excellent fallow culture. Corn is widely used in reseeding processes. It is also used as a protective fence plant [3].

To prevent the defeat of maize seedlings by various diseases and pests (molds, root and stem rot, etc.—what is known to date are more than 100 different diseases affecting this culture) corn seeds are treated with different disinfectants of complex action: fungicides (Vitavaks 200, 2 kg/ton; Vitavaks 200 FF, 2.5–3.0 l/ton; Maxim 025, 1 l/ton; Premis, 1.5 l/ton) and insecticides (Promet 400, 40% mk.s., 25 l/ton or 70% of Gaucho s.p., 5 kg/ton, and other chemical substances for systemic effects) [2].

At the same time, one of the most important methods used to prepare seeds for sowing is called inlaying, which means partial coverage of seeds' surface by various film-forming agents [2]. This method of treatment consists of the following: the surface of seeds is covered by aqueous solutions of polymeric film formers (polyvinyl alcohol, sodium salt of carboxy-methylcellulose, etc.), in which besides disinfectants, various physiologically active substances required for activation of growth processes of corn are introduced. For example, to treat seeds of this culture the following composition is used (for 1 ton of seeds): polyvinyl alcohol (PVA) in the amount of 0.5–11, biologically active substances, and pesticides according to recommended standards per instructions for their use [7].

The environmentally safe nanotechnology method for pretreatment of corn seeds with physiologically active multifunctional and polycomponential nanochips of different composition is developed. This method is based on biopesticides that are derivatives from polysaccharides (chitin, cellulose) and natural minerals utilizing nanotechnological approaches in their development [2].

Preparation of nanochips, coating of the surface of seeds, and assessment of their effectiveness are carried out according to patent USA [2]. Corn seed (Tables 7.11 and 7.12 [1, 2]) precalibrated and presorted by passing through Petkus sieves of different cell diameters (depending on seed dimensions) were fed to an accumulation hopper in the amount of 100 kg, from where the seeds were periodically unloaded in small portions under gravity onto a rotary pelletizer drum. The surfaces of the seeds were coated in the pelletizer drum by spraying the nanochip-containing finely dispersed homogeneous colloidal system (or solution) with a dosing device (based on 30 mL of solution per 1 kg of seeds) for 2 to 3 s. The BANs compositions were prepared in five different variants, as shown in Table 7.11.

To form and fix the BANs on the surfaces of the seeds, the treated seeds were tumbled and mixed for 5 min in the pelletizer and were then unloaded to a feed screw, where during transportation the treated seeds were dried in a flow of air heated to 20 to 30°C. Following this, the seeds were fed to a receiving hopper, packaged, and sent to storage until sowing.

The effect of treating corn seeds with BANs of different compositions on sowing properties and yield is shown in Table 7.12.

It can be seen in Table 7.12 that the contents of BANs as shown in variants 1 to 4 in Table 7.11 provide a higher yield of grain than the control group of untreated seeds. Therefore, the following contents (kg/ton of seeds) can be recommended for BANs intended for presowing treatment of com seeds:

Table 7.11 Concentration of nanochip components and their composition for presowing processing of corn seeds

		Content of components in various preparations				
		Preparation variant				
Component	**Units**	**1**	**2**	**3**	**4**	**5**
Peat[a]	%	3.75	4.00	4.25	4.50	4.71
	kg/ton of seed	10.0	20.0	25.0	40.0	50.0
Polyvinyl-pyrrolidone[b]	%	0.0050	0.0075	0.0100	0.0125	0.0150
	kg/ton of seed	0.01	0.25	0.40	0.80	1.00
Sodium salt of carboxymethyl cellulose[b]	%	0.0050	0.0075	0.0100	0.0125	0.0150
	kg/ton of seed	0.1	0.2	0.4	0.8	1.0
Nicotinic acid[c]	%	0.0050	0.0075	0.0100	0.0125	0.0150
	kg/ton of seed	0.25	0.50	0.70	0.90	1.20
Unum[d]	%	0.0050	0.0075	0.0100	0.0125	0.0150
	kg/ton of seed	0.0005	0.0010	0.0020	0.0050	0.0100
Vitawax[e]	%	0.0050	0.0075	0.0100	0.0125	0.0150
	kg/ton of seed	0.5	1.0	1.5	2.0	5.0
Titus[f]	%	0.0050	0.0075	0.0100	0.0125	0.0150
	kg/ton of seed	0.050	0.150	0.220	0.300	0.400
Water	%	balance	balance	balance	balance	balance
	l/ton of seed	25	27	30	32	34

Source: [1, 2]
[a]Carrier for biologically active components
[b]Water-soluble polymeric binder
[c]Seed-germination stimulator
[d]Biofertilizer; arachidonic acid
[e]Carboxine plus thiram; widely used seed treatment fungicide effective against early season diseases
[f]Herbicide; rimsulfuron

- Peat: 10.0 to 40 kg
- Polyvinyl pyrrolidone: 0.01 to 0.80 kg
- Sodium salt of carboxymethyl cellulose: 0.1 to 0.8 kg
- Nicotinic acid: 0.25 to 0.9 kg
- Unum: 0.005 kg
- Vitavax: 0.5 to 2.0 kg
- Titus: 0.05 to 0.3 kg
- Water: 25 to 35 l

All conducted laboratory and field experiments, including observations, data collections, and analysis, are performed according to generally accepted international standards and methods [3].

Table 7.12 Sowing characteristics, productivity, and biochemical characteristics of corn seeds treated with biologically active nanochips and the effect of the treatment on productivity and nutritive characteristics of food products obtained from the treated corn

Characteristic	Preparation variant					
	1	2	3	4	5	Control
Germination energy, %	90.0	91.5	93.5	92.1	89.0	92.0
Laboratory germination, %	92.5	93.5	95.0	92.5	91.0	90.0
Field germination, %	89.0	91.1	92.5	91.0	90.0	85.0
Yield, ton/ha	62.5	63.6	68.0	67.0	64.1	61.9
Grain yield, %	81.0	82.0	83.5	80.0	79.0	79.5

Source: [1, 2]

During the experiment the seeds treated with nanochips of different compositions were sown on experimental plots of the Agricultural University of Tirana (Albania).

Field and laboratory experiments to study the effectiveness of nanotechnology of presow treatment of seeds with nanochips of various compositions were carried out according to conventional methods. The experiments were laid in four iterations, based on the following general scheme:

- Untreated seeds (control batch)
- Seeds treated with the disinfectant used in production (standard batch)
- Seeds treated with the individual components of nanosystems that are part of nanochips
- Seeds treated by different components with various concentrations and ratios in the physiologically active multifunctional and multicomponent systems
- Seeds treated by different components using various concentrations and ratios in physiologically active multifunctional systems combined with, accepted in production, a disinfectant with a reduced consumption rate

The experiments were laid down per widely accepted methodology [8, 9], in four replications, based on commonly accepted cultivation technology of corn culture in Albania. All fields' experiments were conducted using phenology and phytosanitary observations and registrations in accordance with generally accepted methods and developed recommendations. To protect corn crops from weeds, they were treated with herbicides during the growing season.

During the growing season of corn plants for the experiment the following survey and observations were carried out: a set of phenological observations (start of sprouting) depending on the composition of systems used for presow treatment of seeds, full shoots, density of standing plants (on shoots and before the harvest), appearance of the first true leaf and 6–7 leaves, beginning and full budding, beginning and mass blossoming, accumulation of wet and dry mass, plants' biometrics (plant height, branching height, stem diameter, number of ears per plant, number of seeds per corncob, mass of 1000 seeds, seeds' weight per plant, wax ripeness, and crop capacity.

The population of crops' pests and damage susceptibility was determined by a method. In addition, a structural analysis of corn crop harvest in sheaves, collected from each plot of two nonadjacent repetitions, was conducted from four sites. The biochemical composition of plant samples was studied, and analysis of plants and seeds (after harvest) was conducted, for major contained components, to determine the quality of products in all variations of the experiment. Seed moisture, seed purity, germination energy, germination, oil content, and fatty acid composition of oils were determined by the relevant standards and methodologies per plot.

According to results of field trials in Albania, the following were detected in individual variants of the experiments (with nanotreatment):

- Steady improvement of all growth indicators and development of corn plants compared to the control batch (the maximum increase in plant height by 18.3 cm, the distance from the first cob by 5.0 cm, the length of the cob by 0.6 cm, the weight of the cob by 41.0 g, and the weight of 1000 seeds by 5 g);
- Increase in corn crop capacity to 11.3 dt/ha) depending on the composition of nanochips as compared to the control batch (untreated seeds) according to mathematical processing of data at two levels of reliability/significance of

Table 7.13 Concentration of nanochip components and their composition for presowing processing of tomato seeds

		Content of components in various preparations				
		Preparation variant				
Component	**Units**	**1**	**2**	**3**	**4**	**5**
Diatomite[a]	%	3.75	4.00	4.25	4.50	4.71
	kg/ton of seed	5.0	10.0	20.0	25.0	35.0
Chitosan[b]	%	0.0050	0.0075	0.0100	0.0125	0.0150
	kg/ton of seed	0.075	0.10	0.15	0.20	0.30
Glutamic acid[c]	%	0.0050	0.0075	0.0100	0.0125	0.0150
	kg/ton of seed	0.05	0.10	0.15	0.25	0.40
Succinic acid[d]	%	0.0050	0.0075	0.0100	0.0125	0.0150
	kg/ton of seed	0.25	0.50	0.70	0.9	1.2
Biological humus[e]	%	0.0050	0.0075	0.0100	0.0125	0.0150
	kg/ton of seed	0.025	0.10	0.15	0.20	0.25
Acrobat®[f]	%	0.0050	0.0075	0.0100	0.0125	0.0150
	kg/ton of seed	0.0050	0.0075	0.0100	0.0300	0.0500
Karate[g]	%	0.0050	0.0075	0.0100	0.0125	0.0150
	kg/ton of seed	0.0005	0.0009	0.0010	0.0050	0.0100
Water	%	balance	balance	balance	balance	balance
	l/ton of seed	10	12	15	17	20

Source: [1, 2]

[a] Diatomaceous earth; carrier for biologically active component

[b] Linear polysaccharide composed of randomly distributed β-(1,4)-linked D-glucosamine (deacetylated unit) and N-acetyl-D-glucosamine (acetylated unit); in agriculture, chitosan is used primarily as a natural seed treatment and plant growth enhancer and as a substance that boosts the ability of plants to defend against fungal infections

[c] Growth regulator

[d] Growth stimulator

[e] Biofertilizer

[f] Fungicide that can control various crop diseases; the acting component is dimethomorph

[g] Insecticide; lambda-cyhalothrin

0.05% and 0.01% with a minimum set difference (value of dispersion medium) at 0.01, the productivity was equal to 11.3 dt/ha with 99% accuracy).

7.2.4.7 Composition of BANs for tomato seed preparation for planting

Tomato seeds (Tables 7.13 and 7.14 [1, 2]) precalibrated and presorted by passing through Petkus sieves of different cell diameters (depending on seed dimensions) were fed to an accumulation

Table 7.14 Characteristics of suppression pathogens in tomato grown from biologically active nanochip-treated seeds, development of root system, and effect of nanochip compositions on yield

Characteristic	Preparation variant					
	1	2	3	4	5	Control
Average size of nematode galls (mm^2)	17.4	17.0	13.2	13.8	16.8	19.2
Average size of nematode females (mm^2)	0.295	0.258	0.250	0.254	0.270	0.320
Weight of roots (g)	5.0	6.0	7.5	7.4	5.3	4.4
Yield (ton/ha)	48.0	59.6	62.1	61.4	49.0	50.0

hopper in the amount of 100 kg, from where the seeds were periodically unloaded in small portions under gravity onto a rotary pelletizer drum. The surfaces of the seeds were coated in the pelletizer drum by spraying the nanochip-containing finely dispersed homogeneous colloidal system (or solution) with a dosing device (based on 15 mL of solution per 1 kg of seeds) for 2 to 3 s. The nanochip compositions were prepared in five different variants, as shown in Table 7.13.

To form and fix the biologically active nanochips (BANs) on the surfaces of the seeds, the treated seeds were tumbled and mixed for 5 min in the pelletizer and were then unloaded to a feed screw, where during transportation the treated seeds were dried in a flow of air heated to 20 to 30°C. Following this, the seeds were fed to a receiving hopper, packaged, and sent to storage until sowing.

The effect of treating corn seeds with BANs of different compositions on sowing properties and yield is shown in Table 7.14.

It can be seen in Table 7.14 that the contents of BANs as shown in variants 2, 3, and 4 in Table 7.13 provide a higher yield of tomatoes than the control group of untreated seeds. Therefore, the following contents (kg/ton of seed) can be recommended for BANs for presowing treatment of tomato seeds:

- Diatomite: 10 to 25 kg
- Chitosan: 0.10 to 0.20 kg
- Glutamic acid: 0.10 to 0.25 kg
- Succinic acid: 0.50 to 0.90 kg

- Biological humus: 0.10 to 0.20 kg
- Akrobat®: 0.0075 to 0.03 kg
- Karate: 0.0005 to 0.005 kg
- Water: 10 to 20 l

7.3 Detoxication of Pesticide and Other Toxic Substance in Soil by the Use of Nanomaterials

Environmental safety of the agricultural sector and the quality of food supply are both of great international importance. Constantly growing environmental pollution makes environmental safety an important part of the national safety aspect in general. Among the national safety threats connected with environmental pollution, there are threats connected with water reservoir and drinking water pollution, as there are the threats connected with pesticide presence in the soil and in products. Such threats are the most dangerous threats after radiation pollution and oil spilling.

The Food and Agriculture Organization (FAO) of the UN endorsed the Global Plan of Action to prevent soil degradation [10]. As the FAO deputy director general M. H. Semedo noted: "The soil is the basis for the production of food, feed, fiber and fuel. Without soil, we cannot sustain life on earth, and in the case of loss of soil resources, they cannot be recovered during the lifetime of one generation."

The current escalation of land degradation threatens the potential of future generations to meet their needs. That is why the adoption of the Global Plan of Action for the sustainable use and soil protection is a huge achievement. All this requires the political will and investments to save valuable soil resources. The farmland area contaminated by pesticide residues is above the established norms. Only in Russia this index makes up more than 50 million ha. Their return into the rotation process to obtain clean crop production is one of the most important tasks of our time. The development of agricultural technologies aimed at the rehabilitation of anthropogenically destroyed areas is one way to resolve the situation.

Soil quality is the main aspect in agricultural production; the soil is exposed to great stress as a result of intensive pesticide usage.

It is widely known that to remove pesticide, xenobiotic, or other toxic substance residual quantity from the soil, various sorbates—detoxicants of natural and unnatural origin—are required [11]. This problem becomes even more urgent today: when the ecological crisis is deepening, the need for nature conservation technology is also becoming greater. Application of active carbons (ACs) as high-quality sorbents, and inactive high-porous matrix carriers of active components, is of particular interest for resolving the problems of the agro-industrial sector. It is to be noted that to resolve the problems connected with the environment, more than 10% of activated carbon materials are used; such materials are produced by manufacturers all over the world [12]. Internationally important technologies of AC application for biosphere "ecologization" are shown in Table 7.15 [1, 11]. Besides, one should mention new threats rising all the time, to deal with which ACs are needed.

Owing to their physicochemical properties carbon absorbents (ACs) are unique and ideal adsorption materials, which allow

Table 7.15 Technologies of active carbon application

Part of biosphere	Carbon absorption technology
Atmosphere	Solution recuperation Sanitary end gas purification, including desulfurization Nuclear power plant gas purification system Recovery of transport fuel gas Chemical weapon elimination Solid domestic and sanitary waste elimination Cleaning of the air in living and working space (air conditioning)
Hydrosphere	Drinking water purification Water-borne waste sterilization Liquid radioactive waste treatment Gold nonferrous materials mining
Lithosphere	Soil protection from xenobiotics, including pesticides Soil resuscitation Sanitary control of water source zones
Human	Individual and collective protection means of filter type/ Chemical pharm, vitamins, antibiotic production Enterosorption and hemosorption Environmentally safe manufacturing of products

Source: [1, 11]

resolving of various problems connected with chemical and biological safety of humans, environment, and infrastructure [12]. ACs are high-porous carbon materials with a highly developed internal structure (1000–2000 m^2/g). This property is the result of high internal porosity, which includes macro-, meso-, and micropores. Such porous structure of AC (micro- and mesospore size) absorbs all kinds of organic trace contaminants due to its adsorptive capacity (surface interaction processes). Microspores of carbon absorbents have a slit-like form and are expressed as their half-width $\boldsymbol{x}$. To micropores we refer pores of $\boldsymbol{x} < 0.6–0.7$ nm and also some bigger pores (0.6–0.7 nm $< x <$ 1.5–1.6 nm), called supermicropores. Exactly in these pores the admixture absorption takes place due to van der Waals forces and their nonspecific interactions. It means carbon absorbents are standard nanomaterials.

Certain microspore nanostructure formation is a process taking place during AC obtainment through application of some specific raw material and certain thermal treatment regimes (drying, carbonization, and activation). The macropores and mesopores of ACs and carbon absorbent act as transport links and influence process kinetics. Macropores possess an effective radius (r_{ef}) of more than 100–200 nm, but their surface makes only 0.5–2.0 m^2/g. Mesopores (transitional pores) have efficient radii from 1.5 to 200 nm, and their specific surface is 50–100 m^2/g. Some samples have a surface of 300–400 m^2/g. The parameters of a carbon active porous structure are shown in Table 7.16 [1, 11].

An investigation of kinetics of chlorosulfon extraction from a solution using the AC showed that in 3 days within the etalon experiment 95% of the chlorosulfon turned into the absorbed state of AC [11]. The essence of carbon absorption method of soil detoxification consists of introducing ACs and carbon absorbents alike into the soil with the help of various agricultural equipment at doses of 50–100 kg/ha, followed by their embedding into the soil, 10–15 cm deep.

Choosing certain methods of introducing these materials into the soil one should take into account soil toxicological indices and agro-climatic properties of the regions.

The basic application methods are:

Table 7.16 Porous structure of active carbons

Type of pores	Volume of pores cm^3/g	Size of pores nm	Surface of pores m^2/g
Macropores	0.2–0.6	Pore radius >100 ÷ 200	0.5–2.0
Mesopores (transition pores)	0.1–0.4	Pore radius from 1.5 ÷ 1.6 to 100 ÷ 200	50–200
Micropores:			
• Micropores proper	0.3–0.7	Slit half-width size <0 ÷ 0.7	1200–1500
• Supermicropores	0.1–0.3	Slit half-width size from 0.6 ÷ 0.7 to 0.5 ÷ 1.6	500–800
• Total pore volume	0.7–2.0		1800–2000

Source: [1, 11]

- Surface application (spraying) of polydisperse AC or water suspension made of AC powder onto the contaminated part and then embedding uniformly distributed particles into the soil 5–10 cm deep (seed sowing was accomplished 5–7 days later)
- Introducing into the seed furrows polydisperse AC or its water suspension together with the seeds and embedding them
- Presowing local introduction of polydisperse AC or its water suspension into the seeding furrows and then embedding carbon elements 5–10 cm deep 2–5 days after sowing seeds into the furrows

As raw materials to obtain ACs one can use coal, peat, timber, shell, various plant residues, and other cheap materials.

For resolving environmental problems of the agricultural sector such ACs have certain advantages, for instance, selectivity of organic toxin absorption, absorption properties universality, high absorptive capacity, and hydrophobicity. They are also suitable for preparation of powders, granules, etc.

Below there are the results of various vegetation methods' application and greenhouse and field tests. These data have been obtained from the study of carbon adsorption and soil detoxication when the soil was contaminated with various herbicides. In

Table 7.17 The efficiency of soil fertility restoration when the soils are contaminated with herbicide residues and modified active carbon (application dose, 100 kg/ha)

Herbicide	Herbicide residues in soil, g/ha	Culture	Preserved yield data culture tests % to contaminated level
Chlorosulfon (CSF)	0.2	Cucumber	16–20
		Beet	58–63
		Garden radish	23–28
Terbisil	1.4	Cucumber	23–27
		Beet	64–69
		Garden radish	30–39
Pikloram	2	Cucumber	22–24
Simasin	50	Tomato	22–26
			98–100
Chlorosulfon	0.4	Beet	98–99
		Garden radish	98–100

Source: [1, 11]

particular, the results of vegetation tests on the ash-gray soils are shown in Table 7.17 and Fig. 7.1 [1, 11].

The results of field tests in chernozemic soil are shown in Table 7.18 [1, 11]. In both cases (according to the data from

Figure 7.1 The results of vegetation tests on ash-gray soils [1, 11].

Table 7.18 Agricultural crop yield against the background of herbicides and active carbon (AC) application at doses of 50 kg/ha

Crop	Herbicide and dose background, kg/ha	Yield against the background of herbicides, 100 kg/ha	Yield after AC application	Yield increase, 100 kg/ha (%)
Corn (per grain)	Treflan 1.4	53	78	25 (47)
Tomatoes	Treflan 1.5	333	652	319 (96)
Sugar beet	Treflan 1.5	343	416	73 (21)
Rice	Ronstar 2.0	60	72	12 (20)
Bulb onion	Ramrod 8.5	228	295	67 (29)
Cucumber	Treflan 1.0	85	202	117 (138)
Soya	Dialen 9.0	11	24	13 (118)
Winter wheat*	Unidentified residues	43	49	6 (14)
Corn* (green mass)	Unidentified residues	342	592	250 (73)
Rice	Unidentified residues	63	85	22 (35)

Source: [1, 11]
*Active carbon dose 100 kg/ha

Tables 7.17 and 7.18) the unique effect of herbicide residue "elimination" in various sowing tests is obvious.

One should especially note that in the tests with various agricultural crops cultivated (beetroot, garden radish, cucumber, tomato) in the soils that were not contaminated with herbicides, a positive influence of carbon absorbents on the tested crops has been detected.

Table 7.19 [1, 11] shows the influence of various absorbent brands and doses on beet growth according to the vegetation test conducted at the Institute of Phytopathology (Russia).

The test crop mass increase by 10–15% along with AC dose increase from 100 to 400 kg/ha can probably be explained by certain endogenous phytotoxins, typical for this crop, accumulating in the soil during the natural process of soil exhaustion and then connecting to the absorbent.

The results of comparative research works conducted on the range of agricultural crops being cultivated in the soils

Table 7.19 Beet growth depending on various absorbent brands with the soil contaminated

Contaminating substance dose kg/ha	Plant green mass, in % to the control variant		
	UMD-1[a]	UMD-2[b]	UMD-3[c]
Adsorbent norm 0 kg/ha			
0.1	3	9	16
0.2	2	2	6
Adsorbent norm 100 kg/ha			
0.1	12	46	25
0.2	8	27	16
0	102	105	134
Adsorbent norm 200 kg/ha			
0.1	25	100	34
0.2	21	92	20
0	105	111	201
Adsorbent norm 400 kg/ha			
0.1	40	93	145
0.2	37	101	124
0	107	116	210

Source: [1, 11]
[a]Powder form (active carbon + natural clinoptilolite)
[b]Powder form (active carbon + synthetic zeolite)
[c]Grained form (active carbon + natural clinoptilolite)
In all the absorbent compositions the masses are the same.

contaminated with herbicides like treflan (1 kg/ha) and 2.4D (2,4-dichlorophenoxyacetic acid) (5–10 kg/ha), in accordance with the conventional cultivating technology using carbon adsorbents, have shown that introducing the absorbent into the contaminated soils in an amount up to 100 kg/ha (in the case of barley crop this amount was up to 200 kg/ha) can allow a reduction in (or sometimes even prevent) herbicide accumulation in plant and vegetable products (Table 7.20) [1, 11].

The same results have been obtained from corn cultivation in the soils contaminated with herbicide atrazine (Table 7.21 [1, 11]).

Thus, the application of AC for soil detoxication from fixation herbicide residues has two main aspects: yield increase in contaminated soils by 20–80% and an opportunity to obtain yields brought up to the "dietary" mark.

Table 7.20 Herbicide content in agricultural products

Herbicide and its dose, kg/ha	Active carbon dose, kg/ha	Test crop	Herbicide content in products, g/kg
Treflan - 1	–	Tomatoes	28
Treflan - 1	100	Tomatoes	0.6
Treflan - 1	–	Carrot	95
Treflan - 1	100	Carrot	Not detected
2.4 – D - 5	–	Barley	220
2.4 – D - 5	200	Barley	Not detected
2.4 – D - U	–	Barley	670
2.4 – D - U	200	Barley	Not detected

Source: [1, 11]

Table 7.21 Atrazine accumulation in corn

Active carbon dose, kg/ha	Content of atrazine g/kg		
	In green mass to ensilage at doses of atrazine, kg/ha		
	8	16	32
0	2	5	13
50	Not detected	Not detected	1
100	Not detected	Not detected	Not detected
Active carbon dose, kg/ha	**Content of atrazine g/kg**		
	In grain forms at doses of atrazine, kg/ha		
	8	**16**	**32**
0	7	11	29
50	Not detected	1	2
100	Not detected	Not detected	Not detected

Source: [1, 11]

Both effects are conditional on pesticide absorption in the soils containing AC, which means fixating pesticides within the structure of AC, thus making them unable to penetrate plants and their root systems. It also prevents them from further spreading in the soil. Thus, the negative effect of these toxins on crop yield is being reduced (Table 7.22 [1, 11]).

Besides, it has been noted that the xenobiotics absorbed by ACs are also unavailable for destructive soil microorganisms and

Table 7.22 Herbicide detoxication and crop development levels when using active carbon at doses 100 kg/ha

Crop, herbicide, and the dose g/ha	Herbicide content decrease in soil, %	Green mass increase, %
Garden radish, chlorsulfuron, 0.2	40–45	23–28
Cucumber, picloram, 2	33–37	22–24
Tomato, simazine, 50	34–40	22–26

Source: [1, 11]

also only after desorption of this type of contaminating substances into the aqueous phase, they start being influenced by microflora. However, this process is quite energetically costly (time consuming also), that is, it will take the soil 3–4 years for natural purification. One should note that AC by itself does not affect plant life and soil biota activity. The safety of these components is proved by the fact that they are used as efficient absorbents of various toxins of exo- and endogenous origin in medicine and veterinary for human and animal treatment without any restrictions.

Results of investigations into chlorsulfuron extraction kinetics with the help of active carbon circuit control tester (AC CCT) are shown in Fig. 7.2 [1, 11]. Thus, the mechanism of herbicide fixation consists of herbicide absorption by ACs. As it is obvious from the kinetic curve within three days chlorsulfuron almost disappears from the aqueous phase and becomes an absorbed substance.

A remarkable example, proving these results, has been shown in Ref. [13]. The authors studied the method of biological degradation of benzotiazole in the soil, which is formed as a result of tyre rubber interacting with asphalt roads. Benzotiazole is a very strong carcinogen. These authors used 76 bacteria varieties; two of them decomposed the xenobiotic successfully. In the soil there are 10 times more bacteria and fungi than in any research bank, so it means there is always a destructor of this or that pesticide.

In the case of specific strain destructors, introduced on ACs, carbon pores protect them in the aggressive natural soil bio flora, allowing them to grow and breed, feeding on pesticide, absorbed

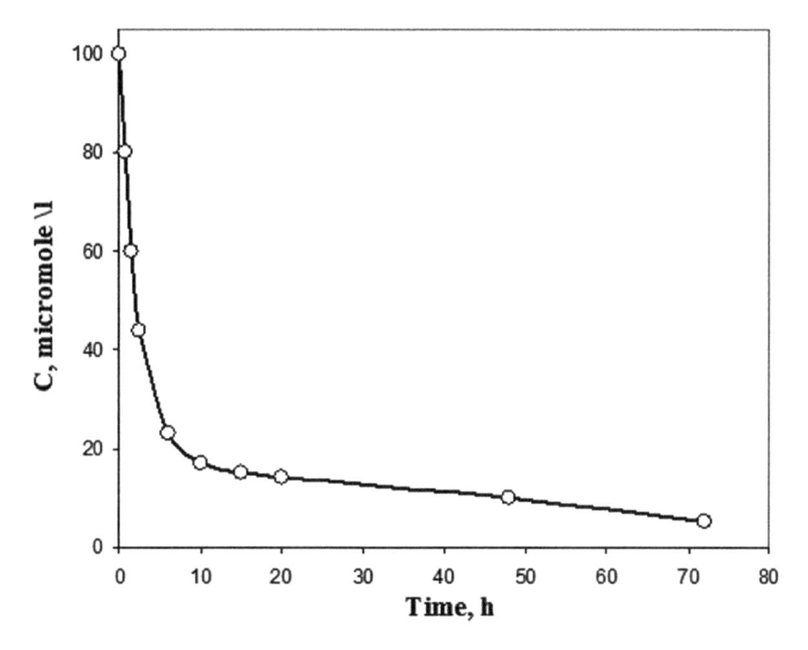

Figure 7.2 Kinetic curve of chlorsulfuron absorption from aqueous solution by active carbon of CKT [1, 11].

from the soil. In this case, purification increasing effect exhibits a synergistic behavior.

Besides, the destruction of pesticides on the surface of ACs may, probably, take place due to oxidation, where oxygen-containing groups take part, catalysis at certain active points, physicochemical influence of absorption potential of micropores, and the other mechanisms.

The efficiency of AC application in agricultural technologies is defined not only by its type (raw material type, absorbing pore size, transport and absorbing pore correlation), but also by application methods (grains and suspension, overall application, application in patches or in certain areas of seed sowing, embedding depth, and other factors).

The newest approach in this area is a sorption-biological soil remediation. This method is connected with the soil cleaning method based on using small doses of absorbent (1–5% of soil

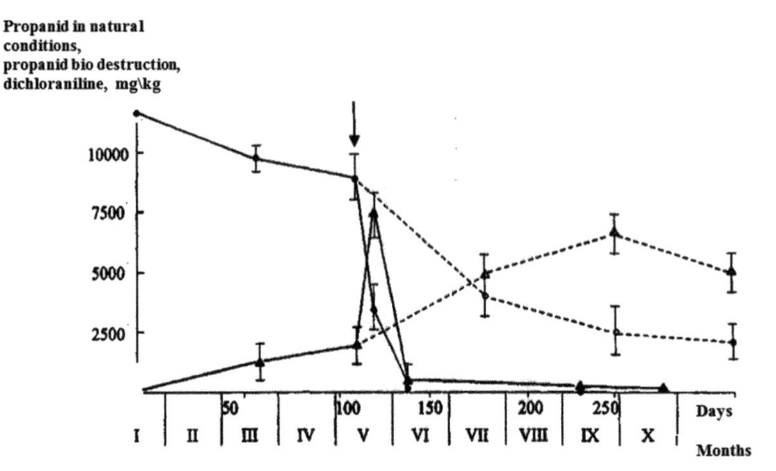

Figure 7.3 The dynamics of propanid concentration changes (1 and 2) and DCHA (3 and 4) in the soil on the site of the accident: solid line, observed; dashed line, the soil without treatment; the arrow shows the beginning of the works aimed at accident consequences' elimination by adding active carbons (50 kg/ha) and (microorganisms for dichloraniline destruction [MFDD]) (10^5 cells/g) into the soil [1, 11].

mass), which is put into the soil together with specifically grown microbe strains—pesticide destructors.

The results of propanid pesticide destruction tests are shown in Fig. 7.3 [1, 11] as an example, where the pesticide got in the soil as a result of a car accident with a truck-mounted tanker having taken place in the Krasnodar region (Russia). This method, apparently, can be adapted to soil cleaning, where the soil is contaminated with various organic chemicals, petrochemicals, and other poisonous products.

Some soil detoxication aspect is connected to defoliant-substance application, causing leaf exfoliation. These substances, used during cotton harvest, were used by the US Army in Vietnam to destroy tree leaves to find out the enemy location. In nature these substances transform into dioxins—extrapoisonous substances for humans.

The preliminary experiments [11] have shown that adding ACs to the soil contaminated with such xenobiotics leads to these absorbents' extraction from the soil and their fixation them in the porous structure. As a result, the negative influence of xenobiotics on the lithosphere and hydrosphere is reduced.

The research works conducted nowadays are aimed at AC obtainment from various agricultural crop residues (straw of oil and grain crops) and their further application as a promising annually renewable source for sorbate development and, thus, for agricultural soil improvement. Properties of the obtained ACs were defined by standard methods using device ASAP 2020.

The sorbates obtained by rape, barley, wheat, and straw processing into ACs (product yield amounts from 8.4–16.0%) have:

- Total pore volume from 3.53 cm^3/g to 4.14 cm^3/g
- Ultimate sorptive volume from 0.28 to 0.73 cm^3/g
- Micropore sizes of 0.6–1.0 nm
- Sorptive power according to iodine 39.0% to 43.0% and according to methylene blue 37.0 mg/g to 87.0 mg/g, which allows using them for soil detoxication from pesticide residues, because they meet the requirements, especially set to sorbates for "soil" improvement

Agricultural sorbates have to hold on to pesticide molecules and their destruction products. Besides, transport porosity of agricultural sorbates should be well developed to provide the necessary absorption speed.

Research (taking sunflower crop as an example) has shown the rapid decrease of metsulfuron-methyl herbicide phytotoxic activity from 73.2% to 4.9% when using agro sorbates based on ACs from rape, barley, and wheat crops, during the sowing period. The average mass of the sunflower plants cultivated (in vegetation tests) changed from 1.1 g when herbicide Zinger® had been put into the soil to 3.9 g when the ACs, made of wheat and barley straw, had been put into the soil together with the herbicide (in the control variant this index amounts to 4.3 g (Fig. 7.4 [1, 11]), which proves a quite high detoxication activity of the sorbates obtained from plant residues of grain crops.

Conducting research works on detoxification of soil with the sorbents of diverse nature, including ACs obtained by processing the straw of different crops (brassica oilseeds, cereals, and others),

Figure 7.4 K, control; 1, Zinger, SP; 2, AC from barley straw; 3, AC of wheat straw; 4, AC of rape straw; 5, AC of antriatsit; and 6, AC of Grosafe brand (Switzerland) [1, 11].

before sowing breeding samples of different cultures to optimize the selection process is currently a very acute approach. Really, the remaining residual amounts of pesticides (herbicides, insecticide, fungicide, bactericide, and other plant protection chemicals) in the soil have an inhibitory effect on the productivity of many species of plants; thereby they significantly distort the results. In breeding practice there has always existed the problem of increasing the objectivity of the field evaluation of selection samples (lines, accessions) by leveling the diversity of soil fertility.

Therefore, a comprehensive approach is required to the preparation of soil and seeds for sowing, and that forms the basis of the research data on the leveling of the harmful effects of various toxicants in the process of breeding.

Taking into consideration that ACs have no adverse effects on plant life and activity of soil biota, as evidenced by numerous studies, and the fact that many preparations containing such carbons are used as sorbents in medical and veterinary practice for the treatment of humans and animals, effectively linking various toxins of exogenous and endogenous origin, the research works were conducted to study the effect of carbon adsorbtion detoxification of soil on growth and development of rape plants in the process of selection.

For this, two approaches were used: the first is to introduce ACs and similar carbon adsorbents into the soil in certain doses together with the seeds (if sowing) using agricultural machinery, followed by their introduction into the predetermined depth, as described above in the cultivation of various crops; and the second is to apply them to the seed surface consisting of (nano)chips for the treatment of seeds according to Ref. [2]. The selection of specific ways of introducing these materials into the soil within this method is carried out taking into account the toxicological indicators of soil and agro-climatic features of the studied areas.

The increase in weight and length of rape sprouts in field trials in the initial period of development of the different varieties of this crop by 11–19%, contributing to the increase of yield up to 10%, was due to, probably, the bondage of carbonaceous adsorbents of "some" endogenous stuff typical only to the given culture of phytotoxicants accumulating in soils under natural soil exhaustion and fixing of the residual amounts of herbicides, pesticides, and other xenobiotics especially dangerous, allowing, in the end, to slightly "smooth" the diversity of soil fertility and get in the future environmentally friendly or safe products.

7.4 Application of Nanochips as Inducers of Disease Resistance during Presowing Seed Treatment

The present level of development of science has led to the emergence of a new method of plant protection, which is based on enhancing the immune capacity of the plant rather than on the destruction of pathogens, as is the case with pesticides. Substances that induce protective responses in plants are called elicitors (inducers of disease resistance). The first biogenic elicitor was obtained in 1968; since then their number has been annually growing. Biogenic elicitors are usually used in very small quantities. Plants are "vaccinated" against the disease, which results in their increased disease resistance [14].

The aim of the studies is to investigate the effect of different compositions of environmentally friendly nanochips based on multi-component polyfunctional nanosystems, including high-sorption-capacity carrier matrices of different natures (water-soluble derivatives of natural polymers, natural minerals, carbonaceous sorbents, biogenic elicitors—inducers of natural disease resistance—based on chitosan, organic acids, and bacterial cultures), as well as the effect of other means of plant protection and their growth and development regulation, used for presowing seed treatment and for the processing of plants during their vegetation with the help of nanotechnology, on the growth, development, yield, and quality of rape seed [15].

Development and obtainment of new sorption-capacity nanochips for presowing treatment, based on carbon-containing nanomaterials (carbon nanotubes, graphene, activated carbons), silicon dioxide, water-soluble polymer and mineral derivatives, as well as the other sorbents, together with different plant protection, are established according to Ref. [2].

Studies [15, 16] have shown that putting tomato seeds, tobacco, sainfoin, rape, wheat, and some other crops into a nutrient solution containing carbon nanotubes leads to stimulation of their germination, plant growth and development improvement, biomass accumulation increase, and also higher yields due to the penetration of nanotubes through the seed coats into various plant organs.

During experiments [14] using for presowing seed treatment new nanoporous nanochips based on carbonaceous sorbents with the different plant protection products of biogenic nature the same effects were identified.

Let us consider in detail the new obtained carbon nanomaterial (CNM).

CNM "Taunit" is known as the one-dimensional nanoscale threadlike formations of polycrystalline graphite in the friable form, resembling powder of black color. CNM granules of micrometer dimensions have the structure of tangled bundles of multitubes (Fig. 7.5) [1, 14].

New types of the CNM Taunit using various catalysts and production modes were obtained—carbon nanotubes Taunit-M and Taunit-MD—having improved morphological and physico-mechanical properties (Fig. 7.6) [1, 14].

Optimal heat treatment of the substances, which preceded the catalyst and its composition, provides a few-layered CNM with small

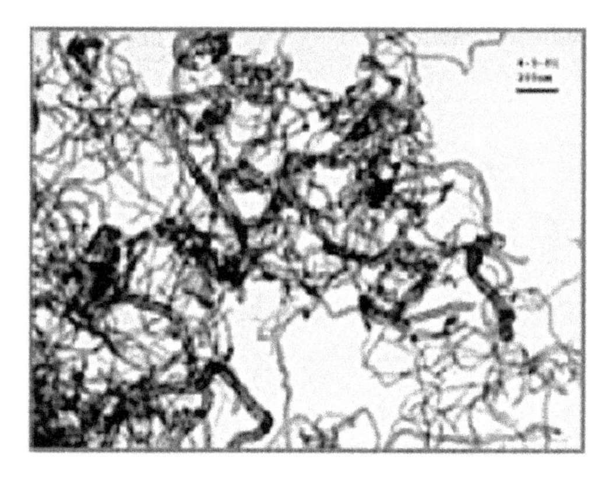

Figure 7.5 Electron micrographs based on the material Taunit [1, 14].

diameter variations. Table 7.23 [1, 14] presents data on various CNM.

ACs were obtained by recycling annually renewable plant waste (the straw of various oilseed brassica crops) and represent a black powder. Powdered recycling activated carbons (RAC) derived from rape, winter cress, mustard, and radish straw (the product yield level

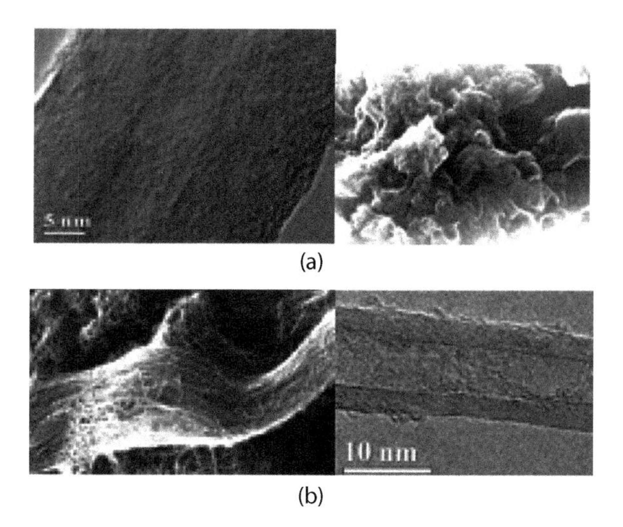

Figure 7.6 Electron micrographs of the CNM Taunit-M (a) and Taunit-MD (b) [1, 14].

Table 7.23 General characteristics of the series of carbon nanomaterials (CNMs): Taunit, Taunit-M, and Taunit-MD

Parameters	Taunit	Taunit-M	Taunit-MD
External diameter, nm	20–70	30–80	8–15
Inner diameter, nm	5–10	10–20	4–8
Length, μm	2 and more	2 and more	2 and more
Total admixtures, % (after purification)	Up to 5 (up to 1)	Up to 5 (up to 1)	Up to 5 (up to 1)
Apparent density, /cm^3	0.4–0.6	0.03–0.05	0.03–0.05
Specific geometric surface area, m^2/g	120–130 and more	180–120	300–320 and more
Thermostability, °C	to 600	to 600	to 600

Source: [1, 14]

is from 5.7% to 17.9%) according to the developed method [17] have the characteristics shown in Tables 7.24 and 7.25 [1, 14]. Electron micrographs of powdered activated carbon from straw rape, winter cress, and mustard are shown in Fig. 7.7 [1, 14].

Thus, well-developed total porosity and relatively vast sorption space allow one to effectively use the resulting adsorption RAC for those technologies where powdered activated carbons are used.

In the trial variants of seed treatment using environmentally friendly nanochips, crop spraying with biopesticides in the phases of 4–6 leaves and budding–flowering was conducted. Phenological and phytosanitary surveillance and surveys were carried out according to the generally accepted methods and the recommendations

Table 7.24 General characteristics of the RAC series of active carbons obtained from the straw of oilseed brassica crops: RAC-rape, RAC-winter cress, RAC-mustard, RAC-false flax, and RAC-radish

No.	Sample	V_Σ cm^3/g	W_s cm^3/g	Moisture %	Δ, g/dm^3	Total %
1	Rape	4.14	0.48	2.4	135	10.3
2	False flax	2.43	0.69	5.7	140	5.7
3	Radish	2.55	0.20	2.2	131	17.9
4	Winter cress	2.28	0.57	1.5	135	8.1
5	White mustard	4.00	0.45	2.8	60.9	11.0
6	Brown mustard	2.81	0.57	3.4	111	7.3

Table 7.24 (*Continued*)

No.	Sample	Sorption capacity with: Iodine %	Methylene-blue MG, mg/g	Total ash, %	Output volatile, %
1	Rape	39	87	16.5	16.0
2	False flax	43	82	15.6	12.6
3	Radish	31	69	26.2	15.3
4	Winter cress	62	73	22.1	12.8
5	White mustard	50	64	24.4	11.2
6	Brown mustard	56	67	27.3	10.6

Source: [1, 14]

Table 7.25 The micropore volume and micropore size for the isotherms of a series of recycling activated carbon (RAC) obtained from the straw of oilseed brassica crops: RAC-rape, RAC-winter cress, RAC-mustard, RAC-false flax, and RAC-radish

Sample	Micropore volume, cm^3/g	Equivalent pore width, nm: Average	Modal	S_{equ}, m^2/g	E_a, kJ/mole
Oil-bearing cabbage plants					
Rape	0.161	1.58	1.42	454	22.7
False flax	0.135	1.66	1.53	380	19.8
Winter cress	0.24	1.53	1.38	350	24.9
White mustard	0.161	1.57	1.39	454	23.2
Brown mustard	0.163	1.50	1.32	459	26.2
Radish	0.116	1.62	1.51	328	21.0

Source: [1, 14]

developed. To protect rape crop from weeds and pests it was treated with herbicides and insecticides during the vegetation period.

The conducted laboratory studies have shown that the treatment of rape seeds before sowing using nanoporous carbonaceous materials (Taunit, Taunit-M, Taunit-MD, graphite, and activated carbons obtained from the straw of different crops with the micronutrient malnutrition included into their composition), as well as the drugs with fungicidal, bactericidal, and eliciting properties, tended to sharply accelerate the process of seed germination. The number of "arisen" seeds on the first germination day, depending

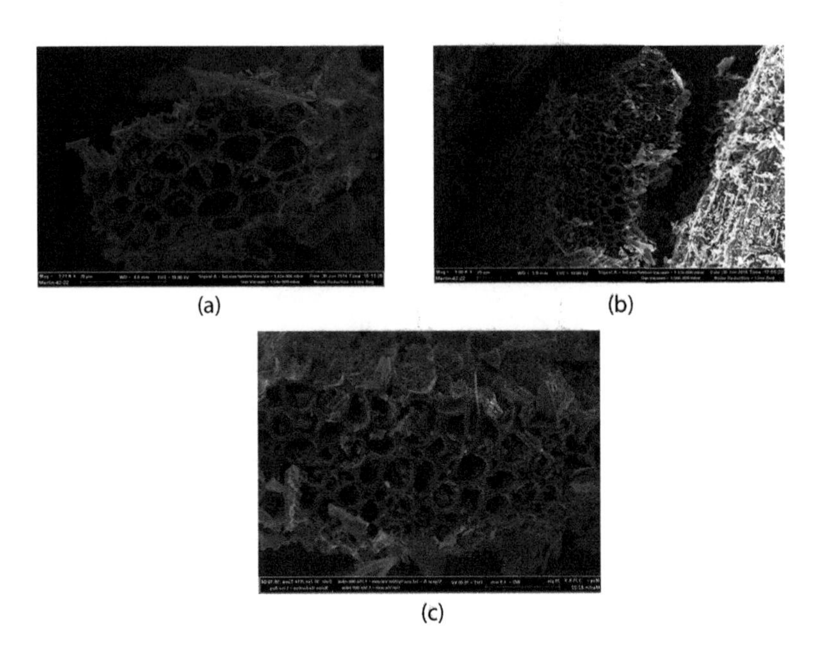

Figure 7.7 Electron micrographs of powdered activated carbon, obtained from rape straw (a), mustard straw (b), and winter cress straw (c) [1, 14].

on the composition, was 20.2–32.5% while in the control variant (where the seeds have not been treated) it was 19.8% and on the second day it was 90.5–93.0%.

It also facilitated the increase of vigor, laboratory germination, seedling length, and fresh weight accumulation, which can be probably explained by the possibility of the idea that physiologically active polyfunctional nanochips based on nanotubes penetrate inside the seed and then into the cells and tissues of seedlings and plants.

The conducted field trials revealed an obvious differentiation between the various trial variants depending on the composition of multifunctional complex nanochips applied to the surface of rape seeds in the process of pretreatment using nanotechnology and in the process of crop spraying with biopesticide having an eliciting activity. Besides, when using nanochips with nanotubes applied as matrix carriers, the yield increase was 14.4–74.0%, depending on their structure; with graphene it was 22.0%; with RAC obtained by rape and camelina straw processing it was 16.1% and 55.8%, respectively; with mineral derivatives it was 5.7–14.8%, depending on their composition; and with water-soluble polymers it amounted to 8.0–12.0%, depending on their nature.

7.5 The Risks Connected with the Use of Polymeric Nanostructures in Technologies of Seed Treatment before Sowing

The creation and use of nanotechnologies is connected with their interaction with ecosystems and living organism. Currently it is actual the problem of a risk use of nanoparticles and nanotechnologies in different sphere of human activity. Influence of nanotechnology on the preparation of seeds before sowing on natural nano-object—a seed of the plants has been studied comprehensively [4].

It was established [18] that in the earliest stages of seed sprouting water inflow in the seed has an oscillating character whereas the polymeric covers disturbing water inflow rhythm regulates some physiological processes, in particular hydrolysis of some substances and velocity of hydrolysis products inclusion in the substances biosynthesis "de novo". The results of these investigations allowed one to consider the seed as a dissipative system (DS) and to elucidate the role of closed space between the rind and the seed germ where this DS is realized [19].

Addressing some similar porous materials (having pores of nanosizes) that are filled by the products of the same nature but lesser size, Ruban et al. [4] have undertaken an attempt to consider the method of the seeds' presowing preparation in the course of seed treatment with the use of polymer covers from the point of view of modern technologies.

Starting from the common concept of the seeds' water regime, in the course of their ripening and transition to the rest state germ dewatering is observed. As a result its volume decreases sharply. It seems that at the same time any closed space is formed between the rind and the germ. The seeds' dewatering and close space formation lead to rind dewatering. Evidently that together with water in the course of dewatering in the closed space the metabolites and part of chemical compounds (or the products of their destruction) that inflow outside and possibly from the rind inside. In the closed space the emanated metabolite solvents, chemical substances, and the products of their hydrolysis are mixed and may create in it aggregates of various structures and sizes, including nanosizes. Starting of relaxation times the probability of rind pores filling with small particles is lesser as compared to larger particles. The last ones form in the closed space aggregates, which provide the seed additive stability due to the limitation of water inflow from outside through the rind in the closed space that provides the seed rest. This phenomenon is connected evidently to the stabilization of smaller particles in the rind pores.

Inside of the closed space there are the ways connecting the germ to the rind inside surface. By these canals (capillaries) the solved metabolites, chemical compounds, and the products of their decomposition move to the rind inside surface and then deeper in it and close its pores, some of which are nanopores. Thus the ripening seed provides itself minimal moisture access inside at its transition into the rest state. The closing of these pores by metabolites, chemical substances, and the produce of their destruction leads to the formation of specific structures where the rind pores are filled with small particles presenting the aggregates previously formed in the closed space and then transferred into the pores. Particles' aggregation in the pores up to nanosizes leads to the formation of their high surface energy, which weakly dissipates due to hypothesis on nanoparticles' stabilization in the pores by agglomerates, formed from the larger particles in the closed space which seem to be "hanging" over the pores. All this in common provides nanoproperties to the system, leading to seed stability.

It seems that the reasoning given for the formation of such nanosystems is that it possesses unique nanoproperties, such as high functionality. High functionality is reflected in the possibility of such nanostructures of the seed energy and mass transfer with the environment on the level that is necessary for the seed in the course of rest conditions and during rest to create seed reliability during environmental parameters' weak fluctuations to keep the necessary low level of metabolism that is characteristic of a resting seed and at last to realize the conditions of functionalizing of closed space between the rind and the germ [19].

In the resting seed the level of metabolite processes is low [4]. In the case of their proper keeping they do not lose their sowing properties and under the definite conditions (moisture, temperature) the seeds get out of the rest state and then sprout. At the higher level of fluctuations the pores extend and the superfluous surface energy of nanoparticles dissipates; as a result the nanostructure is destroyed and water begins to inflow into the seed. When the seed comes in contact with water at optimal temperatures for its sprouting water permeates through the closed pores. It enters through the closed space in the germ and activates the physiological processes lying at the base of sprouting (hydrolysis of stored substances, inclusion of hydrolysis products in metabolism of sprouting seed). As the germ watering increases, its volume increases, the volume of closed space decreases, the pressure in it increases, and larger particles agglomerated in the closed space "hanging" over the pores at the seed rest state "press out" the seed border. Metabolites and the products of destruction as well as the part of chemical substances that closed the pores at the seed transition into the rest state are thrown out, which provides the increase of the rate and quantity of water inflow into the seed.

Is it necessary at such level of safety to increase it more by seed treatment with polymers? Of course, it's necessary, especially for the seeds that have not undergone deep rest and been placed in conditions of often-changing environment parameters in the period of the seeds' sowing. In the extreme conditions of environment, the polymer covers of the seeds that have not undergone deep rest preserve them from unfavorable conditions, providing the additional safety. Under the favorable conditions the previously developed mechanism operates. It consists of hindering the water inflow into the seed and accumulation of the intermediate products of substance exchange, which include the biosynthesis process of the sprouting seed with the dissolution of polymer cover. Owing to the realization of this mechanism the seeds' sowing energy increases as well as their laboratory and field sproutness.

Considering Ruban et al.'s [4] reasoning it is obvious that polymeric covers in the period of polymers' swelling do not hinder the pores' release of small particles but they do not allow the inclusion of the mechanism of the larger particles "pressing out," which formed in the closed space, through the pores and polymeric covers for the time of the cover full solution. Such effects are inherent to water-soluble film-forming polymers possessing high adhesion to the seed surface and high sorption capacity. Besides these polymers have to have high solubility under the creation of optimal conditions of environment for the seeds' sowing. To such polymers providing fine-porous covers belong some water-soluble polymers, vinyl variety polymers, polysaccharide derivatives (cellulose, chitin), their mixtures, etc.

Thus considering the seed's unique adaptive properties and high reliability on environment factors' action, the seed (as a DS) can be considered to have nanostructure possessing nanoproperties. On the mechanism considered realizing in the resting seed and at the outcome from the rest it can be expected that these mechanisms promote the formation of a native nano-object—the seed. The method of seed treatment with polymers that increases the seed reliability to realize the mechanism regulating the velocity of water inflow into the seed and the processes underlying sowing as well as seed sprouting allows to consider it as a nanotechnology and polymeric covers (depending on their nature, molecular mass, concentration, etc.) at the definite conditions the seeds' presowing treatments as nanomaterial which allows to change significantly the properties of the seed as a native nano-object.

For risk estimation of the use of nanotechnology from the preparation of the seeds to the sowing we used some models describing operating the live systems (seed) and the role of a covering in different phases of the seed development [4].

7.5.1 Toward Theory of Water Sorption by Seeds Using the Memory Functions Method

It was elucidated that water inflow in the seed has an oscillating character (Figs. 7.8 and 7.9 [1, 4]).

The oscillation phenomenon appeared in no equilibrium processes of many molecular systems is usually interpreted in analysis of engaging nonlinear equations composed on the basis of a certain

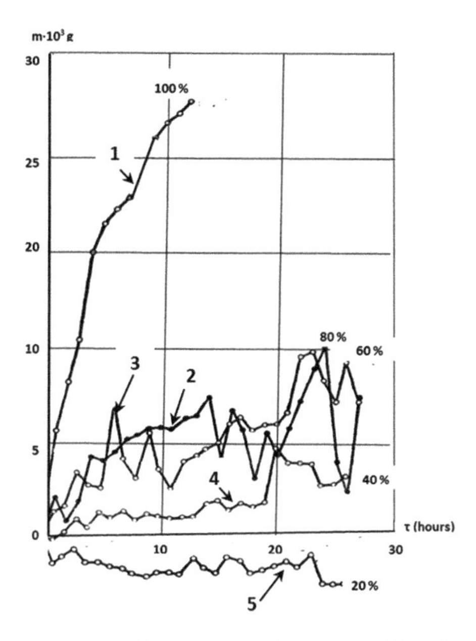

Figure 7.8 Water sorption by cotton seeds at several levels of environmental moisture [1, 4].

hypothesis on the object's internal structure. Besides a synergetic approach the system behavior modeling with a common equation, which solution assumes the oscillation regimes and connects to experimental curves asymptotic behavior is seems interesting. It is essential to use such an approach in relation to compound objects, the internal structure of which may be described from the view of a multitude of hypotheses. A typical example of synergetic effects in such systems is the experiments on oscillatory sorption of water by seeds—system selected as a variable.

In a seed-steam system description the memory function method was applied [20]. In that the integro-differential equation was used:

$$dN/dt = N_0\alpha - \beta \int_0^t N(\tau)P(\tau)d\tau, \tag{7.1}$$

where N and N_0 are water molecules' concentration in the seed and steam respectively, N_0 is accounted from the average water concentration in the seed before sorption starts, $P(\tau)$ is a certain "memory function," and α and β are system parameters.

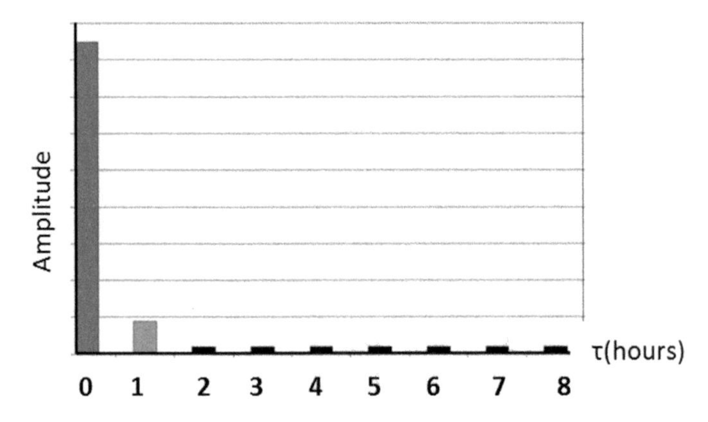

Figure 7.9 Fourier-analysis of the water absorption curve by intact seeds [1, 4].

Equation 7.1 is different from the standard one (which is the special case of Eq. 7.1) in

$$P(\tau) = \delta(t-\tau); \quad dN/dt = N_0\alpha - N/\tau$$

by the presence of integral resorption effects taking into consideration complex processes that happen during $0 \leq \tau \leq t$ time. For $P(\tau)$ the following expression is selected:

$$P(\tau) = f(N_0)\delta(t-\tau) + (1 - f(N_0))\Phi(\tau) \tag{7.2}$$

Function $f(N_0)$ should be equal to $(N_0/N_0^*)^\gamma$, where $\gamma > 0$ (a certain constant) and N_0^* is maximal water molecules' concentration in the steam phase.

Equation 7.1 contains the important special cases related to the limit situations.

Case (a): $N_0 \to 0$ corresponds to the equation:

$$\frac{dN}{dt} = -\beta \int_0^t N(\tau)\,\Phi(\tau)d\tau \tag{7.3}$$

The linear combination of the Wentsel–Kramers–Brilluen (WKB) function is the decision of Eq. 7.20:

$$N(t) = \frac{C_0^\pm}{[k(t)]^{1/2}} \exp\left(\pm i \int k(t)dt\right), \tag{7.4}$$

where $\kappa^2(t) = \beta\Phi(t)$ and $\kappa^2 > 0$. The WKB method's applicability gives the following:

$$\left|\frac{dk(t)/dt}{[k(t)]^{1/2}}\right| \langle\langle \left|\frac{d^2\,[k\,(t)]^{-1/2}}{dt^2}\right| \tag{7.5}$$

So, we have the oscillation regime in accordance with the experiment (Fig. 7.8, curve 1). The $C_0^{\pm}$ constants are selected from initial conditions, and the $\Phi(t)$ function may be concluded as limited both from above and below.

Case (b): $N_0 \rightarrow N_0^*$ and:

$$dN/dt = N_0^* - \alpha - \beta N/2 \tag{7.6}$$

The equation's solution is:

$$N(t) = \frac{2N_0^*\alpha}{\beta}\left[1 - \exp\left(\frac{\beta t}{2}\right)\right] \tag{7.7}$$

For example, the dependence is relaxative (Fig. 7.8, curve 5). In the common case $(0 < N_0 < N_0^*)$ we have

$$N(t) = N_1(t) + N_2(t), \tag{7.8}$$

where

$$N_1(t) = \frac{2N_0\alpha}{f(N_0)\beta}\left[1 - \exp\left(\frac{-f(N_0)\beta t}{2}\right)\right] \tag{7.9}$$

and

$$N_2\,(t) = \exp\left[\frac{-\beta f\,(N_0)\,t}{4}\right] z\,(t)\,, \tag{7.10}$$

where $z(t)$ satisfies the equation

$$\frac{d^2 z\,(t)}{dt^2} + k^2(t)z(t) = -\frac{2N_0\alpha(1 - f(N_0))}{f(N_0)}\left[1 - \exp\left(-\frac{\beta f(N_0)t}{2}\right)\right] \times \exp\left(\frac{\beta f(N_0)t}{4}\right)\Phi(t) \tag{7.11}$$

where

$$\kappa^2(t) = \beta[\Phi(t)(1 - f[N_0]) - \beta f^2(N_0)/16]. \tag{7.12}$$

Term $\kappa^2(t) = 0$ divides the phase plane (N_0, t) into two fields, $\kappa^2(t) > 0$ and $\kappa^2(t) < 0$, that correspond to the different decision types.

Using the WKB method,

$$\left|\frac{i(dk(t)/dt)C^{\pm}(t)}{[k(t)]^{1/2}}\right| \langle\langle \left|\frac{d^2(C^{\pm})(t)\,[k(t)]^{1/2}}{dt^2}\right|, \tag{7.13}$$

where

$$C^{\pm}(t) = C_0^{\pm} \pm i\frac{N_0\alpha(1-f(N_0))}{f(N_0)} \times \int \frac{\left[1-\exp\left(\frac{\sqrt{\beta f(N_0)t}}{2}\right)\right]\exp\left(\frac{\beta f(N_0)t}{4}\right)\Phi(t)}{[k(t)]^{1/2}} \times \exp\left[\pm\int k(t)dt\right]dt \tag{7.14}$$

we obtain the following decisions for the corresponding fields of the phase: in $\kappa^2\,(t) > 0$ the quasi-oscillation regime is realized,

$$N_2(t) = \frac{\exp\left(-\frac{\beta f(N_0)t}{4}\right)}{[k(t)]^{1/2}}\left\{C_1\exp\left(i\int k(t)dt\right) + C_2\exp\left(-i\int k(t)dt\right)\right\} + \frac{2(1-f(N_0))N_0\alpha\exp\left(-\frac{\beta f(N_0)t}{4}\right)}{[k(t)]^{1/2}f(N_0)} \left\{\cos\left(\int k(t)dt\right)\int F(t)\sin\left(\int k(t)dt\right)dt - \sin\left(\int k(t)dt\right)\int F(t)\cos\left(\int k(t)dt\right)dt\right\} \tag{7.15}$$

and in $\kappa^2\,(t) < 0$ the quasi-relaxing regime is realized,

$$N_2(t) = \frac{\exp\left(-\frac{\beta f(N_0)t}{4}\right)}{|k(t)|^{1/2}} \left\{\tilde{C}_1\exp\left(\int |k(t)|\,dt\right) + \tilde{C}_2\exp\left(-\int |k(t)|\,dt\right)\right\} + \frac{2N_0\alpha(1-f(N_0))\exp\left(\frac{\beta f(N_0)t}{4}\right)}{|k(t)|^{1/2}f(N_0)} \left\{ch\left(\int |k(t)|\,dt\right)\int F(t)sh\left(\int |k(t)|\,dt\right)dt - sh\left(\int |k(t)|\,dt\right)\int F(t)ch\left(\int |k(t)|\,dt\right)dt\right\}, \tag{7.16}$$

and

$$F(t) = \frac{\left[1 - \exp\left(-\frac{\beta f(N_0)t}{2}\right)\right] \exp\left(\frac{\beta f(N_0)t}{4}\right)}{|k(t)|^{1/2}} \Phi(t) \tag{7.17}$$

C_1, C_2, $\tilde{C}_1$, and $\tilde{C}_2$ constants are selected for comparison with the experiment.

The following sorption kinetics regimes may be separated for the given N_0 depending on the memory function $\Phi(t)$:

- First regime: $\kappa^2(t) > 0$ in any t that corresponds to the quasi-oscillation regime that doesn't contain the relaxation.
- Second regime: $\kappa^2(t) < 0$ in any t that corresponds to the quasi-oscillation regime that doesn't contain oscillations.
- Third regime: $\kappa^2(t) = 0$ assumes a great number of roots. In this case quasi-oscillation and quasi-relaxation regimes consequentially replace each other [49].

As κ^2 is the function of both t and N_0, at the given memory function $\Phi(t)$ it is possible to pass from one regime to another in an N_0 alteration.

Note that in the developed theoretical schemes (1) and (2) functions $\Phi(t)$ and $f(N_0)$ are determined by a detailed comparison with the experiment.

Returning to the experiment [18] it may be considered that a number of situations in which water sorption by seeds in different experiments varied depending on environmental humidity were described. Significant water sorption oscillations were shown when environmental humidity increased from 40 to 80%, which evidenced the seed functioning character. In 20% humidity the seed is in the rest state and weak fluctuations in water sorption caused by its flow-out to the environment are noted. In 100% humidity sudden water absorption, leading to seed swelling and loss of sowing quality, is noted [22].

So, Eqs. 7.1–7.17 allowed theoretical interpretation of experimental data on oscillation water intake into seeds and the mathematical instrument describing these oscillations was worked out. The relation of environmental humidity and physiological state of living system in vivo was founded, including parameters determining where the seed is in the rest state with insignificant water flow-out, optimal functioning regime, or only water absorption regime. Further the attempts in theoretical base the role of polymer cover in water absorption by seeds, and to this purpose a model describing features of water transport into seeds across polymer cover was suggested.

7.5.2 On Mechanisms of Water Transport into a Seed across the Polymer Cover

To explain the obtained experimental results in a water intake oscillation regime in polymer-covered seeds (Figs. 7.10 and 7.11) [1, 4], let us examine the dynamic system of polymeric closed space between polymer and seed.

The following system of differential equations describes the process of water transfer through a polymer cover and close volume into the seed:

$$dN_o/dt = (Q - \eta + Q - q) \tag{7.18}$$

$$\tau_0 \frac{\partial \theta}{\partial t} = l_o \Delta \theta - q(\theta, \eta, \mathrm{A}) \tag{7.19}$$

in approximation $Q = q_1(dN_0/dt) = (\theta - \eta)$

$$\tau_\eta \frac{\partial \eta}{\partial t} = l^2 \eta \Delta \eta - Q(\theta, \eta, \mathrm{A}), \tag{7.20}$$

where N_o is water concentration within the closed space, Q is water flow into the closed space from the environment through the polymer, dN_o/dt is the velocity of water entering the seed from the closed space, q is the velocity of metabolism products removed from the seed, η is the velocity of water-soluble reaction products removed from the closed space through the polymer into the environment, A is the controlling parameter, τ is time of the process, and l is the specific lengths of process change.

In the analysis of systems, in a certain range of controlling parameters, the following conditions may be fulfilled: $\partial Q/\partial \eta > 0$; $\partial q/\partial Q < 0$. This means that as the velocity of the water entering the closed space increases, the moisture content in the seed also increases. As reaction products accumulate in the closed space, water entering the seed from the closed space will be hindered ($\partial q/\partial Q < 0$). As the velocity of water and metabolism products leaving the closed space becomes greater (η increases), the flow

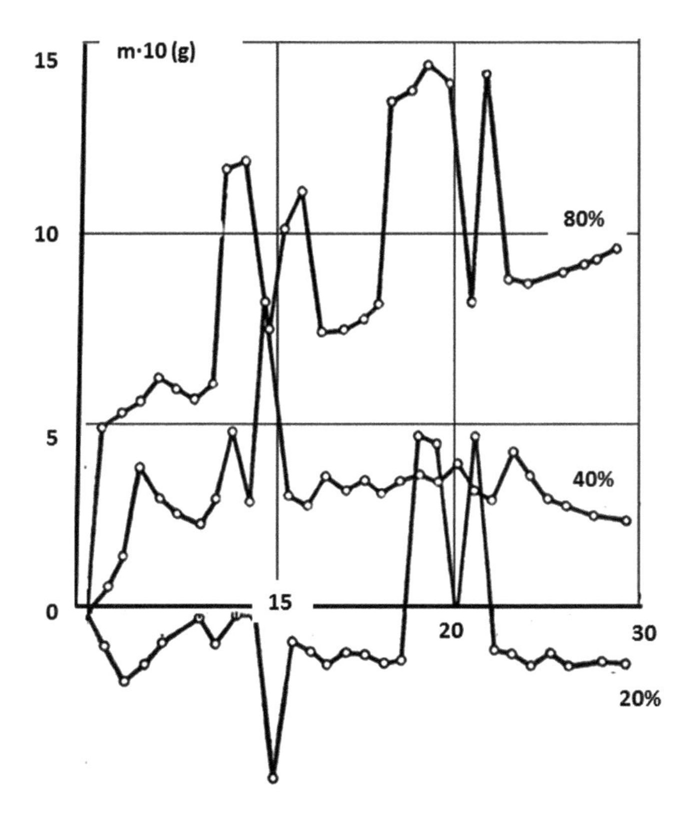

Figure 7.10 Water sorption of seeds covered by a polymer [1, 4].

from the closed space into the environment increases accordingly ($\partial Q/\partial \eta > 0$). Thus, the condition may arise when great instability occurs, shown by great water accumulation within the closed space, because if $\eta = \text{const}$, the number of fluctuations grows. So a greater level of moisture concentration within the closed space is accompanied by a sharp increase in inner pressure and temperature Consequently, quasiperiodic shocks occur in the oscillating regime of water entering the seed. Therefore, we should use the model of the water consumption process in the seed that shows fading oscillations are activated by periodic shocks (Fig. 7.11) [19].

So, Eqs. 7.4–7.17, which describe the complex polymer-skin-closed space between the skin and germ system and are able to determine the role of the polymer cover, which regulates the velocity

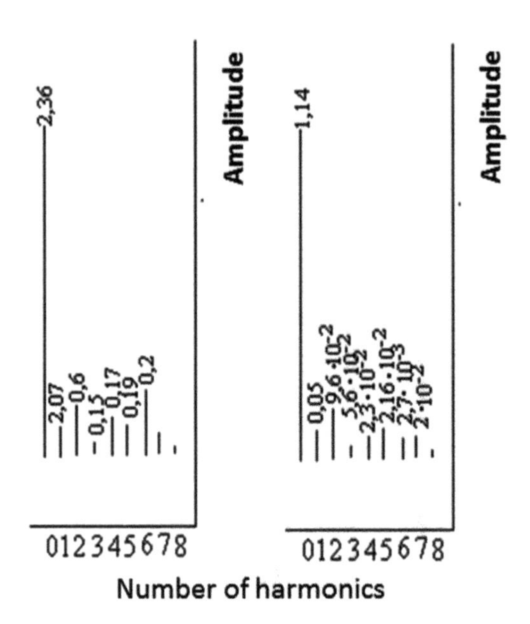

Figure 7.11 Fourier-analysis of water penetration and transport [1, 4].

of water intake into the seed from closed space, as well as velocity of metabolites and water-soluble products removing across skin and polymer, were obtained. Also the processes occurred in closed space (seed outer and inner flow mixing), and also their influence on the curves of water sorption by covered seeds were described.

Starting from the suggested scheme and equations polymer cover adhesive strength and sorption capacity seem important as controlling parameters of water sorption by polymer and seed. At that they may be regulated by polymer type, its molecular mass, and cover thickness alteration, as well as by physiologically active substances included in the polymer matrix. And, finally, it is necessary to take into consideration the nature of the seed skin surface in order to achieve complete covering. The key moment in the investigation is polymer cover sorption capacity study, which is a first stage in the consecutive processes of water intake into the coated seed, from the view of polymer cover action mechanism based on water intake regulation in seeds, as well as the role of physiological-biochemical processes initiating their germination [4].

Taking into consideration the optimal soil temperature range for cotton seed germination, 16–20°C, systematic investigations were made into the sorption capacity of different natural polysaccharide derivatives, vinyl derivatives, and polymer-polymer mixtures on their base with different component ratios in different temperatures [23]. These parameters are so attractive because of the often-coincident stable temperature optimum necessary for seed germination with a water deficit in the soil. Weak oscillations attributed to the water through away process were noted. In water excess (100% relative humidity) water intake is described by a saturation curve with weak oscillations. That's why the selection of polymers with high sorption capacity is essential not only to keeping of seeds in the soil but also to their intensive development until germination, that is, sprouts may be obtained also in water-deficit soil. Following watering allow keeping full sprouts due to seeds polymer cover and creating further the appropriate conditions for plant development.

So, physic-chemical properties of polymer systems used for seeds cover determine the water intake velocity of seeds, and water is a valuable material used in seed processes modeling, especially during the after-rest period. It should be considered that the polymer cover is an external information source that enables regulation of a seed's life-support processes. These processes may be controlled by variations in the cover's physical-chemical properties.

7.5.3 Role of the Polymer Cover in the Seed's Water Sorption

For investigation of complex dynamic systems and especially biological systems the synergistic approach is the most adequate [24]. In this connection interest appears in the translation stationary biological systems in new state with the purpose changes in the quantitative and qualitative characteristics both the biological system as a whole, and it's separate components (subsystems). In this case two ways of system translation in new state may be picked out:

- The biological object turns into the new phase trajectory as a result of the interior structure reorganization,

- The change of the phase volume ΔV at the expense of the initial conditions changes and translates the dynamic system from the phase volume V_1 to the phase volume V_2. This translation is characterized by new ultimate cycles at the minimum of external consumed energy.

The interaction of the germ and rind is of interest from these positions.

The structure of seeds, related to angiosperm seeds, consists of the germ, closed space between the germ and the rind, and the rind. The presence of the rind, which implements protective functions for the germ, and the presence of free space in the volume of the seed permit one to consider the seed as a quasi-closed system [19]. Therefore, one may introduce the notion of energy-mass exchange between the germ and the rind. The transfer of these energy and mass currents take place through the closed space inside a seed.

The closed space between the germ and the rind may be considered as the phase volume of the germ-rind dynamic system. We note that the presence of the final volume and the absence of full germ fixation in the rind give the ground to suppose the space-time and periodic character of the mass and energy currents' passing. In this situation it is worth mentioning the set of waves and harmonics that take part in this process. First, it is necessary to determinate the character of energy and mass currents in the seed, that is, elucidate if they have a linear or nonlinear character. In the case of the linear interaction of the germ and rind it would require the accomplishment of the boundary condition of the next type:

$$\bar{k} \cdot \bar{l} = 2\pi n, \tag{7.21}$$

where $\bar{k}$ is the wave vector and $\bar{l}$ is the distance between germ and rind.

In so doing the distribution of energy-mass currents has to correspond to the reflection regime to the interaction between the germ and rind. But this condition may be implemented only by the deterministic fixation of the germ in the volume of the rind, that is the presence of the identical and constant position of germ in the rind, which cannot be in reality. The germ changes its position in the seed volume the seed, when seed changes its position in the space and in its energy spectrum of energy-mass currents would be changed. Hence, the germ and the rind interact in a nonlinear way. But in such case the processes of interaction between separate harmonics would be observed. The resonance outburst of water flow in the seed wasn't observed. This indicates the absence of such processes of the energy-mass transfer in the volume between the germ and the rind.

The inflow of the water in the seed has a quasi-harmonic character that may correspond with the nonlinear waves' successions, which propagate in the closed space between the germ and the rind [21]. In this connection the conclusion may be drawn that the nonlinear character of the exchange between the germ and the rind has a soliton-like character, when the nonlinearity can be compensated by dispersion. In this case even in conditions of currents overlapping an effective energy exchange doesn't take place and an appearance of resonance term is impossible [25]. The supposition about soliton-like currents, which propagate in the invariable free volume without dependence on germ position, is most acceptable because solitons don't change their form at reflections.

One may suppose that solitons have a finite space area and finite time duration, that is, the dimensionality $2 + 1$, where 2-space and 1-time dimensionalities coordinate. In a soliton's spatio-temporal Fourier-resolution on harmonic components the expression for the soliton has the appearance

$$A(x, y, z, t) = \sum_{i=1}^{N} A_i \exp(-ik_i x_i + i\omega_i t), \qquad (7.22)$$

where i is the number of harmonics, A is the amplitude of harmonics, $k = k_j$, $j = 1, 2, 3(x, y, z)$, and ω_i is the frequency of harmonics.

The concept of the "rest" seed may be introduced in accordance with the accepted determination of soliton-like waves. The two states of the "rest" may separate. Deep rest is described by the single-soliton regime and corresponds to minimum energy-mass exchange between the germ and the rind. Seeds with the thick rind, which prevent germination at weak fluctuations of the outer medium, correspond to such seed types. The seed not in deep rest by energy-mass transfer in closed space character corresponds to many solitons' behavior of nonlinear waves. These are the seeds with the thin rinds, which easily germinate even under weak influences, that render the outer medium. They are subject to the polymer cover's informational impact.

Hence, two possibilities may result on the breach of this physiologic state. The first (for a seed in deep rest) is increase of the energy-mass exchange of the germ with the rind by means of an inside reorganization of germ and rind by means of essential decrease of the rind thickness by mechanical influence on it. This is an energetic method because the withdrawal of the seed from the single-soliton current of the energy exchange is possible only by increase in this exchange energy. The second possibility of the withdrawal of seeds from the state of rest concerns a many-soliton regime of energy-mass exchange in seeds (not in deep rest) and consists of change of boundary conditions, when energy exchange between harmonics and energy redistribution between currents may be possible. It may be realized by roofing of a seed by thin films from water-soluble polymers, which is the technological method of preparation of the sowing material [26].

The change of the amplitude and spectral and space distribution of soliton-like waves takes place by coating seeds by the polymer cover. The distribution amplitude may be changed because of the total or partial dissipation of separate harmonics, when the soliton passes through the polymer. The reason of the change of the spectral structure may be the change of the harmonics dispersion in polymer. The interference of harmonics or the time extensibility takes place in the case of abnormal or normal dispersion, accordingly resulting in compression or broadening of the soliton-like waves. It begins to change its form of envelope, and the soliton-like energy-mass transfer regime breaks.

The space distribution in the soliton is changed because of the change of the reflection conditions. The reflection in a seed with a polymer coating takes place on the boundary "polymer-outer medium" but not on the boundary "rind-outer medium," as in a free polymer. Harmonics have a different refraction in the polymer, that is, displacement of the reflection regime takes place in the seeds that haven't the polymer coating. This effect brings misbalance between dispersion and nonlinearity in the volume of the seed. In this case, energy exchange begins to change by crossing the energy-mass currents, which bring to the emergence resonance and the reorganization the whole energy spectrum in the volume between germ and rind.

The interaction of nonlinear waves in detail is considered in Ref. [27], therefore give the definite precision for the nonlinear process:

$$\frac{\partial n_k}{\partial t} = 18\pi \sum_{k_1 k_2} \frac{|V_{kk_1k_2}|^2}{\omega_k \omega_{k1} \omega_{k2}} [2(n_{k1} n_{k_2} + n_k n_{k2} - n_k n_{k1}) \\ \times \delta(\omega_k + \omega_{k1} - \omega_{k2}) \delta(k + k_1 - k_2) - (n_k n_{k1} + n_k n_{k2} - n_{k1} n_{k2}) \\ \times \delta(\omega_k - \omega_{k1} - \omega_{k2}) \delta(k - k_1 - k_2)], \quad (7.23)$$

where n_{ki} is the number of "quasiparticles" in the k_i harmonic; V_k, k_1, and k_2 are matrix elements of the waves' interaction; k_i is the wave vector, three-dimensional in this case; and ω is the frequency of interacting harmonics. The process that describes Eq. 7.23 corresponds with the resonance:

$$\omega_k = \omega_{k1} + \omega_{k2}, \; k = k_1 + k_2 \quad (7.24)$$

Equation 7.23 is the analog of Boltzmann's kinetic equation:

$$\frac{\partial n_k}{\partial t} = St\{n, n\} \quad (7.25)$$

The considered process has a quadratic nonlinearity because many types of seeds haven't the center of symmetry. In systems with the center of symmetry nonlinearity of the cubic character is possible, that is, interaction of four waves. In this case the equation (nonlinear analog Boltzmann's equation) has the following appearance:

$$\frac{\partial n_k}{\partial t} = St\{n, n, n\} \quad (7.26)$$

The character of the nonlinear waves' interaction process is like the character of hydrodynamic turbulence. The stationary picture arises hydrodynamic turbulence that is characterized by the energy exchange between the nearest local scales. Equation 7.23 describes the interaction of different scales. Therefore, we may say about the hydrodynamic turbulence in a nonlinear wave field [26].

In the seed covered with the polymer, all conditions are realized at the beginning of turbulence in the presence of nonlinear waves' energy-mass transfer. First, the presence of solitons' diffusion in the space because of different harmonics reflection, which corresponds to k change, that is, on the quasi-particle pulse, therefore, conditions of the local instability is performed. The volume itself between the germ and the rind is essentially fractal because of the presence of fibers, films, pores, and other increments, covering the germ and the rind and existing in closed space between the germ and the rind. Secondly, in the energy exchange not all harmonics take part but only the one for which the condition in Eq. 7.24 is performed, that is, the fields exist, or the "islands of stability." Thirdly, both the complicated germ form and the complicated rind form, in conditions of energy exchange between harmonics, promote the formation of spatial chaos, not only time chaos. The concern is that the process of interaction of waves is subordinate to the condition in Eq. 7.24. The fact of new harmonics formation with the sum frequency of interacted waves is expressed in this condition. This process may occur only by performing the second condition (Eq. 7.24), that is, the condition of phase synchronism; in other words, the third wave appears with the total wave vector. Taking into account that the difference between the harmonics frequency and the dispersion between wave vectors is negligible, one may say it allows the formation of both time and space structures of the resonance character. Once again we emphasize that it is applied to conditions of closed seed volume that has a complicated form.

Because of the emergence of space-time chaos in the seed volume the change of the energy exchange inside the germ may occur. The germ must arrange under changes of interaction conditions inside the rind. Therefore, the hierarchy inside the germ arises and transition occurs from a quasi-stationary, many-solitons regime to a regime that possesses energy resonance and space irregularity. This hierarchy must correspond to the release of a germ subsystem that is characterized by different velocities of metabolism. In this case the carrying of seeds in the medium of germination the impulse regime of water entrance in the seed must arise, corresponding to the saturation subsystems' dependence on the velocities of processes proceeded in it. In this case may be possible the breach of hydrolysis velocities, for example, spare albumen's and biosynthesis albumen's de novo, which occurs because different velocities of energy exchange passing subsystems [28].

In conclusion we may state that changing outer conditions, i.e. at forming of cover films from water dissolved polymers on the seed rind essentially changes the process of energy exchange inside the seed. The passage takes place from the quasi-harmonic character of the interaction of the germ and the rind to turbulence. In addition, the hierarchy arises in the seed itself, that is connected velocities change of quasi chemical reactions flow, going in the seed on the early stage of germination.

We may consider the polymer cover as an external information impact that changes the seed's water regime followed by alteration of the physiologic-biochemical parameters attributed to the seed's germination. Because of the influence of polymer nature, molecular mass, and cover thickness on water sorption by seeds, the role of polymer covers was shown as information careers, due to the complex changes in live supporting processes velocity and directions are appeared in the seed.

So, concluding the results of experimental data modeling in the nonmonotonic character of water sorption by seeds, including polymer-coated ones, as well as sorption curves' theoretical analysis, it seems clear that a seed's polymer cover promotes water intake regulation and is followed by the physiologic-biochemical parameters attributed to the seed's germination. It is clear why a seed "warms itself" and combusts if thick and not-readily-soluble covers or high-molecular-mass polymers are used. It leads to the deterioration of seeds' germinating capacity.

Such phenomena, that consider as the risks in seeds presowing treatment, were founded in pellet application (seeds coating by not-readily-soluble covers or insoluble polymers); covers containing many components with particles of variable sizes, including nano-sizes; and colloidal systems (suspensions, emulsions, etc.) [28]. As a result seeds' germination decreased so considerably that the plant's density and yield in many cases also decreased considerably. These losses are related to seeds' self-warming phenomenon, changing velocities, and directions of life support processes down to their death. Besides, risks of seeds' polymer or multicomponent cover nanotechnologies should be attributed to environmental factors, like excessive soil humidity or dryness and soil solutions ion composition.

All these risks modify the water intake velocity of the seed and moisture rhythm and destroy the energy and mass exchange between seed and environment. Such living organisms and environment interactions are attributed to DSs. The alteration of the root system geotropic reaction (the negative geotropism phenomenon, when the root grows the wrong side up) is one example of negative influence of some plant protection means in combination with the polymer cover used in the cotton growth method by seedlings [3]. This amazing phenomenon is connected to hormonal regulation infringements and auxin movement and distribution in cotton seedlings.

7.5.4 The Seed as a Dissipative System

Biological systems are dissipative [4, 29]. Permanent substance and energy and information exchange with environment are attributed to them. At the same time these systems are structured hierarchically. The hierarchical structure of complex living systems, which combine from more simple subsystems, help avoid instability and undesirable dynamics that appear in complex centrally controlled systems. Due to living systems' openness and energy exchange with environment, they are self-organized, that is, they are complex ordered structure formations but at the same time do not contradict laws of thermodynamics.

A system's entropy can decrease in time. The self-organization effect lies in the dissipation by the system of the energy flow entering from an external source. Due to that flow the system is activated, that is, takes the ability for autonomous structure formation. Dissipative structures appear only in systems that may be described using nonlinear equations assumed solution symmetry alteration in definite controlling parameters value. It seems correct applying to biological systems, since quantitative and qualitative alterations in the system occur if the environmental parameters are altered, for instance, the velocity and way of life support reaction in a living organism [4].

Dissipative structures, as noted above, appear in macroscopic systems, that is, in systems containing many components. Due to this fact the synergetic interactions necessary to system reorganization may occur. In synergy the phenomena of space-time structure formation, or space-time self-organization passed in different systems—physical, chemical, biological, social, etc.—is studied [4].

Most basic works in synergy advance three main ideas: two of them are always nonlinearity and openness and the third is complexity/dissipativity. These postulated triads may be used with respect to biological nano-objects, including seeds, because first, biological systems are open; second, nonlinear processes occur at all development stages; third, they are complicated; fourth, they are dissipative; and fifth, individual components' interactions in the whole system scale (metabolism) are observed.

A polymer cover is a major factor that influences a living system's life support processes since it creates the possibility to regulate velocities of water intake into the seed. It is known that water is an important factor for starting metabolism processes; therefore, it is important for all system components' interactions, such as reserve substances hydrolysis in the seed, involvement of hydrolysis products in biosynthesis de novo, as well as all quasi-chemical reactions directed to germinated seed growth and development and compounds exchange with environment [19]. That's why a polymer cover, taking in account temperature and humidity, may be considered as a new factor determining a seed's physiological state on different stages of ontogenesis.

To visualize the physiological processes underlying the basis of the different levels of rest, let us use a modeling method that separates out one variable that has the greatest effect on organism development. That variable is the amount of water absorbed by the seed [19]. However, note that the state of a seed at each stage of its development is the result of many physiological processes.

Such a model should explain the contradictoriness of the generally accepted properties of the seed as an integral organism [4]. Therefore, the "seed system" that is sensitive to fluctuating environmental conditions is protected. Random temperature and humidity fluctuations shouldn't influence the seed that has a protective cover. However, without a cover, a seed's water-exchange mechanism is affected considerably by changes in temperature and humidity. Because the water-exchange process is connected with a specific seed state, it might have a negative influence on seed development. However, different stages of seed development differ from each other by their specific regimes of water consumption, so the system should sufficiently adapt to changing environmental conditions. Consequently, the seed should possess sensitivity to fluctuations. Also, a high seed water requirement during the sprouting stage demands a high seed shell capacity, which naturally decreases the reliability of the system. To understand that contradiction in the given system, one should foresee the possibility of complications if there is not sufficient reconstruction [19].

In examining the seed as a dissipative system (DS), we know from DS theory that seed systems should have certain properties [19]. There are several main properties of the seed system. The seed is an open system, which is far from thermodynamic equilibrium. Its life activity process is connected directly with the existence of inside and outside flows. The existence of a reserved space between the embryo and the shell is essential: in this space mixing of elements takes place. Binding of elements within the space occurs because of transfer processes. Because of this, the space is considered to be a "hotbed." Spontaneous formation of dissipative structures should take place here: that formation is caused by the division into layers of an initially homogeneous state. While water enters the seed, an activator (i.e., the sum of processes that activate the transport of water inside the seed) controls the flow.

The process of increasing water consumption is regulated by an inhibitor (i.e., the sum of processes that inhibit the transport of water into the seed). The presence of an activator and an inhibitor in a seed is undoubted [30]. On the basis of these characteristics, the physiological state of a seed may correspond to a definite dissipative structure—the nonhomogeneous distribution of moisture concentration in the hotbed. The transformation of a seed from one physiological state to another will correspond to the nature of the specific nonhomogeneous moisture distribution in the hotbed. Such transformation may take place as a result of dynamic system reconstruction but not as a result of fluctuations. Because of that, the demand for protection is provided as far as dynamic reconstruction of the DS occurs only at some critical levels of excitation. In other words, fluctuations may not play an essential role in the type of DS formation selected, and therefore in the physiological state. In this case, equilibrium takes place, which is the origination of the genetically predetermined state that indicates that the given type is in equilibrium in spite of the chaotic state of the previous stage (i.e., variability).

If this approach is true, seed dissipation should occur by means of the noneven character of moisture consumption, which was shown by us experimentally (Fig. 7.8) by using cotton seeds [4]. The fluctuation period and amplitude of water entering the seed substantially depend on the moisture concentration of the environment. At concentrations of 20–80% fluctuations were quasi-harmonic, which was indicated by the existence of one frequency with a quick fading of harmonics (Fig. 7.9).

Within the framework of the dissipative approach, it is possible not only to ascertain seed equilibrium mechanics but also to understand the correlation of variability and safety and their role in the processes of growth and development in seeds. Features of safety and variability, being opposite, supplement each other and are in balance. Balance displacement to one or another side causes a sharp appearance of the contrary property.

For example, an increase in seed safety was achieved by creating a seed with a polymeric water-soluble material. This caused strengthening of the seed cover. At the same time, if the fluctuation of water entering into the seed were quasi-harmonic without the cover, then having the cover (with the same water consumption) caused quasi-harmonic fluctuations that were interrupted by oscillations with large amplitudes (Fig. 7.10). In the Furie spectrum, with one main frequency appears a whole group of frequencies having their own maximum (Fig. 7.9): it is characterized by accidental oscillations. Consequently, harmonic oscillations were interrupted by chaotic ones on the seed's water consumption curve when the seed was treated with a polymer.

This is typical for dynamic systems (the seed-polymer-water system presumably may be considered to be this type), in which a change to the chaotic state through alternation is observed [31]. The chaotic state caused by the polymer cover is observed for a considerably long time close to its full solution and to seed sprouting. In the development of these ideas, of interest is the action of the mechanism of seed polymer covers on seed sprouting. All of the above allow us to interpret this as alternation phenomena [30]. Such reliability strengthening caused sudden variability.

7.6 Conclusion

Proposed is a biologically active nanochip (BAN) for treating seeds of agricultural plants in order to improve seed germination conditions and development of plants and for protecting plants from anticipated and averaged adverse conditions. A BAN contains a solid porous carrier, such as mineral, clay, turf, or polymer, the pores of which are intended for accommodating nanoparticles of biologically active substances that penetrate the pores when the substances are applied onto the nanochip surface, for example, by spraying. Also proposed is a method for application of the biologically active substances onto the surfaces of the BANs.

Thus, when applying a nanotechnology while processing the surfaces of seeds and plants, nanosized structures are being formed, providing a high efficiency of mineral nutrition process and plant protection and development and also contributing to the following yield increase. In connection with what is stated above, sorption-capable nanoporous materials can be effectively used to transfer the preparations with elicitor activity, to transfer the necessary plant protection products and their trace elements to accelerate seed germination process, and to improve plant growth and development.

Long-term laboratory and field experiments, aimed at studying the influence of presowing seed treatment using disease resistance inductors and other growth regulating compounds as a part of multicomponent (nano)chips on growth, development, and yield of various agricultural crops, have detected yield increases:

- In rice by 19.9%
- In sugar beet by 14.9%
- In wheat by 11.1%
- In corn by 11.5%
- In soya by 8.9%
- In mung bean by 4.3%
- In tomatoes by 24.4%
- In rape seed by 14.7%
- In cotton by 17.3%

An opportunity for efficient agricultural soil detoxication has been detected, with crops treated with herbicides together with active carbons (ACs) obtained from grain crop straw, an annually renewable natural source.

Analyzing the obtained arguments, the seed may be considered as a dissipative nanosystem, as well as seeds cover methods as nanotechnology. In definite environmental conditions and physical-chemical parameters risks appear that leads to loss of a seed's germinating capacity. To exclude them special approaches in polymer synthesis, their application on the seed surface, and selection of cultures with definite rest type are necessary. Besides, soil-climatic habitats, where seed cover nanotechnology may be applied successfully, are essential.

There are risks related to seed pretreatment nanotechnology. As was considered above, the seed's self-heating processes, realized in definite environmental conditions, and the seed's cover features, at that, as a rule, these processes finished by seeds "self-burning." These processes are based on quasi-chemical reactions' disconnection, which leads to interaction dynamics' alteration on subsystems determining biochemical reactions' coordination and homeostasis.

References

1. Figovsky O., Beilin D. (2016). Nananotecnology in agriculture, *Journal Scientific Israel - Technological Advantages*, **16**(3).
2. Ruban I. N., Voropaeva N. L., Figovsky O., et al. (2012). Biologically active multifunctional nanochips and method of application thereof for production of high-quality seed, Patent US 8,209,902 B2.

3. Voropaeva N., Figovsky O., Ibraliu A., et al. (2012). Innovative nanotechnology for agriculture, *Journal Scientific Israel - Technological Advantages*, **14**(1).
4. Ruban I., Voropaeva N., Sharipov V., Figovsky O. (2010). The risks connected with use of polymeric nanostructures in technologies of seeds treatment before sowing, *Journal Scientific Israel - Technological Advantages*, **12**(1).
5. Voropaeva N., Karpachev V., Varlamov V., Figovsky O. (2014). Influence of improved (nano) systems on cultivated corn growth, development and yield, *International Letters of Chemistry, Physics and Astronomy*, **28**, 1–7.
6. Voropaeva N., Karpachev V., Varlamov V. (2014). Influence of efficient, multicomponent, polyfunctional, physiologically active (nano) chips with herbicide activity on rice crop growth, development, yield and on weed growth inhibition, *Journal Scientific Israel - Technological Advantages*, **16**(1–2).
7. The who recommended classification of pesticides by hazard and guidelines to classification (2009). http://www.who.int/ipcs/publications/pesticides_hazard_2009.pdf
8. Dospekhov B. A. (1983). Method of field experience, Kolos: Moscow, 420 p. (in Russian).
9. Dospekhov B. A. (1972). Planning of field experience and statistical analysis of its data, Kolos: Moscow, 207 p. (in Russian).
10. www.fao.org/news/story/ru/item/239702/icode/
11. Spiridinov Y., Muhin V., Voropaeva N., et al., (2015). Detoxication of pesticide and other toxic substance remains in soil with the help of nanomaterials, *Journal Scientific Israel - Technological Advantages*, **17**(4).
12. Belyaev E. Y. (2000). Obtainment and application of wood active carbons from environmental purposes, *Plant Raw Material*, 2, 5–15.
13. Kovaliova O., Bunescu A., Dragalin I., et al. (2007). Researches on the benzothiazole destruction phenomena occuring in different photobiocatalytic system, *The IInd International Conference of the Chemical Society of the Republic of Moldova "Achievements and Perspectives of Modern Chemistry"*.
14. Karpachev V., Voropaeva N., Tkachev A., et al. (2015). Innovative application technology for challenging inducers of disease resistance in spring rape in nanochips, *Journal Scientific Israel - Technological Advantages*, **17**(2).

15. Gusev A., Akimova O., Zakhrova O., et al. (2014). Morphometric parameters and biochemical status of oilseed rape exposed to fine-dispersed metallurgical sludge, PHMB-stabilized silver nanoparticles and of multi-wall carbon nanotubes, *Advanced Materials Research*, **8**, 212–218.
16. Tripathi S., Sonkar S. K., Sarkar S. (2011). Growth stimulation of gram (Cicerarietinum) plant by water soluble carbon nanotubes, *Nanoscale*, **3**, 176–117.
17. Mukhin V., Voropaeva N., Karpachev V., Figovsky O. (2014). Plant residues of various agricultural crops as renewable raw material for obtaining activated carbon, *Journal Scientific Israel - Technological Advantages*, **16**(2), 186–189.
18. Ruban I. N., Rashidova S. Sh., Voropaeva N. L. (1988). The use of polymeric systems on their base for capsulating cotton seeds, *Physiology and Biochemistry of the Cultural Plants*, **20**(1), 73–78.
19. Voropaeva N. L., Pakxarukov Yu., Rashidova S. Sh., Ruban I. N. (1994). The seed as dissipative system, *The Report of Russian Academy Agricultural Sciences*, 2, 10–12 (in Russian).
20. Repke G. (1990). *Non-equilibrium Statistical Mechanics*, Mir: Moscow.
21. Mathew J., Walker R. (2013). *Mathematical Methods for Physics*, Arfkin Weber and Harris.
22. Abrukina Yu. M., Voropaeva N. L., Oksengendler B. L., et al. (1995). Nature of the oscillations in complex molecular systems in non-equilibrium states-2, *Biophysics*, **40**(3), 511–514.
23. Rashidova S. Sh., Voropaeva N. L. (2006). *Water Soluble Polymer-Polymeric Blends*, FAN: Tashkent (in Russian).
24. Haken H. (1983). *Synergetics. An Introduction: Nonequilibrium Phase Transitions in Physics, Chemistry and Biology*, Springer: Berlin.
25. Ablovitz M., Segur H. (1981). *Solitons and the Inverse Scattering Transform*, SIAM: Philadelphia.
26. Stoneham A. M. (2003). The challenges of nanostructures for theory, *Materials Science and Engineering*, **23**(1–2), 235–241.
27. Sagdeev R. Z., Usikov D. A., Zaslavsky G. M. (1988). *Nonlinear Physics: From the Pendulum to Turbulence and Chaos*, Harwood Academic Publishers: New York.
28. Rashidova S. Sh., Voropaeva N. L., Ruban I. N. (1996). Biological active agricultural polymer (mechanism of action on plants), in: *Synthesis, Properties and Applications//The Polymeric Materials Encyclopedia* (Salamone J. C., ed.), vol. 2, pp. 615–629.

29. Manneville P. (1990). *Dissipative Structures and Weak Turbulence*, Academic Press: London.
30. Prokofiev A. A. (ed.) (1982). *Seeds Physiology*, Science, Moscow, 225 p.
31. Shuster G. H. (2005). *Deterministic Chaos: An Introduction*, Wiley-VCH Verlag GmbH & Co. KGaA: Weinheim.

Index